Water and Wastewater Engineering Hydraulics

T. J. Casey

Professor of Civil Engineering
Centre for Water Resources Research
Department of Civil Engineering
University College, Dublin

Oxford New York Tokyo
OXFORD UNIVERSITY PRESS
1992

Oxford University Press, Walton Street, Oxford OX2 6DP

Oxford New York Toronto
Delhi Bombay Calcutta Madras Karachi
Petaling Jaya Singapore Hong Kong Tokyo
Nairobi Dar es Salaam Cape Town
Melbourne Auckland

and associated companies in
Berlin Ibadan

Oxford is a trade mark of Oxford University Press

Published in the United States
by Oxford University Press, New York

A catalogue record for this book is available from the British Library

Library of Congress Cataloging in Publication Data
Casey, T. J. (Thomas Joseph), 1933–
Water and wastewater engineering hydraulics / T.J. Casey.
p. cm.
Includes bibliographical references and index.
1. Hydraulics. 2. Water-supply engineering. 3. Sewage disposal.
I. Title.
TC163.C32 1992 627—dc20 92-7878
ISBN 0-19-856360-4 (hbk.)
ISBN 0-19-856359-0 (pbk.)

Typeset by Integral Typesetting, Gorleston, Norfolk
Printed in Great Britain by Biddles Ltd, Guildford & King's Lynn

Preface

This book addresses those areas of applied hydraulics of special interest to engineers engaged in the fields of water supply and wastewater disposal. It has been developed from lecture notes prepared by the author for a graduate course in Water/Sanitary Engineering.

The analytical methods employed are developed from first principles with an emphasis on engineering application rather than on mathematical rigour. Parameter correlations are presented in a format appropriate to problem solution by computer. Computer algorithms and programs are written in BASIC to achieve the widest accessibility among readers who have access to microcomputer systems.

Chapter 1 reviews fluid properties which are presented in a database format in a computer program called FLUPROPS. Chapter 2 reviews the basic concepts of fluid flow including the principles of continuity, energy, and momentum. Chapters 3–6 inclusive, deal with closed conduits. Chapters 7, 8, and 10 deal with open channel flow. Chapter 9 is concerned with dimensional analysis and hydraulic modelling. Chapter 11 deals with pumping systems. The accompanying computer programs are described briefly at the end of each chapter and their use is demonstrated by a sample program run. A complete listing of the programs is given in Appendix A and can also be supplied on disk. These programs enable the student or design engineer to tackle computational problems which could not be readily handled by hand calculations. Since the algorithms follow directly from the analytical treatment of the subject matter in the text, users should have little difficulty in following the logic of the computer coding and in adapting programs to meet their own specific needs. The difficulty and required effort will, of course, vary depending on program length and complexity.

While not a design manual as such, the contents and computer-oriented presentation should prove of particular interest to water engineers engaged in hydraulic and treatment process design. The suite of computer programs embraces a wide variety of flow problems including topics such as flow of sewage sludge in pipes and flow distribution in pipe manifolds and networks, which are not generally included in texts on the general theme of fluid flow.

Although the programs have been tested extensively and are believed to be correct, the author and publisher cannot accept responsibility for eventual errors, and do not accept liability for damages or losses that may be caused by use of the programs.

The helpful advice and comments of colleagues, including James Dooge, Aodh Dowley, Patrick Purcell, and the late Rory McPhillips, are gratefully acknowledged. The input of former students, particularly Jim Byrne, Brian Curtis, Donncha McCarthy, and Bashir Khyter to the development of the computer programs for network analysis and flow measurement structures is also gratefully acknowledged. I would also like to acknowledge the help of Larry Clarke who prepared the line diagrams, Myles Christian who prepared the photographic plates, and Mary McNamara for her assistance with the typing.

Dublin T. J. C.
January 1992

Supply of programs on disk

The author will be pleased to supply readers with copies of the program listings on disk. Please apply to the address below stating preferred disk format ($3\frac{1}{2}''$ or $5\frac{1}{4}''$, single or double density, and so on). Please also enclose £5 (Irish or sterling) to cover the costs of disks, postage, etc.

Professor T. J. Casey
University College Dublin
Civil Engineering Department
Earlsfort Terrace
Dublin 2
Eire

Contents

1
Fluid properties

1.1 Introduction

The engineer involved in water supply and sanitation deals with a variety of fluids including clean water, sewage and industrial wastewaters, sludges, gases (including air), biogas, chlorine, oxygen, and so on. Although the physical properties of these fluids and their flow characteristics vary widely, they are all classified as fluids in so far as they flow or continuously deform under the action of any unbalanced external force, no matter how small that force may be. Properties which influence the flow behaviour of all fluids include density, viscosity, and surface tension. Where compressibility effects are significant, as is the case in the flow of gases under certain conditions, thermodynamic properties such as specific heat at constant volume or constant temperature must also be known.

1.2 Viscosity

The viscosity of a fluid is a measure of its resistance to flow under conditions where turbulence is suppressed. A commonly used flow environment for the definition of fluid viscosity is illustrated in Fig. 1.1. Consider the deformation of the fluid layer contained between the moving upper plate and the stationary lower plate. Assuming that there is no relative movement between the fluid and the plate surfaces, the movement of the upper plate at a uniform velocity v_p (m s^{-1}) results in a linear velocity gradient across the fluid. The force required to sustain the movement of the top plate can be expressed as a function of the velocity v_p, the plate area A, and the distance Y between the plates:

$$F \propto \frac{A v_p}{Y}. \tag{1.1}$$

This proportional relationship may be written as an equation by introducing the correlating coefficient μ:

$$\frac{F}{A} = \mu \frac{v_p}{Y} \tag{1.2}$$

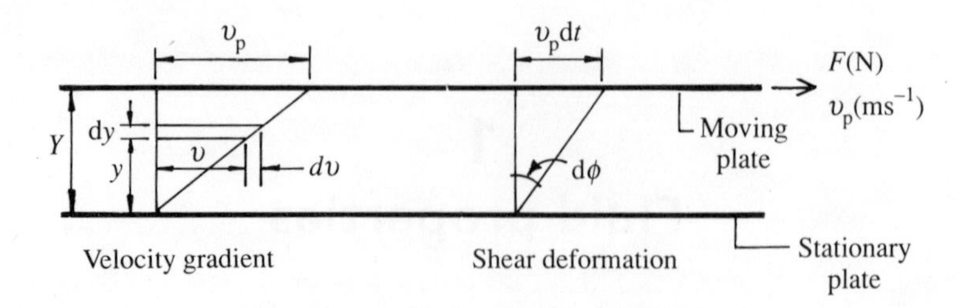

Fig. 1.1 Viscosity definition diagram.

where μ is the coefficient of **dynamic viscosity**. Equation (1.2) may be written in differential form as follows:

$$\tau = \mu \frac{dv}{dy}. \tag{1.3}$$

The units of μ are N s m^{-2}, that is, stress/velocity gradient. Thus, in fluid flow, the maintenance of a velocity gradient requires the application of a shear force.

While the concept of velocity gradient is a very useful one in the general description of fluid flow, particularly in contexts such as mixing, flow can also be represented as a rate of shear deformation, as illustrated on the right-hand side of Fig. 1.1:

$$\frac{dv}{dy} = \frac{d\phi}{dt} \tag{1.4}$$

where ϕ is the angular measure of shear deformation. Thus, the application of a shear stress to a fluid results in a **rate** of shear strain while its application to a solid causes a finite magnitude of strain or deformation.

Those fluids which exhibit the foregoing deformation behaviour are known as Newtonian fluids. They include waters, wastewaters, and gases.

The magnitude of the coefficient of dynamic viscosity μ for liquids decreases with increase in temperature. Its value for water in the temperature range 0 to 100°C is presented in Table 1.1.

The prevailing pressure has only a very minor influence on the dynamic viscosity value for water. At temperatures below 30°C the dynamic viscosity of water slightly decreases with increase in pressure, reaching a minimum value and thereafter increasing with further increase in applied pressure. This minimum disappears at temperatures above 30°C.

The dynamic viscosity of gases increases with increase in temperature. Maitland and Smith (1972) recommended the following empirical correlation

Table 1.1 Physical properties of water. (Source: CRC Handbook of Chemistry and Physics, 67th edn, 1987.)

Temperature (°C)	Density ($kg\,m^{-3}$)	Saturation vapour pressure ($N\,m^{-2} \times 10^{-3}$)	Dynamic viscosity ($N\,s\,m^{-2} \times 10^{3}$)	Surface tension ($N\,m^{-1} \times 10^{3}$)
0	999.87	0.6107	1.787	75.64
5	999.99	0.8721	1.519	74.92
10	999.73	1.2277	1.307	74.22
15	999.13	1.7049	1.139	73.49
20	998.23	2.3378	1.002	72.75
25	997.07	3.1676	0.890	71.97
30	995.68	4.2433	0.798	71.18
35	994.06	5.6237	0.719	70.37
40	992.25	7.3774	0.653	69.56
45	990.24	9.5848	0.596	68.74
50	988.07	12.3380	0.547	67.91
55	985.73	15.7450	0.504	67.05
60	983.24	19.9240	0.467	66.18
65	980.59	25.0130	0.434	65.29
70	977.81	31.1660	0.404	64.40
75	974.89	38.5530	0.378	63.50
80	971.83	47.3640	0.355	62.60
85	968.65	57.8080	0.334	61.68
90	965.34	70.1120	0.315	60.76
95	961.92	84.5280	0.298	59.84
100	958.38	101.3250	0.282	58.90

of gas viscosity with temperature for eleven common gases at low pressure (<2 atm):

$$\ln\left(\frac{\mu}{S}\right) = A \ln \Theta + \frac{B}{\Theta} + \frac{C}{\Theta^2} + D \tag{1.5}$$

where μ is the dynamic viscosity ($N\,s\,m^{-2}$) at temperature Θ (K); S is the dynamic viscosity ($N\,s\,m^{-2}$) at a standard temperature of 293.2 K; and A, B, C, and D are coefficients determined from a least-squares regression analysis. Note that Kelvin temperature K = °C + 273.1.

Recommended values for the foregoing coefficients for air, oxygen, nitrogen, methane, and carbon dioxide are presented in Table 1.2.

The linear correlation of shear stress and velocity gradient, characteristic of Newtonian fluids, prevails only in the absence of turbulence in the flow field. This type of flow environment is described as **laminar** flow and, for Newtonian fluids, is confined to situations where random bulk fluid movement is suppressed as, for example, flow in small-bore pipes or through porous media or very close to solid boundaries. Where turbulence exists in

Table 1.2 Dynamic viscosity coefficients for gases.

Gas	A	B	C	D	S ($\mathrm{N\,s\,m^{-2}} \times 10^7$)
Air	0.63404	−45.6380	380.87	−3.4500	182.0
Oxygen	0.52662	−97.5893	2650.70	−2.6892	203.2
Methane	0.54188	−127.5700	4700.80	−2.6952	109.3
Carbon dioxide	0.44037	−288.4000	19 312.00	−1.7418	146.7

the flow, however, the shear resistance is greatly increased and the associated shear stress can, for convenience, be correlated to the velocity gradient by an expression of the same form as that used to define dynamic viscosity:

$$\tau = \varepsilon \frac{dv}{dy} \tag{1.6}$$

where ε is the coefficient of **eddy viscosity** and is a characteristic of the flow, as distinct from μ which is a property of the fluid. The coefficient of eddy viscosity may be regarded as a coefficient of momentum transfer along the velocity gradient; its magnitude is dependent on the velocity gradient, shear stress, and other factors and is invariably much greater than the dynamic viscosity, μ.

Unlike water and gases, sludges typically exhibit non-Newtonian behaviour, particularly at high concentration. Such behaviour, as illustrated in Fig. 1.2, is characterized by a non-linear relation of shear stress and velocity gradient or rate of shear strain, and, in some fluids, by the existence of a yield stress

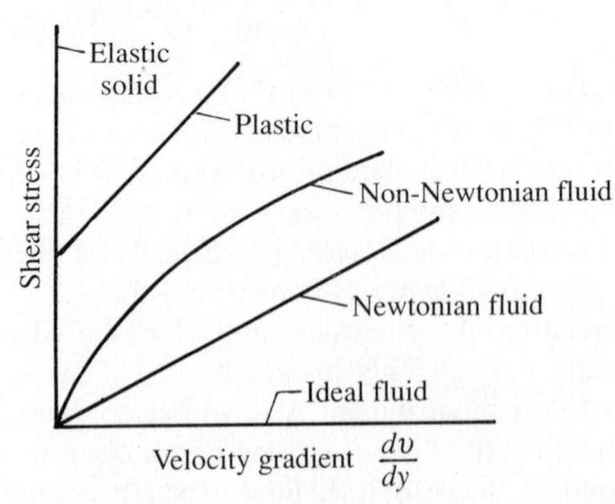

Fig. 1.2 Fluid flow classification.

which must be exceeded for flow to take place. The shear stress/rate of shear strain relation may be expressed in the form

$$\tau = \tau_y + K\left(\frac{dv}{dy}\right)^n \qquad (1.7)$$

where τ_y is the yield stress (N m^{-2}), K is a consistency coefficient, and n is a consistency index. These flow parameters are further discussed in Chapter 2. Newtonian and non-Newtonian flow behaviours are illustrated in Fig. 1.2.

The ratio of fluid viscosity to fluid density, generally known as the **kinematic viscosity**, is a frequently encountered flow parameter in hydraulic computations:

$$\nu = \frac{\mu}{\rho} \qquad (1.8)$$

where ν is the kinematic viscosity ($m^2 s^{-1}$).

1.3 Surface tension

The interfacial liquid at the boundary between a liquid and a gas behaves rather like a membrane which possesses tensile strength. This membrane-like behaviour can be quantified as a strain energy per unit area, that is, N m/m^2 or force per unit length (N m^{-1}), denoted by the symbol σ. The surface tension influence is generally very small in most fluid flow problems encountered by civil engineers. However, in certain applications such as hydraulic modelling, where the model flow depth may be very small, the surface tension influence may be of much greater relative significance in the model than in the prototype and thus distort model flow behaviour. Surface tension is also responsible for the capillary rise above the phreatic surface in fine-grained saturated soils and porous construction materials. When a liquid surface is penetrated by a solid object, surface tension causes the liquid surface in contact with the solid to be raised above the general liquid surface level in the case where the liquid 'wets' the solid surface. On the other hand where the liquid does not wet the solid surface, the liquid surface in contact with the solid is depressed.

The surface tension of water decreases with increase in temperature, as the data presented in Table 1.1 indicate.

1.4 Vapour pressure

When evaporation takes place from the surface of a liquid within an enclosed space or vessel, the partial pressure created by the vapour molecules is called vapour pressure. A liquid may, at any temperature, be considered to be in equilibrium with its own vapour when the rate of molecular transport through the separating gas–liquid interface is the same in both directions. The absolute pressure corresponding to this concentration of gas molecules is defined as the saturation vapour pressure of the liquid. The saturation vapour pressure of every liquid increases with increase in temperature. The temperature at which it reaches a value of 1 atm absolute is the boiling point, which for water is 100°C. Data on the saturation vapour pressure of water in the temperature range 0 to 100°C are presented in Table 1.1.

1.5 Thermodynamic properties

Thermodynamic properties are of particular relevance to gases. The equation of state for the so-called perfect gas is usually written in its general form as follows:

$$PV = mR_u\Theta \tag{1.9}$$

where P is the absolute pressure ($\mathrm{N\,m^{-2}}$), V is the gas volume ($\mathrm{m^3}$), m is the mass of gas (mole), R_u is the universal gas constant ($\mathrm{J\,mole^{-1}\,K^{-1}}$), and Θ is the absolute temperature (K).

The perfect gas has an R_u-value of 8.3144 $\mathrm{J\,mole^{-1}\,K^{-1}}$. The variation from this value for real gases is found to be less than 3 per cent (Daugherty and Ingersoll 1954).

Changing from mole to kg, eqn (1.9) may be written for individual gases in the form

$$\frac{P}{\rho} = R\Theta \tag{1.10}$$

where ρ is the gas density ($\mathrm{kg\,m^{-3}}$) and R is the specific gas constant ($\mathrm{J\,kg^{-1}\,K^{-1}}$), related to R_u as follows:

$$R = \frac{1000R_u}{w} \tag{1.11}$$

where w is the molecular weight.

The constant R can be shown to be the difference between the specific heat capacity of a gas at constant pressure (C_p) and its specific heat capacity at constant volume (C_v). Values for these thermodynamic properties for a number of gases are given in Table 1.3.

The relationship embodied in eqns (1.9) and (1.10) may also be expressed

Table 1.3 Thermodynamic properties of gases. (Source: CRC Handbook of Tables for Applied Engineering Science, 2nd edn, 1976.)

	(25°C and 1 atm)		
Gas	C_p (J kg^{-1} K^{-1})	C_p/C_v	R (J kg^{-1} K^{-1})
Air	1005.0	1.40	287.1
Oxygen	920.0	1.40	262.9
Nitrogen	1040.0	1.40	297.1
Methane	2260.0	1.31	534.8
Carbon dioxide	876.0	1.30	202.2

(K = °C + 273.15)

in the forms

$$PV^{\gamma} = \text{constant} \tag{1.12}$$

or

$$\frac{P}{\rho^{\gamma}} = \text{constant} \tag{1.13}$$

where V is the gas volume (m^3) and γ is the so-called polytropic exponent. The value of γ depends on the process by which the gas undergoes volume change. For adiabatic processes (zero internal energy loss), γ is equal to the specific heat ratio C_p/C_v, whereas for isothermal processes (zero temperature change), γ is equal to unity. Thus, in real situations, the value of γ lies within the range 1.0 to C_p/C_v.

1.6 Compressibility

Compressibility may be defined as the susceptibility of a material to volumetric change on the application of pressure. The coefficient of compressibility K is defined as follows:

$$K = \frac{-\Delta P}{\Delta V/V} \tag{1.14}$$

where K is the bulk modulus or coefficient of compressibility (N m^{-2}), ΔP is the change in pressure (N m^{-2}), and ΔV is the change in volume (m^3) of the original volume V (m^3).

Liquids are highly incompressible; for example, the K-value for water at

10°C is about 21.1×10^8 N m^{-2}. Its value increases marginally with pressure and temperature up to a temperature of about 50°C. Above 50°C, there is a slight decrease with increase in temperature.

Gases are relatively highly compressible, their compressibility depending on temperature and pressure. The coefficient of compressibility K for a gas is given by the relation

$$K = P\gamma \tag{1.15}$$

where γ is the polytropic gas volume exponent, as previously defined, and P is the absolute gas pressure in N m^{-2}. Thus, gas compressibility decreases linearly with increase in pressure. Fluid compressibility has a key influence on the speed of transmission of elastic waves through a fluid and is therefore an important fluid property in the analysis of unsteady flow phenomena such as waterhammer. Compressibility effects may also have to be considered in the steady flow of gases at high velocity.

1.7 Density

The density of a substance is defined as its mass per unit volume (kg m^{-3}). Density is influenced by temperature and pressure. As may be deduced from the preceding data on fluid compressibility, liquids are highly incompressible and thus exhibit negligible change in density with change in pressure. Change in liquid density with variation in temperature is also slight. Density data for water in the temperature range 0 to 100°C are presented in Table 1.1. Gases, on the other hand, are highly compressible and hence subject to significant density change with changing temperature and pressure. Equation (1.10) may be used to compute gas density as a function of temperature and pressure.

1.8 Computer program: FLUPROPS

The computer program FLUPROPS contains an interactively accessible database of fluid properties based on the numerical values presented in Tables 1.1, 1.2, and 1.3. The physical properties of water at a specified temperature are computed by linear interpolation between the relevant tabulated values. Equation (1.5) is used for the computation of gas viscosity and eqn (1.10) for the computation of gas density.

A listing of the program is presented in Appendix A. Program use is illustrated by the following program runs.

Sample program run 1: Program FLUPROPS

```
RUN
    PROGRAM FLUPROPS

    TO ACCESS WATER DATA, ENTER 1
    TO ACCESS GASES DATA, ENTER 2

    ENTER 1 OR 2, AS APPROPRIATE
? 1

ENTER THE WATER TEMPERATURE (deg C) ? 16.3

PHYSICAL PROPERTIES OF WATER AT  16.3  deg C ARE AS FOLLOWS:

DENSITY (kg/m**3) =  998.8963
DYNAMIC VISCOSITY (Ns/m**2) =  1.10338E-03
SURFACE TENSION (N/m) =  73.2976
SATURATION VAPOUR PRESSURE (N/m**2) =  1869.454

PRESS THE SPACE BAR TO CONTINUE

DO YOU WISH TO OBTAIN FURTHER DATA (Y/N)? N
Ok
```

Sample program run 2: Program FLUPROPS

```
RUN
    PROGRAM FLUPROPS

    TO ACCESS WATER DATA, ENTER 1
    TO ACCESS GASES DATA, ENTER 2

    ENTER 1 OR 2, AS APPROPRIATE
? 2
                  1.  AIR
                  2.  OXYGEN
                  3.  NITROGEN
                  4.  METHANE
                  5.  CARBON DIOXIDE

SELECT GAS BY TYPING ITS NUMBER? 2

INPUT GAS TEMPERATURE (deg C)? 16.5
INPUT GAS ABSOLUTE PRESSURE (N/m**2)? 1.03E+5

DENSITY OF OXYGEN (kg/m**3) =  1.352378
DYNAMIC VISCOSITY OF OXYGEN (Ns/m**2) =  2.01349E-05
SPECIFIC HEAT AT CONSTANT PRESSURE (J/kg.K) =  920
SPECIFIC HEAT AT CONSTANT VOLUME (J/kg.K) =  657.1
SPECIFIC GAS CONSTANT (J/kg.K) =  262.9

PRESS THE SPACE BAR TO CONTINUE

DO YOU WISH TO OBTAIN FURTHER DATA (Y/N)? N
Ok
```

References

CRC Handbook of Chemistry and Physics, 67th edn (1987). CRC Press Inc., Boca Raton, Florida.

CRC Handbook of Tables for Applied Engineering Science, 2nd edn (1976). CRC Press Inc., Boca Raton, Florida.

Daugherty, R. L. and Ingersoll, A. C. (1954). *Fluid mechanics*, (5th edn). McGraw-Hill, New York.

Maitland, G. C. and Smith, E. B. (1972). Critical re-assessment of viscosities of 11 common gases. *J. Chem. Eng. Data*, **17**, No. 2, 150–5.

2
Fluid flow

2.1 Introduction

From the macroscopic viewpoint of the engineer concerned with fluid transport it is a convenient idealization to treat fluid flow as that of a continuum, thereby neglecting the complex random motions at molecular level. Flow analysis is concerned with quantifying the flow variables throughout the flow field, as functions of time; these variables are velocity, pressure, and density. This approach may be contrasted with that adopted in solid particle mechanics, where the focus of kinematic analysis is on the motions of individual particles.

Velocity is, of course, a vector, that is, it has magnitude and direction. In a three-dimensional flow field, the components u, v, and w of the velocity vector $\mathbf{V}$, in the x-, y-, and z-directions, respectively, can be written in functional form as follows:

$$u = f_1(x, y, z, t)$$

$$v = f_2(x, y, z, t)$$

$$w = f_3(x, y, z, t).$$

These components define the value of $\mathbf{V}$ in space and time.

A **streamline** is defined as a continuous curve in the flow field that is everywhere tangential to the local velocity vector. It is thus a flow path.

2.2 Flow classification

Flow is described as **steady** at a particular location if the velocity vector at the location does not change with time; it is described as **unsteady** if the velocity vector changes with time. In mathematical terms these definitions are written as follows overleaf.

(1) steady flow

$$\left(\frac{\partial v}{\partial t}\right)_{x_0, y_0, z_0} = 0;$$

(2) unsteady flow

$$\left(\frac{\partial v}{\partial t}\right)_{x_0, y_0, z_0} \neq 0.$$

Flow is said to be **uniform** if the velocity vector is constant along the flow path or streamline. Conversely, flow is described as **non-uniform** if the velocity vector varies along the flow path. These definitions are expressed in mathematical terms as follows:

(1) uniform flow

$$\left(\frac{\partial v}{\partial s}\right)_{t_0} = 0;$$

(2) non-uniform flow

$$\left(\frac{\partial v}{\partial s}\right)_{t_0} \neq 0.$$

The most regulated of the foregoing flow types is steady uniform flow such as that which occurs in pipes of fixed diameter having a constant discharge rate.

An example of steady non-uniform flow is that which occurs upstream of a weir in a river having a steady discharge rate. Examples of unsteady non-uniform flow include estuarine flows (due to variation in channel section and time-variation in flow associated with tides) and flood flows in rivers.

Flow is described as **rotational** if fluid elements undergo a rotation about their centres of mass, and **irrotational** if no such rotation exists. Where there is a spatial velocity gradient in the flow field, as in very many real flow situations, such as in boundary layer flow, there is inevitably some degree of rotation. Flow is obviously rotational where the streamlines are curved.

Flow is described as **compressible** if the fluid undergoes a significant change in density along the flow path, and **incompressible** if there is no significant change in density. Flow of liquid is clearly incompressible flow. Flow of gases at the velocities normally encountered in sanitary engineering practice may also be regarded as incompressible. There may be significant density changes along the flow path in high-velocity gas flows; hence thermodynamic behaviour must be taken into account in analysing such flows.

When fluid flow is confined by solid boundaries, such that random lateral mixing in a direction perpendicular to that flow is suppressed, flow is described as **laminar**, that is, flowing, as it were, in separate layers with minimal lateral momentum transfer between layers. Where there is significant lateral mixing and momentum transfer in a direction normal to the flow direction, flow is classified as **turbulent**. The criteria used to define laminar and turbulent flow conditions are discussed further in Chapter 3.

2.3 Fluid acceleration

The acceleration at any point on a streamline can be expressed in terms of its tangential and normal components. The tangential component $\mathrm{d}v_s/\mathrm{d}t$ may be derived as follows:

$$\mathrm{d}v_s = \frac{\partial v_s}{\partial s}\,\mathrm{d}s + \frac{\partial v_s}{\partial t}\,\mathrm{d}t. \tag{2.1a}$$

Hence

$$\frac{\mathrm{d}v_s}{\mathrm{d}t} = \frac{\partial v_s}{\partial s}\frac{\mathrm{d}s}{\mathrm{d}t} + \frac{\partial v_s}{\partial t} \tag{2.1b}$$

or

$$\frac{\mathrm{d}v_s}{\mathrm{d}t} = v_s\frac{\partial v_s}{\partial s} + \frac{\partial v_s}{\partial t}. \tag{2.1c}$$

Equations (2.1) show the tangential acceleration to be the sum of the spatial or convective acceleration $v_s(\partial v_s/\partial s)$ and the local or temporal acceleration $\partial v_s/\partial t$.

In steady flow $\partial v_s/\partial t$ is zero and hence the steady flow acceleration is

$$\frac{\mathrm{d}v_s}{\mathrm{d}t} = v_s\frac{\delta v_s}{\delta s}. \tag{2.2}$$

The normal or centripetal acceleration $\mathrm{d}v_n/\mathrm{d}t$ of a fluid element moving along a curved path or streamline can similarly be written as the sum of convective and temporal components:

$$\frac{\mathrm{d}v_n}{\mathrm{d}t} = \frac{v_s^2}{R} + \frac{\partial v_n}{\partial t} \tag{2.3}$$

where R is the local radius of curvature of the streamline and v_n is the normal velocity.

2.4 Streamtube and control volume

The concepts of streamtube and control volume are widely used in fluid flow analysis. A streamtube is an elemental flow volume, the end areas of which are normal to the local flow directions and the peripheral surface of which is generated by streamlines. Flow into and out of the streamtube is through its end areas only; there is no flow normal to the peripheral surface since it is generated by streamlines and thus acts as a virtual boundary. The end areas are sufficiently small in extent that any variation in velocity over the cross-section may be neglected.

A bundle of adjacent streamtubes constitutes a control volume. A control volume has the same general characteristics as a streamtube except that there may be a variation in velocity over its end areas. These concepts are illustrated in Fig. 2.1.

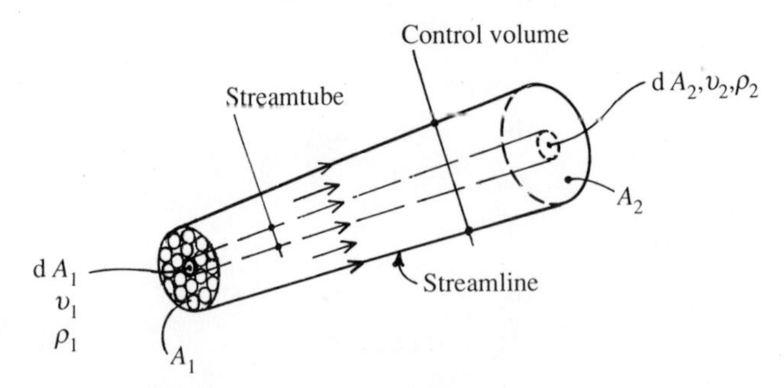

Fig. 2.1 Streamtube and control volume.

2.5 The continuity principle

The concepts of streamtube and control volume facilitate the application of the principle of mass conservation, or the continuity principle as it is known in fluid flow analysis. For example, under steady flow conditions, the mass of fluid contained within a streamtube or control volume does not change with time, hence the rate of mass flow out of such defined zones must equal its rate of inflow:

(1) streamtube

$$\rho_1 \, \mathrm{d}A_1 \, v_1 = \rho_2 \, \mathrm{d}A_2 \, v_2;$$

(2) control volume

$$\bar{\rho}_1 A_1 \bar{v}_1 = \bar{\rho}_2 A_2 \bar{v}_2;$$

where $\bar{\rho}$ and $\bar{v}$ represent the averaged values of these parameters. The mass conservation principle will be applied repeatedly in later chapters to develop the appropriate form of the continuity equation for the problem in hand. In unsteady compressible flow, for example, the fluid mass within the streamtube or control volume varies with time.

2.6 The momentum principle

Newton's second law relates force to the rate of change of momentum:

$$F = \frac{\mathrm{d}}{\mathrm{d}t}(mv).$$

Consider the application of this principle to steady flow through the streamtube illustrated in Fig. 2.2.

At time zero the streamtube contains a mass of fluid between end areas AA and BB. In the following time interval dt this mass moves to the space between A′A′ and B′B′.

$$\text{Initial momentum} = \sum_{\mathrm{BB}}^{\mathrm{AA}} \mathrm{d}m\, v$$

$$\text{Final momentum} = \sum_{\mathrm{B'B'}}^{\mathrm{A'A'}} \mathrm{d}m\, v.$$

Since the flow is steady there is no change in momentum at any point within the streamtubes, that is, the momentum of the fluid in the space A′A′BB, which is common to both streamtubes, remains unaltered. Thus the change

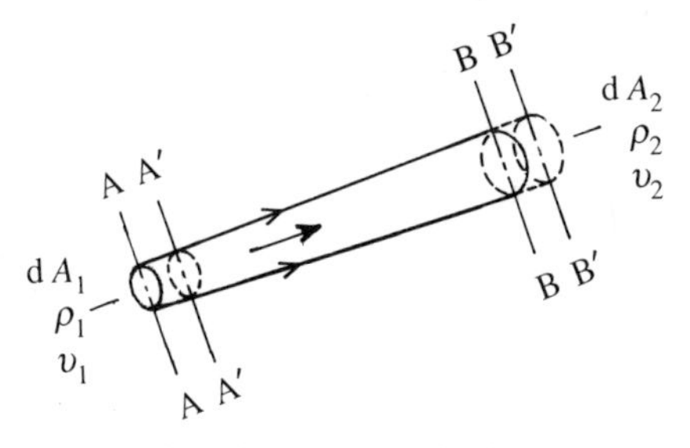

Fig. 2.2 Streamtube flow.

in momentum in the time $\mathrm{d}t$ can be written

$$\text{Momentum change} = \sum_{\mathrm{BB}}^{\mathrm{B'B'}} \mathrm{d}m\, v - \sum_{\mathrm{A'A'}}^{\mathrm{AA}} \mathrm{d}m\, v.$$

Written in terms of ρ, $\mathrm{d}A$, and v, this becomes

$$\text{Momentum change} = (\rho_2\, \mathrm{d}A_2\, v_2\, \mathrm{d}t)v_2 - (\rho_1\, \mathrm{d}A_2\, v_1\, \mathrm{d}t)v_1.$$

The corresponding rate of change of momentum yields the magnitude of the applied force F:

$$F = \rho_2\, \mathrm{d}A_2\, v_2^2 - \rho_1\, \mathrm{d}A_1\, v_1^2 \tag{2.4}$$

The term $\rho_2\, \mathrm{d}A_2\, v_2^2$ represents the rate of outflow of momentum from the streamtube, while $\rho_1\, \mathrm{d}A_1\, v_1^2$ is its rate of inflow. Thus, the applied force corresponds to the difference in momentum flux across the streamtube end areas. It should be noted that this force is the net force applied to the fluid mass within the streamtube by the surrounding bulk fluid.

In unsteady flow, as discussed in Chapters 6 and 10, the change in momentum of the fluid mass throughout the streamtube volume must also be taken into account.

Equation (2.4) may also be applied to a control volume:

$$F = \sum_{A_2} \rho_2\, \mathrm{d}A_2\, v_2^2 - \sum_{A_1} \rho_1\, \mathrm{d}A_1\, v_1^2.$$

Written in terms of the mean velocity $\bar{v}$:

$$F = \beta_2 \rho_2 A_2 \bar{v}_2^2 - \beta_1 \rho_1 A_1 \bar{v}_1^2 \tag{2.5}$$

where β is the momentum correction factor (sometimes called the Boussinesq coefficient), which allows for the use of the mean velocity in the application of the momentum principle to control volumes. Its value is obtained as follows:

$$\beta \rho A \bar{v}^2 = \sum_{A} \rho\, \mathrm{d}A\, v^2;$$

hence

$$\beta = \frac{1}{A} \sum \left(\frac{v}{\bar{v}}\right)^2 \mathrm{d}A. \tag{2.6}$$

In turbulent pipe flow, β is generally less than 1.1; in laminar pipe flow, β is 1.33.

2.7 The energy principle

Consider the idealized flow of an elemental fluid mass along a streamline as depicted in Fig. 2.3.

Applying Newton's second law to this elemental mass:

$$p\,\mathrm{d}A - (p + \mathrm{d}p)\,\mathrm{d}A - \rho g\,\mathrm{d}A\,\mathrm{d}s\cos\theta = \rho\,\mathrm{d}A\,\mathrm{d}s\frac{\mathrm{d}v}{\mathrm{d}t}. \tag{2.7}$$

Since $\mathrm{d}s\cos\theta = \mathrm{d}z$ and $\mathrm{d}v/\mathrm{d}t = v(\mathrm{d}v/\mathrm{d}s)$ in steady flow, eqn (2.7) may be written as follows:

$$\frac{\mathrm{d}p}{\rho} + g\,\mathrm{d}z + v\,\mathrm{d}v = 0. \tag{2.8}$$

This is the **Euler equation**; it relates to steady irrotational flow of a frictionless fluid along a streamline.

Integration of the Euler equation along a streamline yields

$$\int\frac{\mathrm{d}p}{\rho} + gz + \frac{v^2}{2} = \text{constant}. \tag{2.9}$$

If the flow is incompressible, that is, ρ is constant and independent of p, eqn (2.9) becomes

$$\frac{p}{\rho} + gz + \frac{v^2}{2} = \text{constant}, \tag{2.10}$$

which is the **Bernoulli equation**; it relates to steady, irrotational, incompressible flow of a frictionless fluid along a streamline.

When related to liquid flows, the Bernoulli equation is usually written in the form

$$\frac{p}{\rho g} + \frac{v^2}{2g} + z = \text{constant}. \tag{2.11}$$

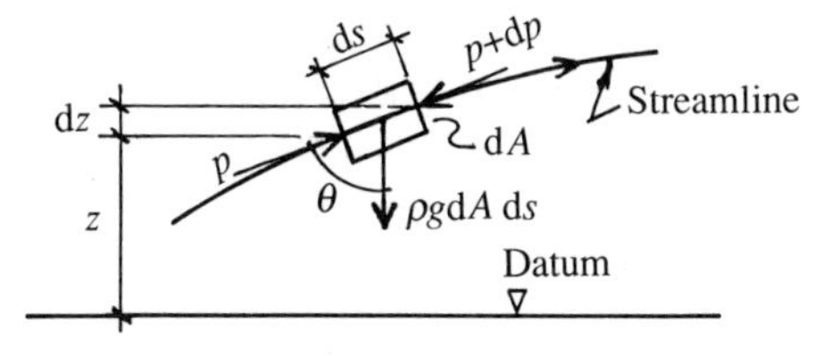

Fig. 2.3 Forces acting on an elemental fluid mass.

Each term in eqn (2.11) has units of length (m) or 'head'. Their sum represents the total head relative to a datum level defined by z. In dealing with incompressible flow the pressure term is conveniently taken as the gauge pressure.

When flow is compressible, integration of the pressure/density term in eqn (2.9) requires reference to the equation of state for gases:

$$\frac{p}{\rho^{\gamma}} = \text{constant}; \tag{1.13}$$

hence

$$\int \frac{\mathrm{d}p}{\rho} = \frac{\gamma}{\gamma - 1} \frac{p}{\rho}$$

or

$$\int \frac{\mathrm{d}p}{\rho} = \frac{\gamma}{\gamma - 1} R\Theta \tag{2.12}$$

which, from the definitions of R and γ (given in Section 1.5), can be written as follows:

$$\int \frac{\mathrm{d}p}{\rho} = C_{\mathrm{p}}\Theta. \tag{2.13}$$

Thus, the integrated form of the Euler equation for steady compressible flow along a streamline becomes

$$C_{\mathrm{p}}\Theta + gz + \frac{v^2}{2} = \text{constant}. \tag{2.14}$$

This is known as the **energy equation** and can also be written in the form

$$C_{\mathrm{v}}\Theta + \frac{p}{\rho} + gz + \frac{v^2}{2} = \text{constant} \tag{2.15}$$

since $R = C_{\mathrm{p}} - C_{\mathrm{v}}$ and $p/\rho = R\Theta$, where P is the absolute pressure. Each term in the energy equations has units of energy per unit mass (J kg^{-1}).

The foregoing Euler, Bernoulli, and energy equations do not take into account the energy loss associated with the flow of all real fluids. Thus, in practice, the sum of the terms on the left-hand side of the Bernoulli and energy equations is not constant but decreases in the downstream direction along a streamline.

When dealing with practical flow situations it is convenient to use the mean velocity $\bar{v}$ and a kinetic energy factor α, where

$$\int \rho \, \mathrm{d}A \, v \cdot v^2 = \alpha \rho A \bar{v} \cdot \bar{v}^2;$$

hence

$$\alpha = \frac{1}{A} \int \left(\frac{v}{\bar{v}}\right)^3 \mathrm{d}A. \tag{2.16}$$

The value of α lies between 1.03 and 1.3 in turbulent flow and has the value of 2.0 in laminar flow.

2.8 Applications of continuity, energy, and momentum principle

2.8.1 Incompressible flow

Many examples of the application of the continuity, energy, and momentum principles are presented in later sections of this book. In this section, the illustrative examples are confined to steady flow problems.

The discharge rate through differential head flow devices such as the Venturi meter and orifice plate meters is derive from application of the Bernoulli and continuity equations. Assuming incompressible flow and neglecting losses, the Bernoulli equation may be applied to the central streamline flow of the Venturi meter illustrated in Fig. 2.4:

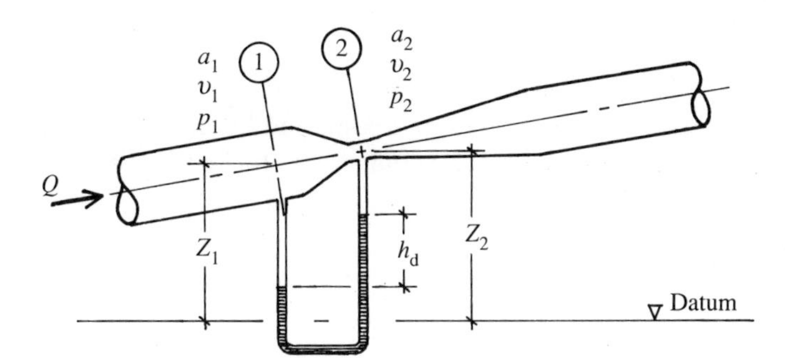

Fig. 2.4 Venturi meter.

$$\frac{v_1^2}{2g} + \frac{p_1}{\rho g} + z_1 = \frac{v_2^2}{2g} + \frac{p_2}{\rho g} + z_2. \tag{2.17}$$

Similarly, the continuity equation may be applied to the flow through sections 1 and 2:

$$Q = a_1 v_1 = a_2 v_2. \tag{2.18}$$

Combining eqns (2.17) and (2.18), the discharge Q can be expressed as a function of the upstream and throat cross-sections and the differential head:

$$Q = a_2 \left[\left(\frac{2g}{1 - (a_2/a_1)^2} \right) \left(\frac{p_1 - p_2}{\rho g} + (z_1 - z_2) \right) \right]^{1/2}. \tag{2.19}$$

Also

$$\frac{p_1 - p_2}{\rho g} + (z_1 - z_2) = h_\mathrm{d} \left(\frac{\rho_\mathrm{m}}{\rho} - 1 \right) \tag{2.20}$$

where h_d is the differential head of manometer fluid, ρ_m is the manometer fluid density, and ρ is the flowing fluid density. Because of flow losses between the upstream and throat pressure tappings, the measured differential head will exceed the theoretical value. This is taken into account in practical applications by introducing a discharge coefficient C_d. The practical discharge equation thus becomes:

$$Q = C_\mathrm{d} a_2 \left[\frac{2gh_\mathrm{d}\left(\frac{\rho_\mathrm{m}}{\rho} - 1 \right)}{(1 - a_2/a_1)^2} \right]^{1/2} \tag{2.21}$$

The value of C_d lies between 0.96 and 0.99 for Venturi meters and in the range 0.60–0.63 for orifice plates. The diameter ratio d_2/d_1 is typically in the range 0.3 to 0.7.

The Pitot tube, illustrated in Fig. 2.5, is a flow measuring device which senses the kinetic streamline head at a point and hence can be used for flow velocity traversing. Writing the Bernoulli terms for the upstream section (1) and the stagnation point section (2):

$$\frac{p_1}{\rho g} + \frac{v_1^2}{2g} = \frac{p_2}{\rho g} + 0.$$

Hence

$$v_1 = \sqrt{2g\left(\frac{p_2 - p_1}{\rho g} \right)} \tag{2.22}$$

or

$$v_1 = \sqrt{2gh}. \tag{2.23}$$

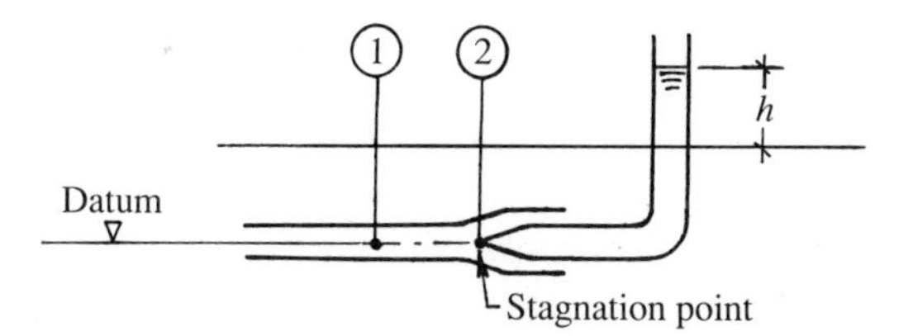

Fig. 2.5 Pitot tube.

The velocity given by the foregoing theoretical derivations is slightly too large and needs to be modified by a correction factor C_v which has a value typically in the range 0.95–1.0. The Bernoulli and continuity equations can also be applied to evaluate the velocity of discharge through a submerged orifice, as in Fig. 2.6:

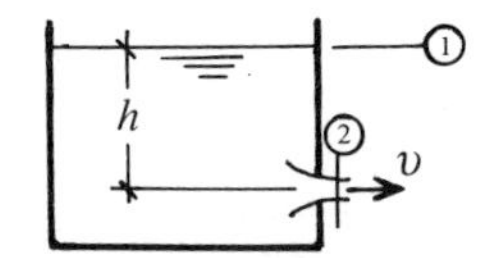

Fig. 2.6 Submerged orifice.

(1) Bernoulli

$$\frac{p_1}{\rho g} + \frac{v_1^2}{2g} + h = \frac{p_2}{\rho g} + \frac{v_2^2}{2g} + 0; \tag{2.24}$$

(2) continuity

$$a_1 v_1 = a_2 v_2. \tag{2.25}$$

Combining eqns (2.24) and (2.25) and assuming that p_1 and p_2 are both equal to atmospheric pressure:

$$v_2 = \left[\frac{2gh}{1 - (a_2/a_1)^2}\right]^{1/2}. \tag{2.26}$$

The actual velocity is found to be somewhat less than the theoretical value. The cross-sectional area of the discharge jet may be effectively equal to the orifice area for a bellmouth orifice, reducing to about 0.6 of the orifice area for a sharp-edged orifice. Assuming that the velocity of approach is negligible, the discharge through a submerged orifice may be written in practical form as follows:

$$Q_0 = C_d a_0 \sqrt{2gh} \tag{2.27}$$

where Q_0 is the **orifice discharge**, a_0 is the orifice area, and C_d is a discharge coefficient, the value of which generally lies in the range 0.50–0.98, depending on orifice geometry, as shown in Fig. 2.7 (Featherstone and Nalluri 1988).

Computation of the forces exerted on pipe fittings such as the taper, illustrated in Fig. 2.8, or the bend, illustrated in Fig. 2.9, requires simultaneous application of the continuity, energy, and momentum principles. Applied to the vertical taper:

(1) continuity

$$Q = v_1 a_1 = v_2 a_2; \tag{2.28}$$

(2) energy

$$\frac{p_1}{\rho g} + \frac{v_1^2}{2g} + 0 = \frac{p_2}{\rho g} + \frac{v_2^2}{2g} + h; \tag{2.29}$$

(3) momentum

$$p_1 a_1 - p_2 a_2 - W + F_T = \rho Q(v_2 - v_1); \tag{2.30}$$

where F_T is the force exerted on the water mass in the taper (positive upward) and W is the weight of water within the taper. Thus, the force applied to the

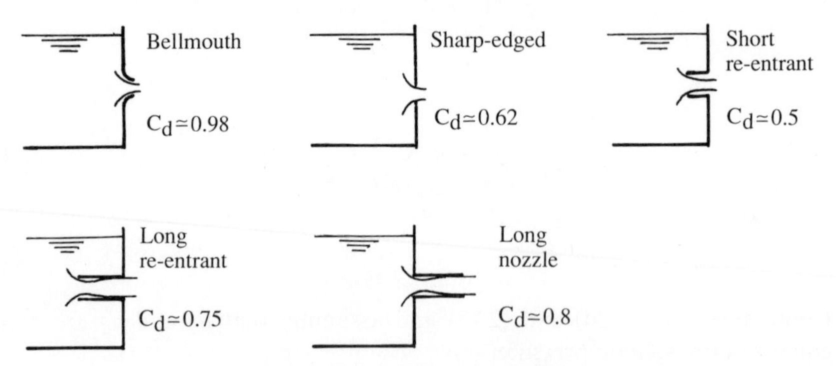

Fig. 2.7 Orifice discharge coefficients.

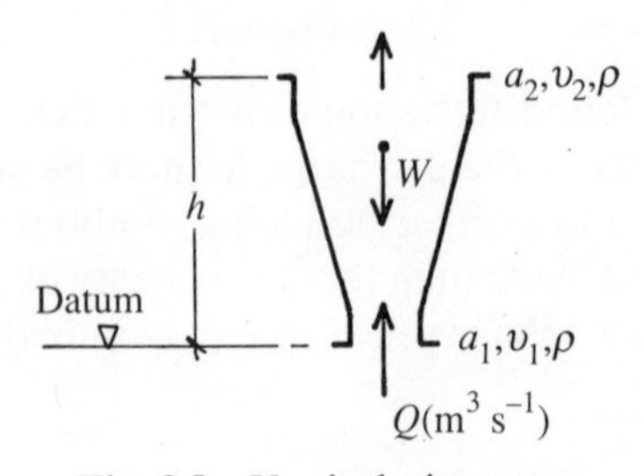

Fig. 2.8 Vertical pipe taper.

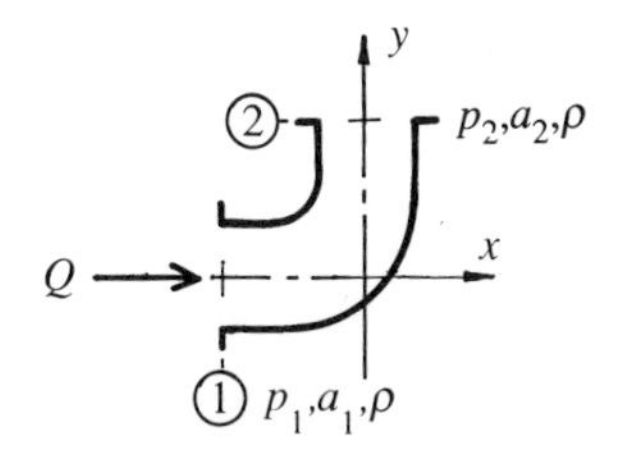

Fig. 2.9 Horizontal pipe bend.

taper, which is the unknown quantity generally sought by the designer, is $-F_T$. When Q and the dimensions of the taper are specified, the magnitude of F can be computed from the foregoing continuity, energy, and momentum equations.

The force exerted by steady flow through a horizontal pipe bend, as illustrated in Fig. 2.9, is also found from application of the continuity, energy, and momentum equations:

(1) continuity

$$Q = v_1 a_1 = v_2 a_2; \tag{2.31}$$

(2) energy

$$\frac{p_1}{\rho g} + \frac{v_1^2}{2g} = \frac{p_2}{\rho g} + \frac{v_2^2}{2g}; \tag{2.32}$$

(3) momentum, x-direction

$$p_1 a_1 + F_x = -\rho Q v_1; \tag{2.33}$$

(4) momentum, y-direction

$$-p_2 a_2 + F_y = \rho Q v_2 \tag{2.34}$$

where F_x and F_y are the forces exerted on the fluid in the x- and y-directions, respectively, and $-F_x$ and $-F_y$ are the corresponding forces exerted by the flow on the bend.

2.8.2 Compressible flow

Compressible flow is marked by variations in fluid density along the flow path. The circumstances in which the related thermodynamic consequences must be taken into account can be examined as follows. The maximum temperature rise along a streamline occurs when the velocity is reduced to zero, that is, at a stagnation point. Applying the energy eqn (2.15) to this flow situation:

$$C_p\Theta + \frac{v^2}{2} = C_p(\Theta + \Delta\Theta)$$

upstream section stagnation point

where $\Delta\Theta$ is the temperature rise. This relationship can be written in the following form:

$$\frac{\Delta\Theta}{\Theta} = \frac{v^2}{2C_p\Theta}. \tag{2.35}$$

For practical design purposes it may be assumed that, where the potential incremental temperature change along the flow path is less than 1 per cent, flow may be regarded as incompressible. Applying this criterion to eqn (2.35), the **limiting value** of v is found to be:

$$v = \sqrt{0.02C_p\Theta} \tag{2.36}$$

giving a limit value for air at 10°C of 75.4 m s^{-1}.

Where, however, thermodynamic changes are significant the flow must be treated as compressible flow. Consider, for example, the steady discharge of a gas from a pipe or reservoir through an orifice or nozzle, as illustrated in Fig. 2.10. The flow is defined by the following correlations:

(1) continuity

$$m = \rho_1 a_1 v_1 = \rho_2 a_2 v_2; \tag{2.37}$$

(2) energy

$$\frac{\gamma}{\gamma - 1}\frac{P_1}{\rho_1} + \frac{v_1^2}{2} = \frac{\gamma}{\gamma - 1}\frac{P_2}{\rho_2} + \frac{v_2^2}{2}; \tag{2.38}$$

(3) pressure/density

$$\frac{P_1}{\rho_1^\gamma} = \frac{P_2}{\rho_2^\gamma}. \tag{2.39}$$

These correlations assume idealized flow with zero energy loss. Using these equations, the mass discharge rate m (kg s^{-1}) can be expressed in terms of known parameter values, P_1, ρ_1, and P_2, and the sectional areas a_1 and a_2:

$$m = a_2\left[\left(\frac{P_1\rho_1}{(P_2/P_1)^{-2/\gamma} - (a_2/a_1)^2}\right)\left(\frac{2\gamma}{\gamma - 1}\right)\left(1 - (P_2/P_1)^{(\gamma-1)/\gamma}\right)\right]^{1/2}. \tag{2.40}$$

The foregoing expression is valid for subsonic flow conditions, that is, up to the pressure ratio at which the velocity at the orifice or nozzle throat reaches

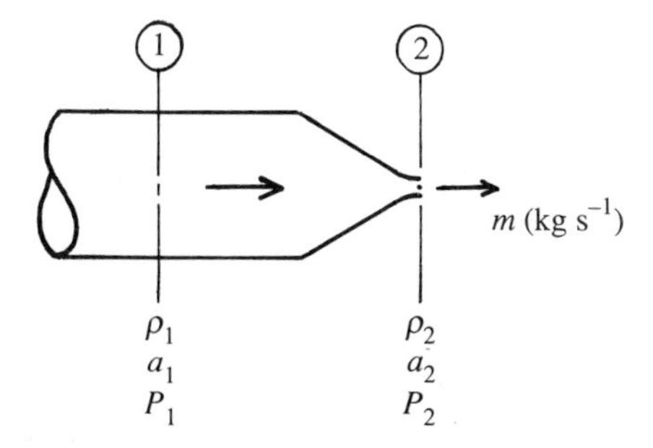

Fig. 2.10 Gas discharge through an orifice.

the sonic value. The sonic velocity α is the velocity of transmission of a weak pressure wave through the gas. Its magnitude (Benedict 1983) is

$$\alpha = \sqrt{\gamma P_2/\rho_2}. \tag{2.41}$$

When the critical pressure ratio is exceeded, flow control changes from a dependency on the pressure ratio P_2/P_1 to a dependency on the upstream pressure P_1, the upstream flow remains subsonic, and the velocity through the throat remains at sonic value. The critical pressure ratio may be found by applying the energy equation to the upstream and throat sections at the onset of sonic flow at the throat section:

$$\left(\frac{\gamma}{\gamma-1}\right)\frac{P_1}{\rho_1}+\frac{v_1^2}{2}=\left(\frac{\gamma}{\gamma-1}\right)\frac{P_2}{\rho_2}+\frac{\alpha^2}{2}; \tag{2.42}$$

Combining eqns (2.41), (2.42), and (2.39) and neglecting v_1 (this is equivalent to assuming p_1 as the upstream stagnation pressure), the **critical pressure ratio** is found to be

$$\frac{P_2}{P_1}=\left(\frac{2}{\gamma+1}\right)^{\gamma/(\gamma-1)}. \tag{2.43}$$

The critical pressure ratio for air ($\gamma = 1.4$) is found from eqn (2.43) to be 0.528.

When the pressure ratio exceeds the above limiting value, the rate of discharge through the orifice or nozzle is found by inserting the limiting value of P_2/P_1 in the discharge eqn (2.40). By omitting the area ratio term, a_2/a_1, the resulting discharge expression is found to be

$$m = a_2\left[P_1\rho_1\gamma\left(\frac{2}{\gamma+1}\right)^{(\gamma+1)/(\gamma-1)}\right]^{1/2}. \tag{2.44}$$

Equations (2.40) and (2.44) are theoretical expressions for the mass discharge rate of a compressible fluid through an orifice or nozzle. In practical computation these expressions are modified by the introduction of a discharge coefficient which allows for variations from ideal behaviour and in particular for flow contraction effects associated with the geometry of the orifice or nozzle.

The admission of air to a pipeline through an air valve consequent on a drop in pipeline pressure below atmospheric pressure provides a practical example of the foregoing compressible flow behaviour from the water engineering field, as illustrated in Fig. 2.11. In this instance the external pressure (P_1) remains constant at atmospheric pressure. The inflow of air through the valve orifice increases with the decrease in the internal pipeline pressure (P_2) in accordance with flow eqn (2.40) until the critical pressure ratio (eqn (2.43)) is reached. Therefore, any further drop in internal pressure does not cause a corresponding increase in the rate of inflow since the latter now depends only on P_1, that is, atmospheric pressure, which is constant. An orifice operating under such conditions is sometimes described, for obvious reasons, as being 'choked'.

2.9 Resistance to fluid flow

The resistance to fluid flow arises primarily from the drag forces exerted on flowing fluids by the solid boundary surfaces of flow conduits. This drag results from the fact that there is zero slippage or relative movement at the contact interface between a flowing fluid and a solid surface, resulting in high shear rates in the adjacent boundary fluid layer. This shear deformation is manifested as a spatial velocity gradient in a direction normal to the boundary surface, decreasing in magnitude with distance from the boundary.

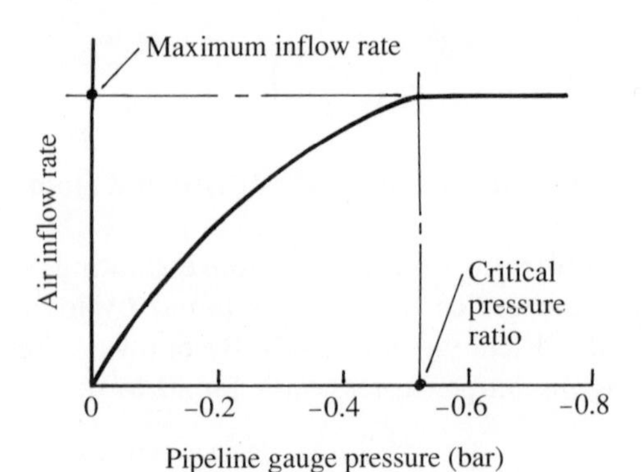

Fig. 2.11 Air entry rate for an air valve.

The existence of a velocity gradient implies a causative shear stress, which is essential to maintain flow and which is a measure of the resistance to flow. Where the flow is laminar, that is, where there is no turbulence in the flow, the local shear stress/velocity gradient ratio is a constant. This constant is by definition the fluid viscosity μ.

Where, however, flow conditions are turbulent, as is generally the case in civil engineering hydraulics, the correlation of shear stress and velocity gradient becomes more complex, being a flow property rather than a fluid property. The nature of turbulent flow and flow resistance in turbulent boundary layers is discussed in Chapter 3.

References

Benedict, R. P. (1983). *Fundamentals of Gas Dynamics.* John Wiley, New York.

Featherstone, R. E. and Nalluri, C. (1988). *Civil Engineering Hydraulics*, (2nd edn). Edward Arnold, London.

Related reading

Francis, J. R. D. and Minton, P. (1984). *Civil Engineering Hydraulics*, (5th edn). Edward Arnold, London.

Shames, I. H. (1983). *Mechanics of Fluids*, (2nd edn). McGraw-Hill, New York.

Streeter, V. L. (1966). *Fluid Mechanics*, (4th edn). McGraw-Hill, New York.

3
Steady flow in pipes

3.1 Introduction

Pipes are the most frequently used conduits for the conveyance of fluids, both gases and liquids. They are produced in a variety of materials, including steel, cast iron, ductile iron, concrete, asbestos cement, plastics, glass, and non-ferrous metals. In their new condition, the finished internal wall surfaces of these materials vary considerably in roughness from the very smooth glass or plastic surface to the relatively rough concrete surface. Also, depending on the fluid transported and the pipe material, the condition of the pipe wall may vary with time, either due to corrosion, as in steel pipes, or deposition, as in hard water areas (see Fig. 3.5).

The fluids of interest in the present context are water, wastewater, sewage sludges, and gases such as air, oxygen, and biogas. As will be seen later, the flow of water, wastewater, and gases in pipes is invariably turbulent. The flow of sludges, however, may well be laminar at practical design velocity values. It is self-evident that fluid density and viscosity are key fluid properties in pipe flow analysis, both obviously having an influence on the power input required to induce flow.

The design engineer is primarily interested in being able to predict accurately the discharge capacity of pipe systems. To do this, it is essential to know the relation between head and loss and mean flow velocity.

3.2 Categorization of pipe flow by Reynolds number

Turbulence in fluid flow is characterized by random local motions which transfer momentum and dissipate energy. This random motion increases with increase in the mean velocity and is suppressed by solid boundaries. Reynolds (1885) carried out extensive pipe flow tests from which he was able to define the flow regime as being either laminar, transitional, or turbulent. The non-dimensional flow index which he developed is known as the Reynolds number R_e, which for pipes, is defined as

$$R_e = \frac{v\,d\rho}{\mu} \tag{3.1}$$

where d is the pipe diameter and v is the mean flow velocity. The ranges of values of R_e are as follows:

(1) laminar flow: $R_e \leq 2300$;

(2) transitional flow: $2300 \leq R_e \leq 4000$;

(3) turbulent flow: $R_e \geq 4000$.

3.3 Hydraulic and energy grade lines

The flow of real fluids through pipes results in a loss of energy or head along the direction of flow. Referring to Fig. 3.1, the Bernoulli equation can be applied as follows:

$$\frac{p_1}{\rho g} + \frac{\alpha v_1^2}{2g} + z_1 = \frac{p^2}{\rho g} + \frac{\alpha v_2^2}{2g} + z_2 + h_f \tag{3.2}$$

where h_f is the head loss over the pipe length L and α is the kinetic energy factor, as defined in Chapter 2 (eqn (2.16)). There is always an energy loss associated with flow. It is graphically represented as a gradient in piezometric head, that is, an hydraulic grade line (HGL), or a gradient in energy or 'total head', that is, an energy grade line (EGL). In steady uniform flow, as depicted in Fig. 3.1, these lines are parallel. The slope of the energy line or friction slope S_f is

$$S_f = \frac{h_f}{L}. \tag{3.3}$$

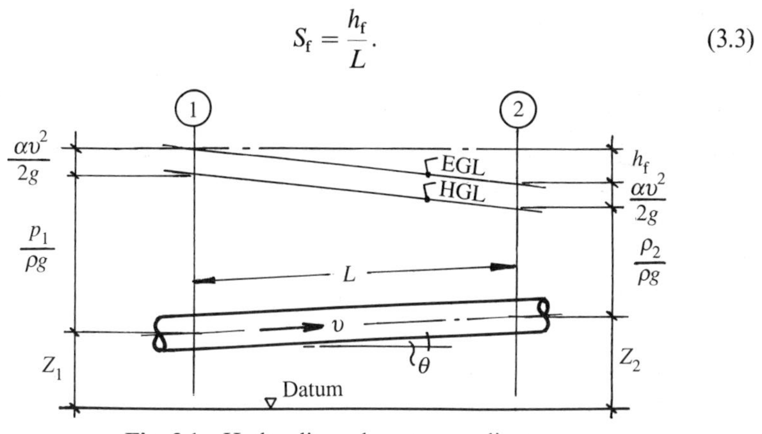

Fig. 3.1 Hydraulic and energy gradients.

3.4 Shear stress distribution

The radial variation of shear stress, under conditions of steady uniform flow, is derived from consideration of the forces acting on the flowing fluid. Since there is no acceleration, the net force acting on the flowing mass of fluid must be zero. The forces acting on the fluid mass between sections 1 and 2 on Fig. 3.1 are as follows:

$$p_1 a - \rho g a L \sin\theta - p_2 a - \tau_0 P L = 0 \tag{3.4}$$

where a is the pipe cross-sectional area, τ_0 is the wall shear stress, and P is the section perimeter length. Dividing by $\rho g a$ and rearranging:

$$\frac{p_1 - p_2}{\rho g} + (z_1 - z_2) = \frac{\tau_0 P L}{\rho g a}.$$

As may be seen from Fig. 3.1, the left-hand side of this equation is equal to the head loss h_f. Therefore

$$h_f = \frac{\tau_0 P L}{\rho g a}.$$

Hence

$$\tau_0 = \rho g \frac{a}{P}\frac{h_f}{L}$$

or

$$\tau_0 = \rho g R_h S_f \tag{3.5}$$

where R_h is the hydraulic radius, that is, the ratio of flow area to perimeter length, and S_f is the friction slope. Thus, in steady uniform flow, the fluid shear stress at the wall is linearly related to the friction slope, which is a readily measureable flow parameter.

The foregoing analysis may also be applied to any concentric cylindrical volume of fluid of smaller diameter than that of the pipe, to give the local fluid shear stress τ_y:

$$\tau_y = \rho g R_y S_f \tag{3.6}$$

where τ_y is the shear stress at a distance y from the pipe wall and R_y is the corresponding hydraulic radius. Thus, the shear stress in pipe flow varies linearly from a maximum value at the pipe wall to zero at its centre.

3.5 Laminar pipe flow

Of primary interest in the hydraulic design of pipe systems is the relationship between carrying capacity and head loss or friction slope. Under laminar flow conditions, the spatial variation in velocity is governed by the fluid viscosity μ and applied shear stress τ:

$$\tau = \mu \frac{\mathrm{d}v}{\mathrm{d}y}.$$

Combining this relationship with eqn (3.6):

$$\rho g \frac{(D-2y)}{4} S_\mathrm{f} = \mu \frac{\mathrm{d}v}{\mathrm{d}y}. \tag{3.7}$$

Equation (3.7) can be integrated subject to the boundary conditions that $v = 0$ at $y = 0$, to give a parabolic velocity distribution for laminar flow (illustrated in Fig. 3.2):

$$v_y = \frac{\rho g S_\mathrm{f}}{4\mu}(Dy - y^2). \tag{3.8}$$

The maximum velocity at the pipe axis is

$$v_{\mathrm{max}} = \frac{\rho g S_\mathrm{f} D^2}{16\mu}. \tag{3.9}$$

The mean velocity v is found by integration over the flow cross-section:

$$v = \frac{\int_0^{D/2} v_y \pi(D-2y)\,\mathrm{d}y}{\pi D^2/4} \tag{3.10}$$

giving the result:

$$\text{Laminar flow:} \qquad v = \frac{\rho g S_\mathrm{f} D^2}{32\mu}. \tag{3.11}$$

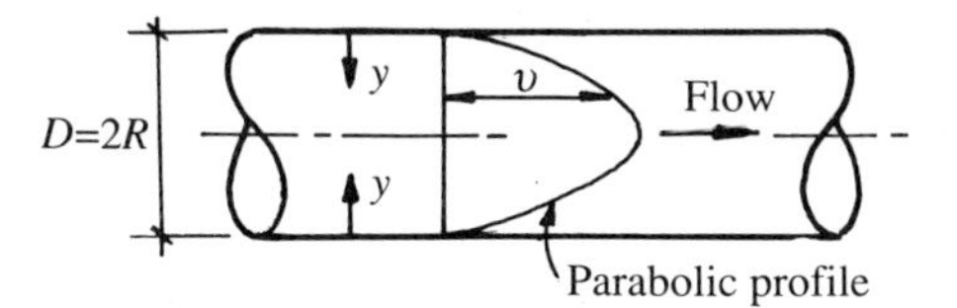

Fig. 3.2 Laminar velocity distribution.

3.6 Turbulent flow in pipes

The random component in turbulent flow renders exact mathematical analysis impossible. However, through a combination of experiment and theoretical reasoning, the magnitude of the resistance to flow of Newtonian fluids under turbulent conditions in pipes has been modelled in mathematical terms, allowing the reliable prediction of head loss for a very wide range of flow and conduit surface conditions. The research works of Nikuradse (1932, 1933), Prandtl (1933), von Karman (1930), and Colebrook and White (1938), among others, have contributed greatly to this development.

As in laminar flow, the starting point is the velocity distribution over the flow cross-section, which may be expressed in the following form as proposed by Prandtl:

$$\tau = \rho L^2 \left(\frac{\mathrm{d}v}{\mathrm{d}y}\right)^2 \tag{3.12}$$

where L is the so-called 'mixing length', which is not a physical dimension of the system but may be thought of as a measure of the random displacement of fluid elements characteristic of turbulent flow. The value of L has been found to be proportional to y, the distance from the flow boundary:

$$L = Ky \tag{3.13}$$

where K is a numerical constant having a value of approximately 0.4. On insertion of this value for K, eqn (3.12) may be rewritten in the form

$$\frac{\mathrm{d}v}{\mathrm{d}y} = \sqrt{\frac{\tau}{\rho}}\,\frac{2.5}{y}. \tag{3.14}$$

It is assumed that the shear stress τ is constant over the flow cross-section under turbulent flow conditions. The term $\sqrt{\tau/\rho}$ has the dimensions of velocity and is sometimes known as the 'shear velocity', denoted by v_*. The velocity distribution over the pipe cross-section is found by integration of eqn (3.14):

$$v_y = 2.5 v_* \ln y + \text{constant}. \tag{3.15}$$

This logarithmic velocity distribution clearly cannot be valid at the pipe wall, since $\ln y$ has an infinite negative value when y is zero. We may, however, assume that eqn (3.15) is valid down to very small values of y, that is, very close to the pipe wall. This condition is satisfied by defining a wall distance y_1, at which the velocity has a zero value. Using this boundary condition, eqn (3.15) becomes

$$v_y = 2.5 v_* \ln\left(\frac{y}{y_1}\right). \tag{3.16}$$

The value of y_1, which may be regarded as effectively defining a new hydraulic boundary inside the actual physical boundary, is determined by flow conditions at the wall.

The turbulent velocity distribution, as represented by eqn (3.16), is thus a logarithmic one in which the velocity magnitude varies from a maximum at the centre to a zero value at the virtual boundary, as shown in Fig. 3.3. The mean velocity v is found by integrating the velocity distribution over the cross-section:

$$v = \int_0^{R-y_1} \frac{2\pi r v_y \, \mathrm{d}r}{\pi R^2},$$

which, on substitution of the right-hand side of eqn (3.16) for v_y (noting that $y = R - r$) and integration, becomes

$$v = 2.5v_*\left(\ln\frac{R}{y_1} - 1.5 + 2\frac{y_1}{R} - \frac{y_1^2}{2R^2}\right).$$

Neglecting terms in y_1/R, which are very small, this expression simplifies to

$$v = 2.5v_* \ln\frac{0.112D}{y_1}. \tag{3.17}$$

Thus, the mean velocity is numerically equal to the local velocity at $y = 0.112D$.

Where the pipe wall is smooth, as, for example, with glass, plastics, and similar surfaces, the flow adjacent to the wall is laminar and fluid drag is exerted on the boundary surface solely by viscous shear. Under such conditions, the magnitude of y_1 is governed by wall shear stress and fluid viscosity and its value has been experimentally found to be

$$y_1 = \frac{0.1\nu}{v_*}. \tag{3.18}$$

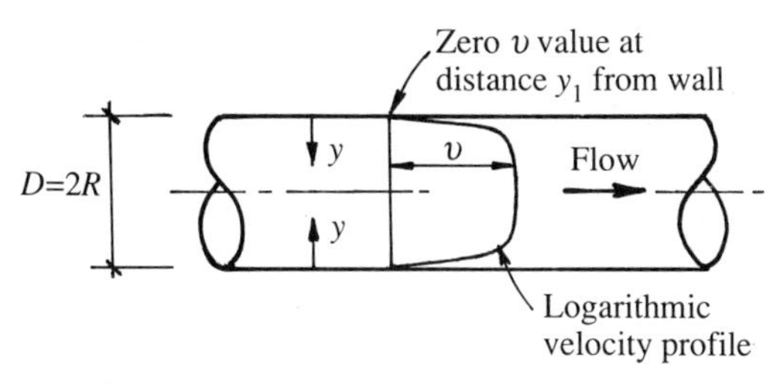

Fig. 3.3 Turbulent velocity distribution.

Insertion of this value for y_1 in eqn (3.17) gives the following value for the mean velocity in **smooth turbulent flow**:

$$\text{Smooth:} \qquad v = 2.5v_* \ln\left(\frac{1.12v_* D}{\nu}\right). \tag{3.19}$$

The roughness of the internal surfaces of pipes is measured in terms of the 'equivalent sand roughness' k (m). This measure of roughness follows from the work of Nikuradse (1932, 1933) who used layers of uniform size sand, glued to the internal surface of his experimental pipes, to provide well-defined rough surfaces. Viscosity is found to have a negligible influence on flow when the wall roughness is such that the ratio $k/(\nu/v_*)$ exceeds about 60. Flow under such conditions is described as rough turbulent flow, for which the value of y_1 has been experimentally determined as $k/33$. Inserting this value for y_1 in eqn (3.17) gives the following value for the average velocity in **rough turbulent flow**:

$$\text{Rough:} \qquad v = 2.5v_* \ln\left(\frac{3.7D}{k}\right). \tag{3.20}$$

When the ratio $k/(\nu/v_*)$ is less than 60 and greater than 3, the flow is categorized as being in the **transition** region between smooth turbulent flow and rough turbulent flow. Flow of water in commercial pipes at conventional velocities is typically in this zone. It is clear that fluid viscosity and wall roughness both influence flow resistance in this transition region between smooth and rough turbulent flow. Colebrook (1939) proposed that the effective wall displacement in transition flow be taken as the sum of the wall displacements for smooth and rough flow:

$$y_1 = \frac{0.1\nu}{v_*} + \frac{k}{33}. \tag{3.21}$$

Insertion of this value for y_1 in eqn (3.17) gives the following expression for the mean velocity in **transition turbulent flow**:

$$\text{Transition:} \qquad v = -2.5v_* \ln\left(\frac{\nu}{1.12v_* D} + \frac{k}{3.7D}\right). \tag{3.22}$$

It is clear that the transition expression can be applied over the full regime of turbulent pipe flow. When the pipe roughness is very small, the transition expression approaches the smooth law and, likewise, when the wall displacement associated with the laminar sublayer is small compared with that due to wall roughness, the transition expression approaches the rough law.

3.7 Practical pipe flow computation

3.7.1 The Darcy–Weisbach and Colebrook–White equations

Although the foregoing correlations between mean velocity v and shear velocity v_* may be used directly in pipe flow computation, they are not commonly used in engineering practice. Instead, the findings are incorporated in the Darcy–Weisbach (Darcy 1858, Weisbach 1842) equation, which has the form

$$S_f = \frac{fv^2}{2gD}. \tag{3.23}$$

This empirical formula, which predated the development of the foregoing turbulent pipe flow analysis, has the computational advantage that it incorporates a non-dimensional friction factor or resistance coefficient f. The Darcy–Weisbach equation can be used for all pipe flow categories by treating f as a flow variable, using the previously developed flow resistance equations to determine its value. The advantage of non-dimensionality is retained by expressing viscosity in terms of Reynolds number and pipe roughness k in terms of relative roughness k/D.

The value of f under laminar flow conditions is found by combining eqns (3.11) and (3.23):

$$\text{Laminar flow:} \qquad f = \frac{64}{R_e}. \tag{3.24}$$

As already noted, the transition eqn (3.22) can be used to model flow resistance over the full range of turbulent flow. The corresponding expression for the friction factor f is found by combining eqns (3.22) and (3.23) and using the relation

$$v_* = \sqrt{\frac{\tau}{\rho}} = \sqrt{\frac{gDS_f}{4}}$$

resulting in the following expression which is generally known as the Colebrook–White (1937) equation:

$$\text{Turbulent flow:} \qquad \frac{1}{\sqrt{f}} = -0.88 \ln\left(\frac{k}{3.7D} + \frac{2.5}{R_e\sqrt{f}}\right) \tag{3.25}$$

or

$$\frac{1}{\sqrt{f}} = -2.0 \log\left(\frac{k}{3.7D} + \frac{2.5}{R_e\sqrt{f}}\right). \tag{3.25a}$$

Thus, the friction factor f is a function of Reynolds number and pipe relative roughness. This functional relationship is illustrated graphically in Fig. 3.4, often known as the Moody diagram.

Equations (3.24) and (3.25) together cover the entire spectrum of pipe flow conditions. Pipe flow computation typically involves the calculation of head loss when the velocity and other relevant parameters are known, or the calculation of velocity when the head loss and other relevant parameters are known. Direct computation of f for turbulent flow conditions, using eqn (3.25) is not feasible because of the non-explicit form of the equation. An iterative method of solution must therefore be used. The computer program FRICTF, presented in Section 3.10, uses the interval-halving iterative procedure (see Appendix B) to calculate f.

Direct computation of velocity is, however, feasible, using the relation

$$f = \frac{2gDS_f}{v^2}.$$

Insertion of this value for f in eqns (3.25) and (3.25a) gives the following explicit expressions for the velocity v:

$$\text{ln function:} \quad v = -0.88\sqrt{2gDS_f}\,\ln\left(\frac{k}{3.7D} + \frac{2.5\nu}{D\sqrt{2gDS_f}}\right) \tag{3.26}$$

$$\text{log function:} \quad v = -2.0\sqrt{2gDS_f}\,\log\left(\frac{k}{3.7D} + \frac{2.5\nu}{D\sqrt{2gDS_f}}\right). \tag{3.26a}$$

With the exception of the conveyance of sewage sludge in pipes, which is dealt with in Section 3.8, fluid flow in sanitary engineering is invariably turbulent. The Colebrook–White equation provides the most soundly based correlation of head loss and mean flow velocity in pipes and its adoption is therefore recommended in preference to empirical exponential equations. A number of explicit exponential approximations of the Colebrook–White equation are to be found in the literature (Barr 1975). These may be used where access to a computer is not available.

3.7.2 Design values for pipe roughness

The pipes used for conveying waters and wastewaters are made from a variety of materials with surface finishes varying from the very smooth to the moderately rough. The effective surface roughness or k-value in the 'as-new' condition can be readily evaluated from Table 3.1. The prediction of roughness increase with age is much more problematic (Colebrook and White 1938; Perkins and Gardiner 1982). Chemically unstable water or

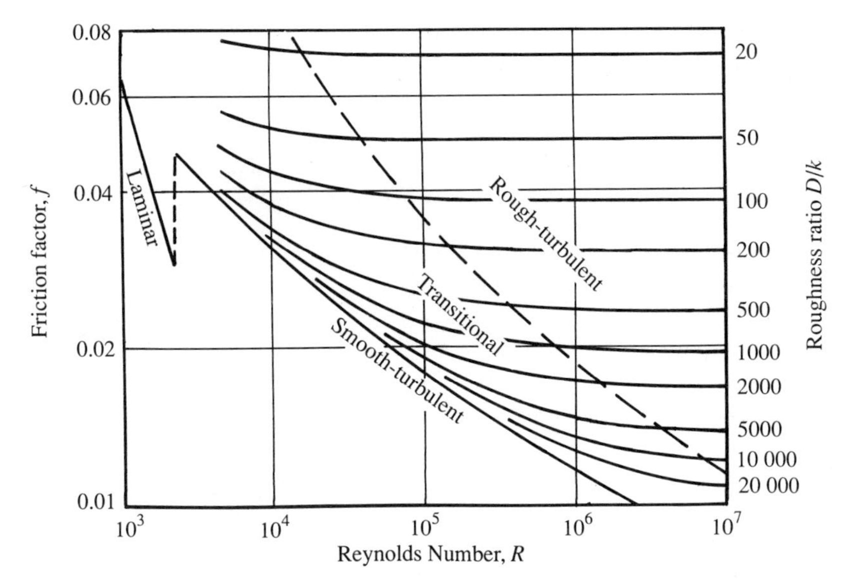

Fig. 3.4 Correlation of friction factor with Reynolds number and relative roughness.

wastewater may cause corrosion of metal pipes resulting in surface tuberculation or may give rise to scale deposition which may not only cause increased surface roughness over a period of time but may also significantly reduce the effective pipe bore. Figure 3.5 illustrates both of these phenomena. Waters, such as sewage which contain biodegradable organics, give rise to the growth of a biological slime layer on the inner surfaces of the pipes in which they are conveyed. This is a more serious problem in part-filled conduits (discussed in Chapter 7) than in pipes flowing full. The degree of sliming is inversely related to the flow velocity. Table 3.1 gives recommended equivalent k-values for a range of flow velocities in pipes flowing full.

3.7.3 Other pipe flow equations

The Hazen–Williams formula, published in New York in 1920, is still widely used by water supply engineers. In its original FPS units, it was expressed in the form:

$$v = CR_{\mathrm{h}}^{0.63}S_{\mathrm{f}}^{0.54} \tag{3.27}$$

where C is the Hazen–Williams coefficient. C is a dimensional coefficient and its numerical value should therefore change on conversion to metric units. As this would be confusing for users, the C-value is treated as a pipe

Table 3.1 Recommended values for the surface roughness parameter k (Hydraulics Research, Wallingford, 1990).

Classification (assumed clean and new unless otherwise stated)	Suitable values of k (mm)		
	Good	Normal	Poor
Smooth materials			
Drawn non-ferrous pipes of aluminium, brass, copper, lead, etc. and non-metallic pipes of Alkathene, glass, perspex, etc.	–	0.003	–
Asbestos–cement	0.015	0.03	–
Metal			
Spun bitumen or concrete lined	–	0.03	–
Wrought iron	0.03	0.06	0.15
Rusy wrought iron	0.15	0.6	3.0
Uncoated steel	0.015	0.03	0.06
Coated steel	0.03	0.06	0.15
Galvanized iron, coated cast iron	0.06	0.15	0.3
Uncoated cast iron	0.15	0.3	0.6
Tate relined pipes	0.15	0.3	0.6
Old tuberculated water mains with the following degrees of attack:			
Slight	0.6	1.5	3.0
Moderate	1.5	3.0	6.0
Appreciable	6.0	15	30
Severe	15	30	60
(Good: Up to 20 years use; Normal: 40–50 years use; Poor: 80–100 years use)			
Wood			
Wood stave pipes, planned plank conduits	0.3	0.6	1.5
Concrete			
Precast concrete pipes with 'O' ring joints	0.06	0.15	0.6
Spun precast concrete pipes with 'O' ring joints	0.06	0.15	0.3
Monolithic construction against steel forms	0.3	0.6	1.5
Monolithic construction against rough forms	0.6	1.5	–
Clayware			
Glazed or unglazed pipe:			
With sleeve joints	0.03	0.06	0.15
With spigot and socket joints and 'O' ring seals: diameter <150 mm	–	0.03	–
With spigot and socket joints and 'O' ring seals: diameter >150 mm	–	0.06	–
Pitch fibre (lower value refers to full bore flow)	0.003	0.03	–
Glass fibre	–	0.06	–

(*continued*)

Table 3.1 (*continued*)

Classification (assumed clean and new unless otherwise stated)	Suitable values of k (mm)		
	Good	Normal	Poor
uPVC			
With chemically cemented joints	–	0.03	–
With spigot and socket joints, 'O' ring seals at 6–9 metre intervals	–	0.06	–
Brickwork			
Glazed	0.6	1.5	3.0
Well pointed	1.5	3.0	6.0
Old, in need to pointing	–	15	30
Sewer rising mains: all materials, operating as follows			
Mean velocity 1 m s^{-1}	0.15	0.3	0.6
Mean velocity 1.5 m s^{-1}	0.06	0.15	0.30
Mean velocity 2 m s^{-1}	0.03	0.06	0.15
Unlined rock tunnels			
Granite and other homogeneous rocks	60	150	300
Diagonally bedded slates	–	300	600
(values to be used with design diameter)			

constant; conversion to SI units is achieved by introducing a numerical conversion coefficient. The SI form of the Hazen–Williams formula may be written in the form:

$$v = 0.355CD^{0.63}S_f^{0.54} \tag{3.28}$$

Typical design values for the pipe coefficient C are given in Table 3.2.

The range of applicability of the Hazen–Williams formula may be examined by correlation of the Hazen–Williams C-value and the friction factor f. This correlation is found to be

$$f = \frac{13.62gv^{-0.148}}{C^{1.852}D^{0.167}} \tag{3.29}$$

or, expressed in terms of the Reynolds number,

$$f = 13.62gC^{-1.852}R_e^{-0.148}D^{-0.019}\nu^{-0.148} \tag{3.30}$$

which, for given values of C, D, and ν may be written in the form

$$\log f = \log C_* - 0.148 \log R_e \tag{3.31}$$

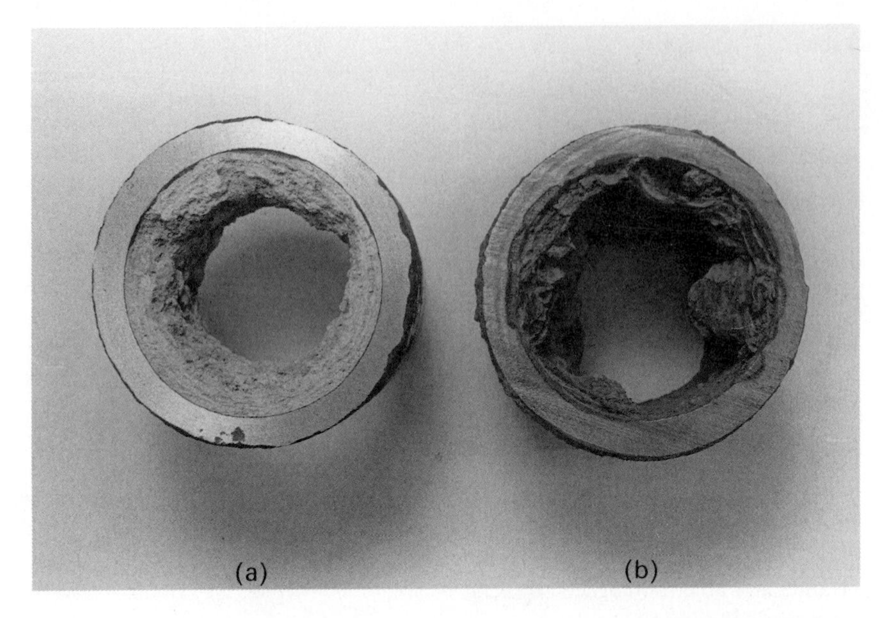

Fig. 3.5 (a) Example of calcium carbonate deposition in an old cast iron water main (courtesy of Dr P. O'Connor); (b) example of severe surface tuberculation in an old cast iron water main (courtesy of P. O'Reilly).

where C_* is a constant. This correlation of f and the Reynolds number would plot as a straight line of negative slope in Fig. 3.4, showing that the Hazen–Williams formula is valid in the transition turbulent flow zone. It should therefore not be applied in the rough turbulent flow region, that is, when the estimated C-value is less than about 100.

The **Manning** formula (1891, 1895) is widely applied to flow in partly filled conduits. It is generally expressed in the form

$$v = \frac{1}{n} R_{\mathrm{h}}^{0.67} S_{\mathrm{f}}^{0.5}. \tag{3.32}$$

Table 3.2 Hazen–Williams C-values for new pipes.

Type of pipe	C-value range
Smooth: plastics, glass, copper, lead, asbestos-cement	135–150
Uncoated cast iron	125–130
Coated metal pipes	135–140
Prestressed concrete ($D > 0.5$ m)	135–145

Table 3.3 Values of Manning's n for various types of conduit surface.

Surface	Manning n value
Smooth metal	0.010
Smooth concrete	0.012
Rough concrete	0.017
Cut earthen channel	0.025–0.035

Expressed in terms of the pipe diameter D, this becomes

$$v = \frac{0.397}{n} D^{0.67} S_f^{0.5}. \tag{3.33}$$

Typical values for the Manning n roughness coefficient are given in Table 3.3.

The region of applicability of the Manning equation may be examined through its correlation with the friction factor f. This correlation is found from eqns (3.23) and (3.33):

$$f = 12.7 g n^2 D^{-0.33} \tag{3.34}$$

Thus, for given values of D and n, f has a constant value, indicating that the Manning equation is valid for rough turbulent flow and hence should not be used for flow computations relating to flow in the smooth and transitional regions.

3.8 Flow of sewage sludge in pipes

The extent to which the head loss associated with sludge flow in pipes exceeds that for water flow is dependent on both the concentration and nature of the suspended solids. The primary rheological parameter affected by the presence of suspended solids is the fluid viscosity, while the change in density is of lesser significance. Both parameters increase with increase in suspended solids. The flow of sludge in pipes may fall in the laminar, transitional, or turbulent flow categories. While laminar flow is not encountered in the practical design flow range used in water distribution, the flow of sludge in pipes is frequently in the laminar range.

The linear correlation of shear stress and shear rate, characteristic of Newtonian fluids, does not apply to sewage sludges at suspended solids concentrations above certain threshold levels. The viscosity of sludge is difficult to measure because of the problem of solids separation and typical thixotropic behaviour, that is, a decrease in viscosity following previous agitation. For many such highly viscous fluids and suspensions, the relation of shear stress and shear rate is non-linear and may be represented as follows:

$$\tau = \tau_y + K\left(\frac{dv}{dy}\right)^n \qquad (3.35)$$

where τ_y is the yield stress, K is a consistency coefficient (N s^n m^{-2}), and n is a non-dimensional consistency index. Equation (3.35) is generally known as the Herschel–Bulkley model of fluid flow and is also referred to as the generalized Bingham model of fluid flow. The Herschel–Bulkley model simplifies to the Bingham model when $n = 1$ and to the so-called power law model when the yield stress $\tau_y = 0$. The values of τ_y, K, and n depend on the sludge type and the sludge concentration. Recommended guideline values for various types of sewage sludge, based on data reported by Frost (1983), are given in Table 3.4.

3.8.1 Laminar sludge flow in pipes

The flow of sludge in pipes can be classified in laminar, transitional, or turbulent categories using a modified Reynolds number criterion, which for a fluid having power law flow characteristics, has the form (Frost 1982):

$$R_e = \frac{\rho v D}{K\{(3n+1)/4n\}^n (8v/D)^{n-1}}. \qquad (3.36)$$

The term $8v/D$ is a measure of the wall shear stress (it represents the velocity gradient at the wall in Newtonian fluid flow). The flow is categorized by Reynolds number as follows:

(1) laminar flow: $R_e < 2300$;

(2) transitional flow: $2300 < R_e < 4000$;

(3) turbulent flow: $R_e > 4000$.

Table 3.4 Guidelines values for the rheological parameters K, n, and τ_y. C is the sludge solids concentration (kg m^{-3}).

Sludge type	K	n	τ_y
Primary	$5.0 \times 10^{-5}C^{2.82}$	$0.79C^{-0.17}$	$1.3 \times 10^{-4}C^{2.72}$
Activated	$9.0 \times 10^{-5}C^{3.00}$	$1.70C^{-0.45}$	$1.3 \times 10^{-4}C^{3.00}$
Anaerobically digested	$6.0 \times 10^{-6}C^{3.50}$	$0.90C^{-0.24}$	$1.4 \times 10^{-5}C^{3.37}$
Humus	$2.0 \times 10^{-5}C^{3.00}$	$1.90C^{-0.45}$	$1.6 \times 10^{-5}C^{3.00}$

As in all fluid flow, the wall shear stress τ_w is related to the friction slope S_f as follows:

$$\tau_w = \rho g R_h S_f \tag{3.37}$$

where R_h is the hydraulic radius. In laminar flow, the shear stress varies linearly from its maximum value at the pipe wall to a zero value at the pipe centre:

$$\tau_r = \tau_w \frac{2r}{D}. \tag{3.38}$$

Combining eqns (3.35), (3.37), and (3.38), and noting that $\mathrm{d}v/\mathrm{d}y = -\mathrm{d}v/\mathrm{d}r$:

$$-\frac{\mathrm{d}v}{\mathrm{d}r} = \left[\frac{0.5\rho g r S_f - \tau_y}{K}\right]^{1/n}. \tag{3.39}$$

Clearly, a velocity gradient will exist only where the imposed shear stress $\rho g r S_f/2$ is greater than the yield stress τ_y. Where this is not the case, that is, near the centre of the pipe, there is a region of plug flow. The discharge Q is found by integrating the velocity distribution over the flow cross-section:

$$Q = \int_0^R 2\pi r v_r \,\mathrm{d}r \tag{3.40}$$

where v_r is a function of r, as defined by eqn (3.39). Integration of eqn (3.40) yields (Frost 1982)

$$Q = \frac{\pi D^3}{8}\left(\frac{n}{3n+1}\right)\left(\frac{\tau_w - \tau_y}{K}\right)^{1/n}\left\{1 - \frac{\tau_y/\tau_w}{2n+1}\left[1 + \frac{2n}{n+1}\left(\frac{\tau_y}{\tau_w}\right)\left(1 + \frac{n\tau_y}{\tau_w}\right)\right]\right\} \tag{3.41}$$

The wall shear stress τ_w and hence the friction slope S_f, for a given discharge rate Q, can be found by solution of eqn (3.41). This requires knowledge of the flow parameters K, τ_y, and n, guidelines values for which are given in Table 3.4. The computer program, SSFLO, a listing of which is given at the end of this chapter, uses an interval-halving procedure to solve eqn (3.41) for τ_w.

3.8.2 Turbulent sludge flow in pipes

It has been observed that the head loss in turbulent flow of sludges in pipes can be reliably related to the corresponding head loss for clean water at the same velocity and temperature. The relation is expressed (Frost 1982) in the form of head loss ratio (HLR) factors as follows, where C is the sludge solids concentration (kg m^{-3}):

(1) primary sludge: HLR = 1.5;

(2) activated sludge: HLR = $0.88 + 0.024C$ for $C > 5$ kg m^{-3};

(3) anaerobically digested sludge: HLR = $0.80 + 0.016C$ for $C > 15$ kg m^{-3};

(4) humus sludge: HLR = $0.84 + 0.020C$ for $C > 10$ kg m^{-3}.

It must be emphasized that the foregoing methods of computation only provide a desk estimate of the head loss due to sludge flow in pipes, which can be used for design purposes in the absence of field measurements of the sludge rheological parameters. It is important to note that some sludges may have significantly higher apparent viscosities than those determined by the above methods of computation. These include activated sludges which have been mechanically thickened by centrifuge or dissolved air flotation processes, and also gravity-thickened activated sludge to which polyelectrolyte has been added.

A measurement procedure for the experimental determination of the sludge rheological parameters K, τ_y, and n is presented by Frost (1983).

3.9 Head loss in pipe fittings

The total head loss in pipe flow comprises the distributed energy loss over straight pipe lengths plus the local losses at bends, tees, valves, and so on. These local losses may constitute the major part of the total flow resistance in the interconnecting pipework in water and wastewater treatment plants and in the pipework within pumping stations. Poor joint alignment and internal projections associated with welding or gaskets may also contribute significantly to the overall resistance to flow.

The head loss in fittings is conveniently expressed in terms of the equivalent length of straight pipe or in terms of the velocity head $v^2/2g$. In the latter form, it is expressed as follows:

$$h = K\frac{v^2}{2g} \tag{3.42}$$

where h is the head loss (m), v is the mean pipe velocity (m s^{-1}), and K is a numerical coefficient. The overall head loss for a pipe of length L (m) and diameter D (m) can therefore be expressed as follows:

$$h = \sum K\frac{v^2}{2g} + \frac{fLv^2}{2gD} \tag{3.43}$$

or

$$h = \frac{v^2}{2g}\left(\sum K + \frac{fL}{D}\right) \qquad (3.43a)$$

where the summation relates to the K-values for all local losses in the pipe system.

3.9.1 Head losses in valves

A variety of valve types is used in water and wastewater engineering practice. They include gate or sluice valves, butterfly valves, float valves, non-return valves, diaphragm valves, ball valves, and pressure-reducing valves. The head loss in flow through these devices depends on the operational position of the device element regulating flow, which may vary from the fully open to the fully closed position. Head loss is also influenced by the detailed design of the device, which varies from one manufacturer to another. Typical K-values for gate valves, butterfly valves, and float valves, over their full operational range (fully open to fully closed) are given in Table 3.5.

3.9.2 Other pipe fittings

In this context, the designation pipe fitting includes pipe tapers, entries, bends, and T-junctions. Recommended K-values for these fittings are

Table 3.5 Typical K-values for valves. (These K-values are for use in eqn (3.42), where v is the computed velocity based on a fully open valve.)

% Open	Butterfly	Gate	Float
100	0.3	0.1	4.2
90	0.5	0.2	4.8
80	0.9	0.4	5.5
70	2.5	0.8	6.6
60	6.3	1.7	8.5
50	14.5	3.3	11.8
40	32.6	5.8	19.0
30	80.0	10.0	41.0
20	220.0	23.0	171.0
10	1000.0	80.0	2500.0
0	(valve closed, zero flow)		

presented in Table 3.6. It should be noted that where there is a velocity change in flow through a fitting, the fitting K-value must be related to a defined velocity. For example, at a T-junction:

1 | 2 3

$$h_{13} = \left(\frac{v_1^2}{2g} + h_1\right) - \left(\frac{v_3^2}{2g} + h_3\right) = K\left(\frac{v_3^2}{2g}\right)$$

where h_{13} is the branch flow head loss in flow from 1 to 3.

3.9.3 Head loss in flow of sludge through fittings

The hydraulic resistance to the flow of sludge through pipe fittings, such as bends, tees, and so on, can be correlated with pipe velocity in the same manner as outlined for clean water:

$$h = K_s \frac{v^2}{2g}$$

where K_s is the sludge head loss coefficient, which can be related (Frost 1982, 1983) to the corresponding K-value for clean water (Table 3.6):

$$K_s = K\left(1 + \frac{2000}{R_e}\right) \tag{3.44}$$

where R_e is the sludge Reynolds number, as defined by eqn (3.36). Thus, at low Reynolds numbers, that is, under laminar flow conditions, the hydraulic resistance to flow through pipework fittings considerably exceeds that for clean water.

3.10 Computer programs: FRICTF, PIPFLO, SSFLO

This chapter includes three programs relating to the material covered. They are FRICTF, PIPFLO, and SSFLO, which are concerned with friction factor computation, steady water flow in pipes, and flow of sludges in pipes, respectively.

FRICTF

Program FRICTF computes the friction factor f from supplied input data, using eqns (3.24) or (3.25), as appropriate. The latter equation is solved iteratively by the method of interval-halving (described in Appendix B). The program outputs the computed values for the friction factor and Reynolds

Table 3.6 Head losses in pipe fittings: K-factors. Head loss (m) $= K(v^2/2g)$, where v is the mean pipeline velocity (m s^{-1}).

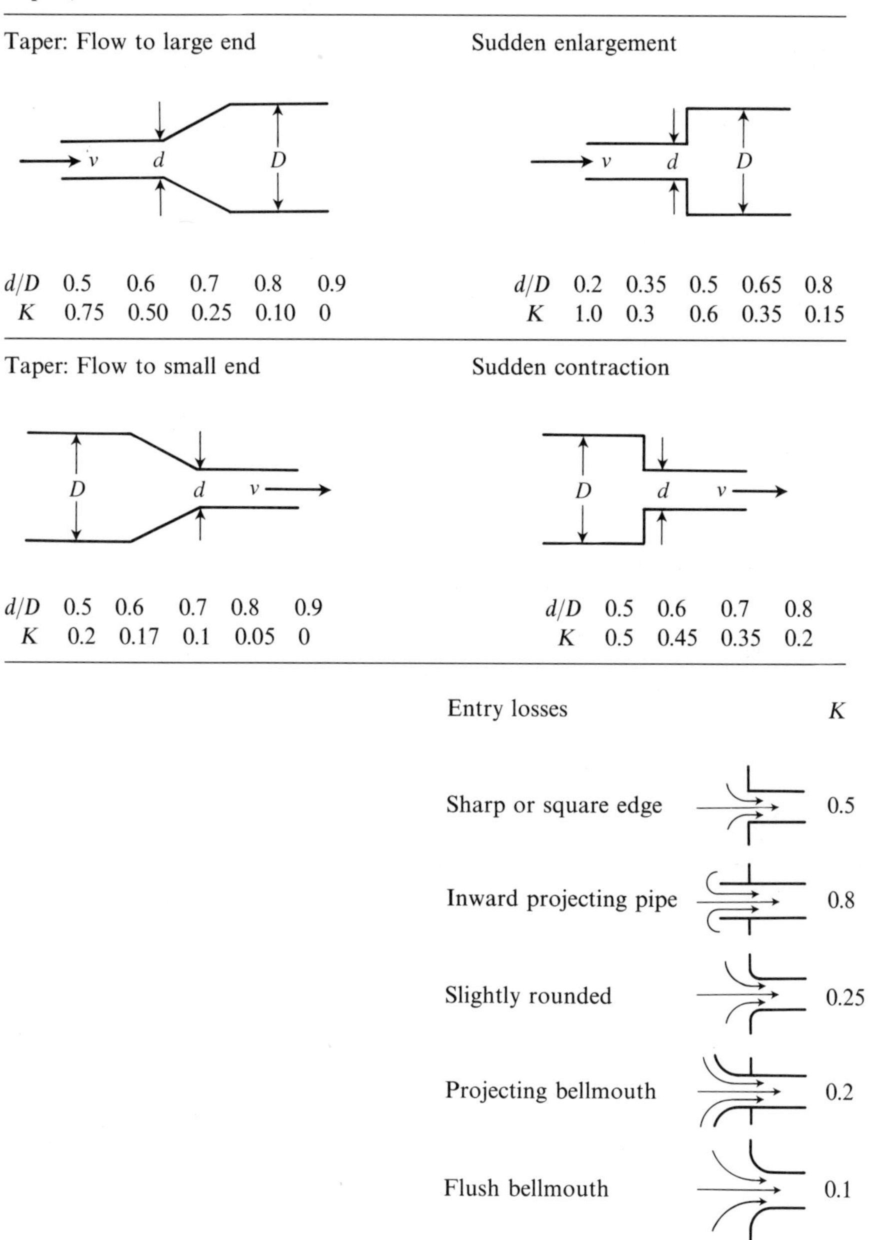

Tapers, entries

Taper: Flow to large end

d/D	0.5	0.6	0.7	0.8	0.9
K	0.75	0.50	0.25	0.10	0

Sudden enlargement

d/D	0.2	0.35	0.5	0.65	0.8
K	1.0	0.3	0.6	0.35	0.15

Taper: Flow to small end

d/D	0.5	0.6	0.7	0.8	0.9
K	0.2	0.17	0.1	0.05	0

Sudden contraction

d/D	0.5	0.6	0.7	0.8
K	0.5	0.45	0.35	0.2

Entry losses	K
Sharp or square edge	0.5
Inward projecting pipe	0.8
Slightly rounded	0.25
Projecting bellmouth	0.2
Flush bellmouth	0.1

Table 3.6 (*continued*)

Bends and elbows

Bends and elbows	Diameter range	K
Cast iron:		
90° D/F bends	50–1200	0.40
45° D/F bends	50–1200	0.20 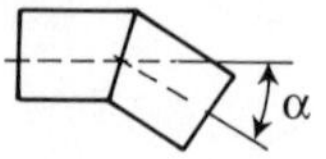
Steel welding bends:		
90° Short radius	50–400	0.40
45° Short radius	50–400	0.20
90° Long radius	50–400	0.35
45° Long radius	50–400	0.17
PVC/ABS:		
90° Elbows	$\frac{1}{2}''$–8″	1.25
45° Elbows	$\frac{1}{2}''$–8″	0.50 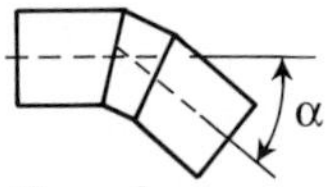
90° Long radius bends	$\frac{1}{2}''$–4″	0.45
90° Long radius bends	150–600	0.30
45° Long radius bends	$\frac{1}{2}''$–4″	0.25
45° Long radius bends	150–600	0.15
$22\frac{1}{2}^\circ$ Long radius bends	150–600	0.10 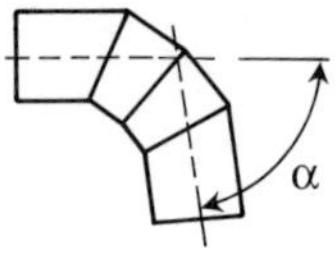
$11\frac{1}{4}^\circ$ Long radius bends	150–600	0.05
Screwed steel:		
90° Elbows	$\frac{1}{2}''$–6″	1.25
45° Elbows	$\frac{1}{2}''$–6″	0.50

Mitre bends	α	K
Type 1	90°	1.20
	80°	1.00
	70°	0.80
	60°	0.60
	50°	0.40
	40°	0.30
	30°	0.15
	20°	0.10
	10°	0.05
Type 2	60°	0.25
	45°	0.20
	30°	0.15
Type 3	90°	0.30
	75°	0.25
	60°	0.20

(*continued*)

Square-edged tees

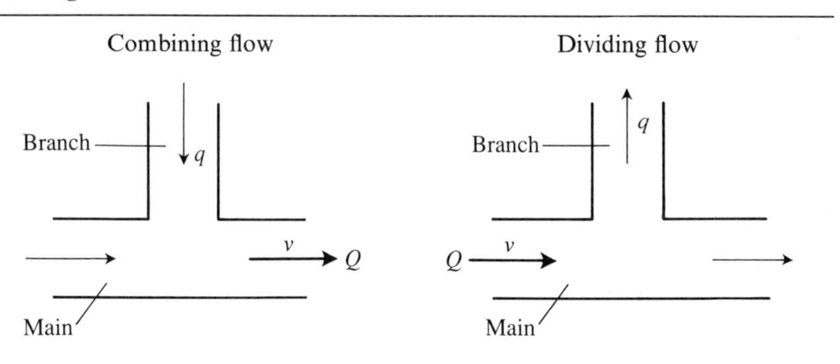

Flow ratio q/Q	Diameter ratio (branch/main) 0.5	0.75	1.0	Flow ratio q/Q	Diameter ratio (branch/main) 0.5	0.75	1.0
	Headloss in line				Headloss in line		
0	0.1	0.1	0.1	0	0.1	0.1	0.1
0.25	0.4	0.4	0.4	0.25	0	0	0
0.50	0.7	0.6	0.5	0.50	0	0	0
0.75	1.0	0.8	0.6	0.75	0.2	0.2	0.2
	Headloss: branch to main				Headloss: main to branch		
0.25	0.7	0	−0.2	0.25	2.2	1.0	0.9
0.50	3.5	0.9	0.5	0.50	6.5	1.3	0.9
0.75	7.0	2.0	0.9	0.75	11.0	1.7	1.1
1.00	11.0	3.0	1.2	1.00	14.0	2.3	1.3

Combining equal flows
Diameter ratio = 1
(branch/main); $K = 0.7$

Dividing flow equally
Diameter ratio = 1
(branch/main); $K = 1.2$

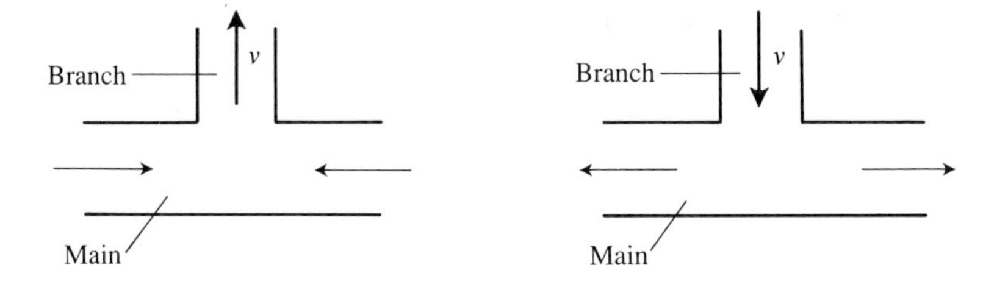

Table 3.6 (*continued*)

Radiused tees

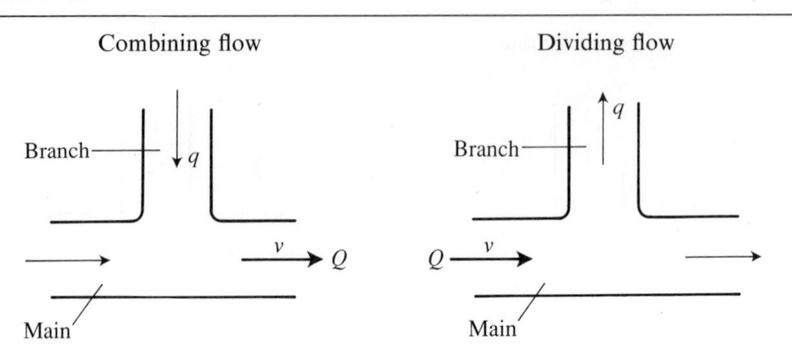

Combining flow

Flow ratio q/Q	Diameter ratio (branch/main) 0.5	0.75	1.0
	Headloss in line		
0	0.1	0.1	0.1
0.25	0.3	0.3	0.3
0.50	0.4	0.3	0.3
0.75	0.2	0.1	0.1
	Headloss: branch to main		
0.25	0.7	0	−0.2
0.50	1.4	0.4	0.2
0.75	3.5	0.7	0.4
1.00	8.3	2.0	0.7

Dividing flow

Flow ratio q/Q	Diameter ratio (branch/main) 0.5	0.75	1.0
	Headloss in line		
0	0.1	0.1	0.1
0.25	0	0	0
0.50	0	0	0
0.75	0.2	0.2	0.2
	Headloss: main to branch		
0.25	1.5	0.8	0.4
0.50	2.8	0.8	0.6
0.75	3.9	0.8	0.6
1.00	4.9	1.0	0.7

Combining equal flows
Diameter ratio = 1
(branch/main); $K = 0.4$

Dividing flow equally
Diameter ratio = 1
(branch/main); $K = 1.8$

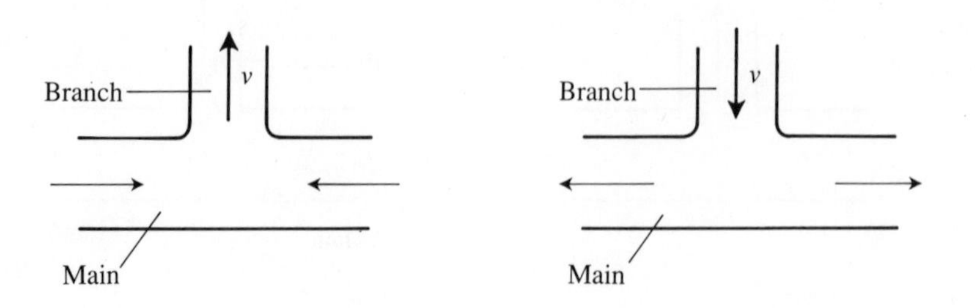

number. A program listing is given in Appendix A. The use of the program is illustrated by the following sample program run.

Sample program run: program FRICTF

```
RUN
Program FRICTF.BAS

THIS PROGRAM COMPUTES THE FRICTION FACTOR F
FOR FLUID FLOW IN PIPES


ENTER FLUID DENSITY (kg/m**3)? 1000

ENTER FLUID DYNAMIC VISCOSITY (Ns/m**2)? 1.0E-03

ENTER PIPE VELOCITY (m/s)? 1.2

ENTER PIPE INTERNAL DIAMETER (mm)? 250

ENTER PIPE WALL ROUGHNESS (mm)? 0.1

    Friction factor f =  1.763916E-02
      Reynolds number =  300000

DO YOU WISH TO COMPUTE ANOTHER VALUE (Y/N)? N
Ok
```

PIPFLO

Program PIPFLO relates to steady flow of clean water in pipes. It computes the hydraulic gradient for a specified flow rate or the flow rate for a specified hydraulic gradient. It offers the user a choice of flow formula, Colebrook–White (3.25), Manning (3.33), or Hazen–Williams (3.28). It incorporates FRICTF as a subroutine for the computation of the friction factor f, using a fixed value of $1 \times 10^{-6}\ \mathrm{m^2\ s^{-1}}$ for kinematic viscosity. A program listing is given in Appendix A. The use of the program is illustrated by the following sample program run.

Sample program run: program PIPFLO

```
RUN
Program PIPFLO.BAS

THIS PROGRAM RELATES TO FLOW OF CLEAN WATER
(TEMP 10 deg C) IN PIPES.  IT COMPUTES:

                   1. HEAD LOSS FOR GIVEN FLOW
                   2. FLOW AT GIVEN HEAD LOSS

ENTER 1 OR 2, AS APPROPRIATE? 1
```

```
ENTER PIPE DISCHARGE RATE (m**3/h)? 40

WHICH OF THE FOLLOWING FORMULAE DO YOU WISH TO USE ?

                   1. COLEBROOK-WHITE
                   2. HAZEN-WILLIAMS
                   3. MANNING

ENTER YOUR SELECTED FORMULA NUMBER? 1

ENTER VALUE OF PIPE ROUGHNESS, K (m)? 0.0001
ENTER PIPE INTERNAL DIAMETER (m)? 0.150

       Friction factor =  2.191162E-02
    Hydraulic gradient =  2.94642E-03
       Reynolds number =  72197.3

DO YOU WISH TO MAKE ANOTHER COMPUTATION Y/N? N
Ok
```

SSFLO

Program SSFLO computes the head loss associated with steady flow of sludges in pipes, including primary, activated, digested, and humus sludges. Computation is based on the headloss/flow correlations presented in this chapter. The program outputs hydraulic gradient, Reynolds number, and velocity. A program listing is given in Appendix A. The use of the program is illustrated by the following sample program run.

Sample program run: program SSFLO

```
RUN
Program SSFLO.BAS

THIS PROGRAM COMPUTES THE HEAD LOSS DUE TO FLOW
OF PRIMARY, ACTIVATED, DIGESTED AND HUMUS SLUDGES
IN PIPES.

        1. PRIMARY
        2. ACTIVATED
        3. ANAEROBICALLY DIGESTED
        4. HUMUS

ENTER YOUR SELECTED NUMBER? 1

INPUT THE SUSPENDED SOLIDS CONC. (kg/m**3)? 30.0
INPUT THE PIPE INTERNAL DIAMETER (m)? 0.1
INPUT THE PIPE ROUGHNESS (m)? 0.0001
INPUT THE SLUDGE DISCHARGE RATE (m**3/h)? 72.0

         Velocity (m/s) =  2.547771
         Reynolds number =  5783.295

HENCE, FLOW IS TURBULENT

         Hydraulic gradient =  .1032874

DO YOU WISH TO MAKE ANOTHER COMPUTATION (Y/N)? N
Ok
```

References

Barr, D. I. H. (1975). Two additional methods of direct solution of the Colebrook–White function, TN 128. *Proc. Inst. Civ. Eng.*, Part 2, **3**, P827.

Colebrook, C. F. and White, C. M. (1938). The reduction in the carrying capacity of pipes with age. *J. Inst. Civ. Eng.*, **7**, 99–118.

Colebrook, C. F. (1939). Turbulent flow in pipes, with particular reference to the transition between the smooth and rough pipe laws. *J. Inst. Civ. Eng.*, **8**, 133–56.

Darcy, H. (1858). Recherches experimentales relative au mouvement de l'eau dans les tuyaux. *Mem. Acad. Sci.*, Paris.

Frost, R. C. (1982). Prediction of friction losses for the flow of sewage sludge in straight pipes, TR 175. Water Research Centre, Stevenage.

Frost, R. C. (1983). How to design sewage sludge pumping systems, TR 185, Water Research Centre, Stevenage.

Hazen, A. and Williams, G. S. (1920). *Hydraulic Tables*, Wiley, New York.

Hydraulics Research, Wallingford (1990). *Charts for the hydraulic design of channels and pipes*, (6th edn). Thomas Telford Ltd., London.

Manning, R. (1891). On the flow of water in channels and pipes. *Proc. Inst. Civ. Eng. of Irl.*, **20**, 161.

Manning, R. (1895). On the flow of water in channels and pipes. *Proc. Inst. Civ. Eng. of Irl.*, **24**, 179.

Nikuradse, J. (1932). Gesetzmassigkeiten der turbulenten stromung in glatten rohren. *Verein Deutscher Ingenieure*, 356.

Nikuradse, J. (1933). Stromungesetzein rauhen rohren. *Verein Deutscher Ingenieure*, 361.

Perkins, J. A. and Gardiner, A. M. (1982). The effect of sewage slime on the hydraulic roughness of pipes, IT 218, Hydraulics Research Station, Wallingford.

Prandtl, L. (1933). Neue ergebnisse der turbulenzforshung. *Verein Deutscher Ingenieure*, 5.

Reynolds, O. (1885). An experimental investigation of the circumstances which determine whether the motion of water shall be direct or sinuous, and the law of resistance of parallel channels. *Phil. Trans.*, **174**, 935.

von Karman, T. (1930). *Mechanische ahnlichkeit und turbulenz*, Nachr. Gess. Wiss., Gottingen.

Weisbach, J. (1850). *Lehrbuch der Ingenieure und Maschinen-Mechanik* (2nd edn). Braunschweig.

Related reading

Ito, H. and Imai, K. (1973). Energy losses at 90° junctions. *J. Hyd. Div. ASCE*, **99**, HY9, 1353–68.

Miller, D. S. (1971). *Internal flow—a guide to losses in pipe and duct systems, Br. Hydromech. Res. Assoc.*, Cranfield, England.

4
Flow in pipe manifolds

4.1 Introduction

The problem of achieving a uniform distribution or collection of fluid over an area is a commonly encountered design task in many areas of fluids engineering. In the field of water and wastewater engineering, manifolds are important components in several treatment processes, including biofilters, sand filters, and fluidized bed clarifiers. Manifolds are used in surface and subsurface irrigation systems for the distribution of water and/or effluents over land areas. They are also employed for the dispersal of effluent into large bodies of water, as in lake and sea outfalls.

The designer needs to be able to compute the variation in discharge reliably over the length of a manifold in order to size conduits efficiently to meet specified tolerances in flow variation.

The mechanics of manifold flow have been reviewed and discussed in papers by McNown (1954), Rawn *et al.* (1961), French (1972), and Hudson *et al.* (1979), among others. In general, theoretical descriptions of manifold flow have been found to be insufficiently accurate for practical design purposes. However, with the aid of experimental measurements on manifold systems, reliable semi-empirical methods of analysis of flow distribution have been established.

4.2 Orifice-type pipe manifold

The simplest type of manifold is a pipe with orifices spaced along its length, as illustrated in Fig. 4. 1. The discharge through such orifices is a function of the differential pressure across the orifice and the velocity of flow in the pipe. Rawn *et al.* (1961) expressed the orifice discharge q_0 as a function of the total differential head E as follows:

$$q_0 = C_D A_0 \sqrt{2gE} \tag{4.1}$$

where C_D is the coefficient of discharge for the orifice, and the total differential head E (shown in Fig. 4.1) is the sum of the differential pressure head (h)

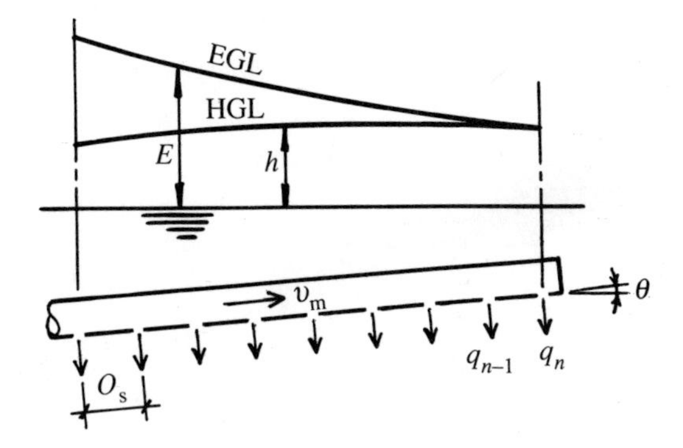

Fig. 4.1 Orifice-type pipe manifold.

across the orifice and the velocity head in the manifold location, $v_m^2/2g$. The discharge coefficient C_D varies with flow conditions and has been found to be a function of the ratio of the manifold velocity head and the total differential head, that is, $(v_m^2/2g)/E$.

For smooth bellmouthed ports (nozzle area contraction 4:1 or more) and for manifold Reynolds numbers exceeding 20 000, French (1972) recommended the following expression for C_D:

$$\text{Bellmouth ports:} \qquad C_D = 0.975\left(1 - \frac{v_m^2/2g}{E}\right)^{0.375}. \tag{4.2}$$

Experimental data for sharp-edged pipe orifices for water and air (Fitzpatrick 1988) are presented on Figs 4.2 and 4.3, respectively. These findings yield the following empirical expressions for C_D for sharp-edged orifices:

$$\text{Water:} \qquad C_D = 0.66 - 0.75\,\frac{v_m^2/2g}{E} \tag{4.3}$$

$$\text{Air:} \qquad C_D = 0.76 - 1.88\,\frac{v_m^2/2g}{E}. \tag{4.4}$$

It should be noted that Fitzpatrick's water manifold data related to orifice discharge to air, while his air manifold data related to air orifice discharge under water.

Using eqn (4.1) and values for C_D given by eqns (4.2), (4.3), or (4.4), as appropriate, the orifice discharges may be calculated in turn, starting from the 'dead' end, that is, orifice n. As an initial approximation, the manifold velocity v_m at orifice n may be assumed to be zero, thus enabling an initial approximate value of C_D to be calculated and hence an initial approximate value for the orifice discharge q_n. By simple iteration, more precise values

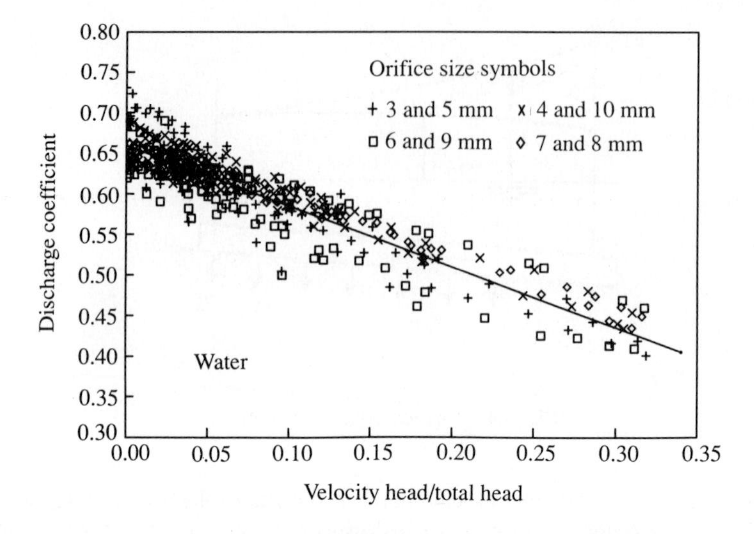

Fig. 4.2 Experimental data for an orifice-type pipe manifold discharging water to air. Manifold ID 43 mm, PVC; R_e range 4000–70 000 (Fitzpatrick 1988).

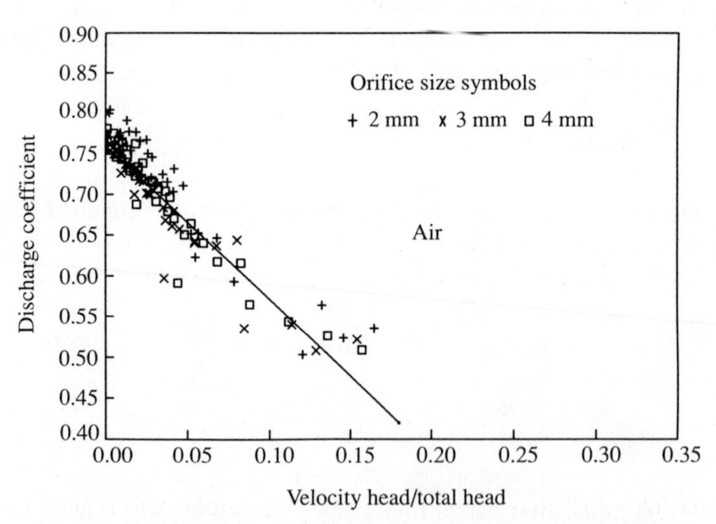

Fig. 4.3 Experimental data for an orifice-type pipe manifold discharging air under water. Manifold ID 27 mm, PVC; R_e range 4000–30 000 (Fitzpatrick 1988).

for v_m and q_n are found. Proceeding to the next upstream orifice, $n - 1$, the value of E_{n-1} is obtained as follows:

$$E_{n-1} = E_n + O_s\left\{S_f + S_0\left(1 - \frac{\rho}{\rho_m}\right)\right\} \tag{4.5}$$

where O_s is the orifice spacing (m), S_f is the manifold pipe friction slope ($S_f = fv_m^2/2gD_m$), S_0 is the manifold slope, as defined in Fig. 4.1 ($S_0 = \sin\theta$), ρ_m is the manifold fluid density, and ρ is the external fluid density. (Note that the differential head is expressed in terms of head of the manifold fluid.) As before, the initial approximation of C_D is found using the known value of v_m downstream of orifice $n - 1$. Using this value for C_D, a first approximation of q_{n-1} is calculated, enabling an improved estimate of v_m upstream of orifice $n - 1$ to be computed: $v_m = (q_n + q_{n-1})/A_m$. Thus, by simple iteration a precise value for q_{n-1} is determined.

Computation proceeds in this manner, orifice by orifice, back along the pipe manifold to the supply end. This computation procedure assumes zero head loss in the manifold pipe due the lateral orifice discharges. McNown (1954) has shown this to be a reasonable assumption.

4.3 Pipe manifold with pipe laterals

The analysis of flow distribution into the individual laterals of a manifold pipe/pipe lateral system is carried out using the same type of iterative computation procedure as that described for an orifice manifold system. The discharge q_L into an individual lateral pipe may be written as follows:

$$q_L = C_L\sqrt{E_m - h_e} \tag{4.6}$$

where C_L is the lateral discharge coefficient, E_m is the total differential head in the manifold at the manifold/lateral junction, and h_e is the lateral entry head loss. The coefficient C_L correlates flow into the lateral to the total head in the lateral on the downstream side of its junction with the manifold. Typically, the lateral may be a submanifold pipe with orifices, in which case the discharge coefficient C_L is calculated in the manner described above for an orifice-type pipe manifold.

Hudson *et al.* (1979) reviewed published data for entry head loss in dividing flow manifolds with square-edged laterals (as shown in Fig. 4.4). From these data, they derived an empirical relationship between the junction entry head loss and the manifold/lateral velocity ratio v_m/v_L, distinguishing between 'short' laterals (less than three lateral diameters) and long laterals. The proposed expressions are as follows:

Long laterals: $$h_e = \frac{v_L^2}{2g}\left(0.9\left(\frac{v_m}{v_L}\right)^2 + 0.4\right) \tag{4.7}$$

Short laterals: $$h_e = \frac{v_L^2}{2g}\left(1.67\left(\frac{v_m}{v_L}\right)^2 + 0.7\right) \tag{4.8}$$

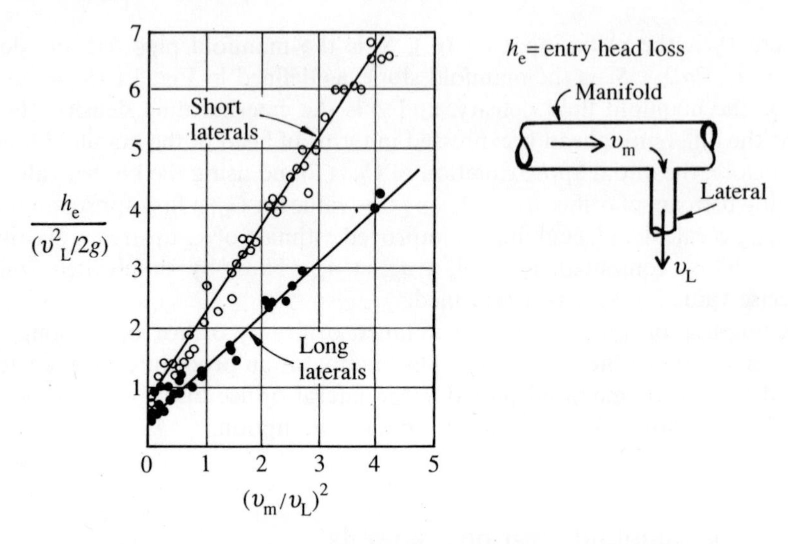

Fig. 4.4 Lateral entry head loss, h_e (Hudson *et al.* 1979).

As in the case of an orifice manifold, computation starts at the dead end (lateral n) and proceeds, lateral by lateral, towards the supply end of the manifold. The total differential head in the manifold at a lateral junction is related to its value at the next downstream lateral junction as follows:

$$E_{n-1} = E_n + L_s S_f + L_s S_0 \left(1 - \frac{\rho}{\rho_m}\right) \tag{4.9}$$

where L_s is the lateral spacing (m) and the remaining terms are as previously defined for orifice laterals.

Thus, the inflow into each lateral can be determined using eqn (4.6), refining the corresponding estimate of the entry head loss h_e by iterative calculation to obtain the desired computational precision.

4.4 Design of manifold systems

Generally, the primary objective in manifold design is the achievement of a nearly uniform discharge rate through the outlets of the manifold system. This design criterion is conveniently specified in terms of the ratio of the maximum outlet discharge to the minimum outlet discharge.

As shown in the foregoing analysis of manifold flow, the rate of discharge through an outlet is influenced by a number of system variables. One of these variables is the manifold pipe slope S_0; S_0 influences manifold performance only where the manifold fluid density is different from the

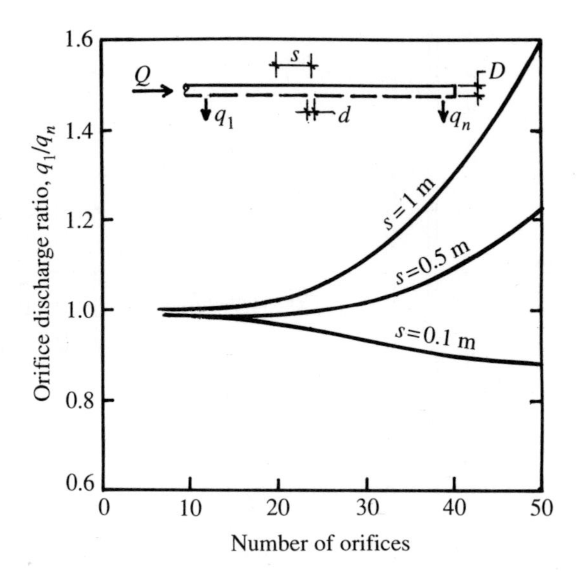

Fig. 4.5 Influence of system dimensions on orifice-type manifold performance: $D = 50$ mm; $d = 6$ mm; $Q = 2$ l.p.s.; $\rho = \rho_m$.

external fluid density, for example, an air manifold discharging under water or a sea outfall manifold discharging sewage. Clearly, in the case of the air manifold, where the difference in the fluid densities is very large, the parameter S_0 has a major influence on outlet flow variation. The other system variables which influence flow distribution are the friction slope S_f and the orifice discharge coefficient C_D.

In general, the objective of uniform discharge is satisfied by ensuring that the ratio of total head variation in the manifold system to the head loss across individual outlets is kept low. This is influenced by the ratio of manifold cross-sectional area to the sum of the outlet cross-sectional areas and the spacing of the outlets. Figure 4.5 illustrates the influences of these system dimensions on flow distribution for an orifice-type pipe manifold.

4.5 Computer program: MANIFLO

The computer program MANIFLO computes the flow distribution through a pipe manifold system of the type illustrated in Fig. 4.6, consisting of a trunk main with lateral pipes on one or both sides, the lateral pipes having uniformly spaced sharp-edged orifices. The program computations are based on the foregoing equations; the Colebrook–White equation is used for the calculation of distributed head loss. A program listing is given in Appendix A.

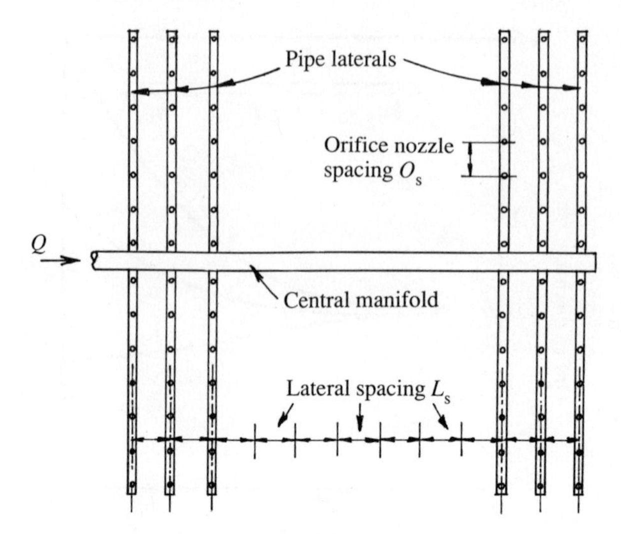

Fig. 4.6 Pipe manifold with laterals.

The use of the program is illustrated by the following sample program run.

Sample program run: program MANIFLO

```
RUN
Program MANIFLO.BAS

THIS PROGRAM COMPUTES THE DISTRIBUTION OF FLOW IN A
PIPE MANIFOLD SYSTEM WITH SQUARE-EDGED LATERAL PIPES
HAVING UNIFORMLY SPACED SHARP-EDGED ORIFICES

INPUT THE FOLLOWING DATA IN THE SPECIFIED UNITS:
(pipes sloping upwards in flow dir. have pos. slope)

Enter orifice diameter (mm)? 15
Enter orifice spacing (m)? 0.5
Enter number of orifices per lateral? 10
Enter lateral diameter (mm)? 100
Enter lateral slope (sin theta)? 0
Enter lateral wall roughness (mm)? 0.1
Enter lateral spacing (m)? 1

Enter manifold diameter (mm)? 500
Enter manifold slope (sin theta)? 0
Enter manifold wall roughness (mm)? 0.1
Enter number of laterals (total both sides)? 20
Are laterals on both sides of manifold (Y/N)? Y

Enter flowrate (m**3/h)? 600
Enter density of manifold fluid (kg/m**3)? 1000
Enter viscosity of manifold fluid (Ns/m**2)? 0.001
Enter density of external fluid (kg/m**3)? 1000
```

```
     COMPUTATION IN PROGRESS
                       ------- please wait -------
                        Manifold Hydraulics

Orifice  Orifice   No per    Lateral   Lateral   No of     Manifold  Discharge
Diameter Spacing   lateral   diameter  spacing   laterals  diameter
(mm)     (m)                 (mm)      (m)                 (mm)      (m3/h)
 15       .5        10        100       1         20        500       600

DISCHARGE RELATIVE TO DEAD END DISCHARGE:

IN LATERALS: 1.00 1.00 1.00 1.00 0.99 0.99 0.99 0.99 0.98 0.98

IN MANIFOLD: 1.00 1.00 1.00 1.00 1.00 1.00 1.00 1.00 1.00 0.99

Min orifice discharge/max orifice discharge =  .9739636

System headloss (m) =    2.655819

Manifold inlet end velocity (m/s) =  .8403074
Ok
```

References

Fitzpatrick, J. M. (1988). Hydraulics of manifold flow. MEngSc thesis, Dept. of Civil Engineering, University College, Dublin.

French, J. A. (1972). Internal hydraulics of multiport diffusers. *J. Water Pollution Cont. Fed.*, **44**, 782–95.

Hudson, H. E., Uhler, M., and Bailey, R. W. (1979). Dividing flow manifolds with square-edged laterals. *J. Env. Div. ASCE*, **105**, 745–55.

McNown, J. S. (1954). Mechanics of manifold flow. *Trans. ASCE*, **119**, 1103–18.

Rawn, A. M., Bowerman, A. N. and Brooks, N. H. (1961). Diffusers for the disposal of sewage in sea-water. *Trans. ASCE*, **126**, Part 3, 344–84.

5
Steady flow in pipe networks

5.1 Water pipe networks

A pipe network can be described as a system of interconnected pipes, forming one or more closed loops. A loop may be defined as a connected set of pipes and their end nodes, every node of which is an end node of exactly two pipes of the set. The term 'node' is applied to any point at which water enters or leaves the network or to any pipe junction within the network. Thus each pipe is defined by a pair of end nodes.

The distribution of steady flow within such a pipe system is determined by the following factors:

(1) the head–discharge relationship for each pipe;

(2) the governing network flow equations;

(3) the boundary conditions of the system.

5.2 Head–discharge relationships for pipes

The head loss h in a pipe of given length can be related to the discharge Q by a correlation of the general form

$$h = rQ^{m} \tag{5.1}$$

where r is a pipe resistance coefficient and the exponent m is a constant. The values to be assigned to r and m depend on the flow equation being used.

The preferred flow equation for pipes flowing full, as discussed in Chapter 3, is the Darcy–Weisbach eqn (3.23):

$$h = \frac{fLv^2}{2gD}$$

where the friction factor f is a function of Reynolds number and pipe relative roughness, as defined by the Colebrook–White equation (3.25). For hand calculation the appropriate value of f may be read from tabulated value sets for pipes. The program PIPFLO (Chapter 3) uses an iterative procedure for the computation of f.

The Hazen–Williams equation (3.27) is also widely used for water network computation purposes. It can be written in the form

$$h = 6.818C^{-1.852}Lv^{1.852}D^{-1.167}. \tag{5.2}$$

When expressed in the form of eqn (5.1), these equations become:

$$\text{Darcy–Weisbach:} \qquad h = \frac{8fL}{\pi^2 g D^5} Q^2 \tag{5.3}$$

$$\text{Hazen–Williams:} \qquad h = \frac{10.704L}{C^{1.852}D^{4.871}} Q^{1.852} \tag{5.4}$$

where h is expressed in metres and Q in $m^3\ s^{-1}$.

5.3 Network analysis

In any pipe network the number of unknown flows corresponds to the number of pipes in the network and their evaluation involves the solution of an equal number of simultaneous equations. These governing equations are of two types.

1. Continuity equations: the algebraic sum of the flows at any node must be zero (flows into and away from each node must balance).
2. Loop equations: the integrated head loss around any loop must be zero.

Consider a pipe network having P pipes and N nodes. The continuity or node equations are of the general form

$$\sum Q_{ij} + E_i = 0 \qquad \text{for any node } i$$

where Q_{ij} refers to flow from node i to node j, the subscript j representing nodes connected to node i. E_i is the external supply or demand at node i. The sign convention adopted in this text is that flow towards a node is considered as positive, flow away from a node is assigned a negative value. The maximum number of independent node equations is $N - 1$.

The loop equations are of the general form

$$\sum h_{ij} = 0 \qquad \text{for each loop}$$

where $h_{ij} = r_{ij}Q_{ij}^m$ is the head loss in the pipe connecting nodes i and j and the summation covers all pipes which comprise the loop. The sign convention adopted here is that head loss associated with clockwise flow is considered positive and head loss associated with anticlockwise flow is considered negative. The maximum number of independent loop equations is $P-(N-1)$. Thus the total number of independent equations, derived from the continuity and loop conditions, equals P, the number of pipes in the network.

The network equations may be expressed in terms of the flow variable Q, as above, or in terms of node piezometric head H, using the correlation

$$h_{ij} = H_i - H_j = r_{ij} Q_{ij}^m$$

The simple network shown in Fig. 5.1 has ten pipes and eight nodes and the supply and demand values are shown in Table 5.1. Its network equations are therefore made up of seven node equations and three loop equations, which are shown in Tables 5.2 and 5.3. It may be noted that the node equation for node 8 is not an additional independent node equation since it could have been obtained by taking the negative sum of node equations 1 to 7, inclusive. It should also be noted that this set of three loops, though an independent set, is not a unique set; any other independent set might have been chosen.

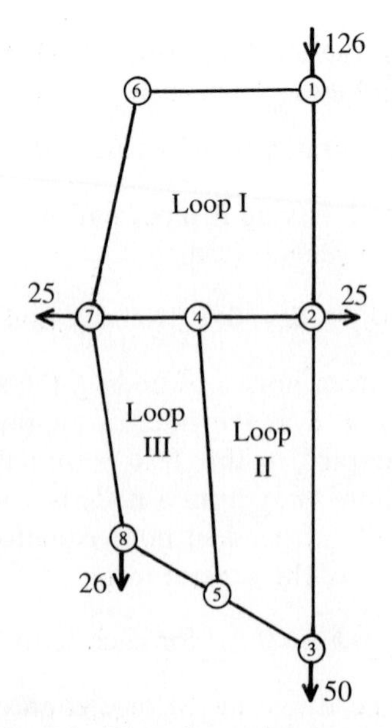

Fig. 5.1 Three-loop pipe network.

Table 5.1 Indicated supply and demand values in l.p.s.

	Pipe data		
Pipe	Length (m)	Diameter (mm)	*C*-value
1–2	300	300	100
2–3	450	300	100
2–4	150	200	100
3–5	150	180	100
4–5	360	180	100
4–7	150	200	100
5–8	150	180	100
1–6	240	200	100
6–7	300	180	100
7–8	300	180	100

Table 5.2

	Node	Equation
Node equations	1	$126 + Q_{12} + Q_{16} = 0$
	2	$-25 + Q_{21} + Q_{23} + Q_{24} = 0$
	3	$-50 + Q_{35} + Q_{32} = 0$
	4	$Q_{45} + Q_{47} + Q_{42} = 0$
	5	$Q_{58} + Q_{54} + Q_{53} = 0$
	6	$Q_{67} + Q_{61} = 0$
	7	$-25 + Q_{76} + Q_{74} + Q_{78} = 0$

Table 5.3

	Loop	Equation
Loop equations	1	$h_{12} + h_{24} + h_{47} + h_{76} + h_{61} = 0$
	2	$h_{23} + h_{35} + h_{54} + h_{42} = 0$
	3	$h_{45} + h_{58} + h_{87} + h_{74} = 0$

5.4 Boundary conditions

The boundary conditions must be sufficient to define flow distribution. Boundary conditions for water networks include: supplies (inflows to the system) and demands (outflows from the system); nodes having constant head, for example, service reservoirs; and flow control devices, such as non-return valves, pressure-reducing valves, and pumps.

Typical examples of sets of boundary conditions sufficient to define flow and pressure distribution in water networks are:

(1) magnitude of supplies and demands are known; one nodal pressure is known;

(2) magnitudes of pressures at supply nodes and magnitudes of demands are known.

5.5 Solution of network equations

As already outlined, the network equations consist of a set of $N - 1$ node equations which are linear in Q and a set of $P - (N - 1)$ loop equations which are linear in H. When the latter are expressed in terms of Q, they convert to a set of non-linear equations of the general form:

$$\sum r_{ij} Q_{ij}^{m} = 0.$$

Alternatively, if the node piezometric head H is taken as the computational variable instead of Q, the network equations convert to a set of $P - (N - 1)$ linear equations in H and a set of $N - 1$ non-linear node equations in H. Since Q is almost invariably the computational variable used in practice, the following discussion is confined to the solution of the network equations expressed in terms of Q.

Direct solution of systems of non-linear simultaneous equations is not feasible; hence it is necessary to use iterative solution methods. In general, these methods start with an estimated solution which is iteratively refined by repeated corrections until the deviation from the true solution is reduced to an acceptable tolerance value. Three such iterative solution techniques, which have been applied to pipe network analysis, are reviewed. This review is followed by a manually worked illustrative example using the Hardy Cross method of analysis. The latter method is also used as the basis of the computer program PNA, which analyses flow in pipe networks. The selected three iterative computational procedures are based on:

(1) loop by loop flow correction (Hardy Cross);

(2) simultaneous loop flow correction;

(3) linearization of the network loop equations.

5.5.1 Hardy Cross method

The most widely used technique for water network analysis is based on a loop by loop iterative computational procedure first described by Hardy Cross (1936). Its use may be illustrated by reference to the three-loop network in Fig. 5.1. As already shown, the set of network equations for this network comprises seven node equations and three loop equations. As a first step, an arbitrary flow distribution, which satisfies flow continuity at nodes, that is, complies with the node equations, is assumed. Inevitably this assumed flow distribution will not satisfy the loop equations. Let the out-of-balance head in loop 1 be h_1, where

$$h_1 = r_{12}Q_{12}^m + r_{24}Q_{24}^m + r_{47}Q_{47}^m + r_{76}Q_{76}^m + r_{61}Q_{61}^m$$

The first-order estimate of the loop flow correction Δq_1, which would reduce h_1 to zero, is found by the Newton–Raphson approximation (described in Appendix B):

$$-h_1 = \frac{\mathrm{d}h_1}{\mathrm{d}q_1}\Delta q_1 \tag{5.5}$$

where the numerical values of h_1 and its differential are based on the current Q values in loop 1 pipes. The loop flow correction Δq_1 is therefore

$$\text{Loop 1:} \qquad \Delta q_1 = -\frac{h_1}{\mathrm{d}h_1/\mathrm{d}q_1}.$$

On the same basis, the loop flow corrections for loops 2 and 3 are

$$\text{Loop 2:} \qquad \Delta q_2 = -\frac{h_2}{\mathrm{d}h_2/\mathrm{d}q_2}$$

$$\text{Loop 3:} \qquad \Delta q_3 = -\frac{h_3}{\mathrm{d}h_3/\mathrm{d}q_3}.$$

Because these flow corrections are loop flows, that is, clockwise or anti-clockwise flows applied to all pipes in the loop, they do not invalidate the initially satisfied nodal continuity equations. Flow corrections are repeated on a loop by loop basis until the maximum out of balance loop head is reduced to a specified tolerance value.

5.5.2 Simultaneous loop flow correction

The Newton–Raphson method has also been used to compute simultaneous flow corrections for all loops (Martin and Peters 1963; Epp and Fowler

1970). This computational procedure takes into account the interactive influence of flow corrections between loops which have common pipes. As in the Hardy Cross method, an initial flow distribution, which satisfies flow continuity at nodes, is assumed. The network of Fig. 5.1 is again used as an illustrative example. The first-order estimates of the loop flow corrections, which would reduce the loop out-of-balance heads to zero, are found from the following Newton–Raphson approximations:

$$-h_1 = \frac{\partial h_1}{\partial q_1}\Delta q_1 + \frac{\partial h_1}{\partial q_2}\Delta q_2 + \frac{\partial h_1}{\partial q_3}\Delta q_3$$

$$-h_2 = \frac{\partial h_2}{\partial q_1}\Delta q_1 + \frac{\partial h_2}{\partial q_2}\Delta q_2 + \frac{\partial h_2}{\partial q_3}\Delta q_3$$

$$-h_3 = \frac{\partial h_3}{\partial q_1}\Delta q_1 + \frac{\partial h_3}{\partial q_2}\Delta q_2 + \frac{\partial h_3}{\partial q_3}\Delta q_3 .$$

This set of linear simultaneous equations in Δq can be written in matrix form as follows:

$$\begin{bmatrix} \frac{\partial h_1}{\partial q_1} & \frac{\partial h_1}{\partial q_2} & \frac{\partial h_1}{\partial q_3} \\ \frac{\partial h_2}{\partial q_1} & \frac{\partial h_2}{\partial q_2} & \frac{\partial h_2}{\partial q_3} \\ \frac{\partial h_3}{\partial q_1} & \frac{\partial h_3}{\partial q_2} & \frac{\partial h_3}{\partial q_3} \end{bmatrix} \begin{bmatrix} \Delta q_1 \\ \Delta q_2 \\ \Delta q_3 \end{bmatrix} = \begin{bmatrix} -h_1 \\ -h_2 \\ -h_3 \end{bmatrix} .$$

This set of equations is conveniently solved by matrix inversion, thus yielding the required set of loop flow corrections, Δq_1, Δq_2, and Δq_3.

The illustrative example used has three loops, resulting in a 3×3 matrix of differentials. Thus, a network having L loops would result in an $L \times L$ matrix of differentials. It is clear that the matrix elements will have non-zero values only where they connect loops having common pipes, that is, dh_x/dq_y will only have a non-zero value if loops x and y share one or more common pipes.

It is reasonable to expect that the simultaneous loop flow correction method should converge more rapidly to the true solution than the loop by loop Hardy Cross method.

5.5.3 Linearization of network equations

Wood and Charles (1972) used a linearization of the loop equations, thus enabling the direct solution of the full set of network equations. Each loop equation is of the general form

$$\sum r_{ij}Q_{ij}^{m} = 0$$

which can be written in the modified form

$$\sum r_{ij}Q_{ij}^{m-1}Q_{ij} = 0$$

or

$$\sum r_{ij}^{*}Q_{ij} = 0$$

where $r_{ij}^{*} = r_{ij}Q_{ij}^{m-1}$ is treated as a numerical coefficient based on the current value of Q_{ij}.

To start this method it is also necessary to assume an initial flow distribution. After each solution of the full set of linearized network equations, a new set of r^{*} pipe coefficients is calculated for use in the next iteration.

5.5.4 Convergence of methods

It has been demonstrated (Wood and Charles 1972) that, as might be expected, of the foregoing methods of analysis, the Hardy Cross method provides the least rapid convergence. This is because each loop flow correction is made in isolation from the rest of the network, whereas the other two methods incorporate full network interaction in each iteration.

The Hardy Cross method, however, is not to be dismissed as a useful method of network analysis as the computer time required may not be of great economic significance in the overall task of network analysis and design. The method has the advantage over the other methods of requiring less computer memory as it does not involve the solution of large sets of linear simultaneous equations.

Convergence of all iterative methods is aided by judicious selection of the initial flow distribution, as discussed in Section 5.7.

5.6 Worked example: Hardy Cross

The calculation procedure for the computation of the flow distribution for the simple three-loop pipe network shown in Fig. 5.1, based on the Hardy Cross computation method, is outlined in Table 5.4. The first column of Q_{ij} values shows the assumed initial flow distribution, which, as required, satisfies the node continuity condition at all nodes. The loop flow correction Δq is based on eqn (5.5), which for any loop x can be written as

$$\Delta q_x = -\left(\frac{h}{\mathrm{d}h/\mathrm{d}q}\right)_x .$$

Table 5.4 Analysis of network shown in Fig. 5.1 by iterative solution of the loop equations.

Loop	Pipe i–j	r_{ij}	Q_{ij}	H_{ij}	H_{ij}/Q_{ij}	ΔQ		Q_{ij}	H_{ij}	H_{ij}/Q_{ij}	ΔQ	
I	1–2	223.7	+0.090	+2.600	28.89	+0.011	+0.000	+0.101	+3.218	31.85	−0.000	+0.000
	2–4	806.0	+0.010	+0.161	16.10	+0.011	+0.007	+0.028	+1.080	38.57	−0.000	−0.004
	4–7	806.0	+0.005	+0.045	9.00	+0.011	−0.003	+0.013	+0.261	20.08	−0.000	+0.002
	7–6	2693.0	−0.036	−5.746	159.61	+0.011		−0.025	−2.927	117.08	−0.000	
	6–1	1289.5	−0.036	−2.752	76.44	+0.011		−0.025	−1.402	56.08	−0.000	
			Σ	−5.692	290.04			Σ	0.230	263.67		
			$\Delta Q = 5.692/1.85 \times 290.04 = 0.011$					$\Delta Q = -0.230/1.85 \times 263.67 = -0.000$				
II	2–3	335.5	+0.055	+1.568	28.51	−0.007		+0.048	+1.219	25.40	+0.004	
	3–5	1346.5	+0.005	+0.075	15.00	−0.007		−0.002	−0.014	7.00	+0.004	
	5–4	3231.5	−0.005	−0.179	35.80	−0.007	−0.003	−0.015	−1.365	91.00	+0.004	+0.002
	4–2	806.0	−0.010	−0.161	16.10	−0.007	−0.011	−0.028	−1.080	38.57	+0.004	+0.000
			Σ	+1.303	95.41			Σ	−1.240	161.97		
			$\Delta Q = -1.303/1.85 \times 95.41 = -0.007$					$\Delta Q = 1.240/1.85 \times 161.97 = 0.004$				
III	4–5	3231.5	+0.005	+0.179	35.80	+0.003	+0.007	+0.015	+1.356	91.00	−0.002	−0.004
	5–8	1346.5	+0.010	+0.269	26.90	+0.003		+0.013	+0.437	33.62	−0.002	
	8–7	2693.0	−0.016	−1.282	80.13	+0.003		−0.013	−0.873	67.12	−0.002	
	7–4	806.0	−0.005	−0.045	9.00	+0.003	−0.011	−0.013	−0.261	20.08	−0.002	+0.000
			Σ	−0.879	151.83			Σ	+0.668	211.85		
			$\Delta Q = 0.879/1.85 \times 151.83 = 0.003$					$\Delta Q = -0.668/1.85 \times 211.85 = -0.002$				

Hence

$$\Delta q_x = -\left(\frac{\sum r_{ij} Q_{ij}^m}{m \sum r_{ij} Q_{ij}^{m-1}}\right)_x$$

or

$$\Delta q_x = -\left(\frac{\sum h_{ij}}{m \sum (h/Q)_{ij}}\right)_x. \tag{5.6}$$

In general, the iterative correction process is continued until the integrated head loss around all loops in the network is reduced to a specified limiting value.

5.7 Loop selection

As shown earlier, the number of independent loop equations for a network is $P - (N - 1)$. Even in simple networks such as that shown in Fig. 5.1, the total number of possible loops is greatly in excess of the number of independent loops, which for this network is three, as shown in Table 5.5. It will be evident that the loop equation based on loop 4 could be obtained by combining the loop equations for loops 1, 2, and 3, while the loop equation for loop 5 could be obtained by combining the loop equations for loops 2 and 3, and so on.

For simple networks, such as that in Fig. 5.1, a set of independent loops is immediately obvious by inspection. This is not usually the case for more complex networks and hence a defined procedure is required for selecting an independent set of loops. The optimal set of independent loops is that set which leads to the most rapid rate of convergence of the iterative calculation process.

It has been shown (Travers 1967) that convergence is most rapid if the chosen set of independent loops comprises those loops for which the

Table 5.5

								$\sum r$
Loop 1	P_{12}	P_{24}	P_{47}	P_{76}	P_{61}			5817.5
Loop 2	P_{23}	P_{35}	P_{54}	P_{42}				5719.5
Loop 3	P_{45}	P_{58}	P_{87}	P_{74}				8077.0
Loop 4	P_{12}	P_{23}	P_{35}	P_{58}	P_{87}	P_{76}	P_{61}	9927.7
Loop 5	P_{23}	P_{35}	P_{58}	P_{87}	P_{74}	P_{42}		7333.5
Loop 6	P_{12}	P_{23}	P_{35}	P_{54}	P_{47}	P_{76}	P_{61}	9925.7
Loop 7	P_{12}	P_{24}	P_{45}	P_{58}	P_{87}	P_{76}	P_{61}	12 283.0

summed resistances (r-values) of their constituent pipes are minimized. Obviously, such loops are based on the low-resistance pipes of the system, that is, the low-resistance pipes are included in as many loops as possible while the high-resistance pipes are included in as few loops as possible.

Ideally, therefore, a loop selection procedure should be able to define the optimal set of independent loops so as to reduce computational time to a minimum. The following loop selection algorithm, described by Travers, approximately satisfies this requirement. The method includes the following steps, as illustrated in Fig. 5.2.

1 All pipes in the system are ranked in order of increasing resistance (r-value), the pipe of lowest resistance becoming pipe number 1, the pipe of next lowest resistance becoming pipe number 2, and so on.
2. A **tree** structure is constructed within the network, consisting of the pipes taken in order from the ranked list. A tree consists of a connected set of pipes and their end nodes, no subset of which comprises a loop. When this procedure is complete, all N nodes of the network will be included and the number of pipes forming the tree structure will be $N - 1$. If the total number of pipes is P, then the number of non-tree pipes is $P - (N - 1)$. These non-tree pipes form the closing pipes of the required set of independent loops.
3. When the tree structure is complete there is then a unique path from each node back to the origin of the tree (one of the end nodes of the pipe of lowest resistance). The loop designated by any of the closing pipes is found as follows. The end nodes of each closing pipe are assigned a level number which corresponds to the number of pipes in the path from that node back

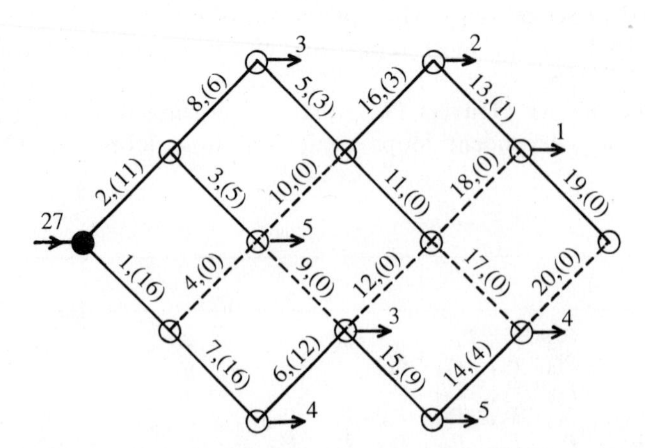

Fig. 5.2 Loop selection and flow initialization: $x(y)$ are pipe rank and initial flow, respectively; ● is the origin of the 'tree' structure; —○——○— indicates a tree pipe; —○---○— indicates a closing pipe.

Table 5.6 Selected set of loops

Loop number	Pipes
1	4, 1, 2, 3
2	10, 3, 8, 5
3	9, 3, 2, 1, 7, 6
4	12, 6, 7, 1, 2, 8, 5, 11
5	17, 14, 15, 6, 7, 1, 2, 8, 5, 11
6	18, 11, 16, 13
7	20, 14, 15, 6, 7, 1, 2, 8, 5, 16, 13, 19

to the origin. Taking, as an example, a particular closing pipe with end nodes *a* and *b* and, starting from the nodes of higher level (node *a*), proceed towards the origin of the tree until a node having the same level as *b* is reached. This node may in fact coincide with node *b*, in which case the required loop is defined. If this is not the case, then proceed along both paths towards the origin, adding in successive pipes in each path in turn until the paths cross before or at the origin, and a closed loop is formed.

This loop selection procedure is outlined in Fig. 5.2 and in Table 5.6.

5.8 Initial flow distribution

Before the iterative process can be started, an initial distribution of flow must be assumed. Where supplies and demands are specified, this initial flow distribution is uniquely defined by the network tree if all closing pipes are assumed to have zero flow. Since the pipes in the tree are mainly the lower-resistance pipes, this initial assignment of flows may be assumed to be nearer the final solution than an arbitrarily assumed distribution. An example of the above procedure is shown in Fig. 5.2.

In cases where fixed heads are specified rather than supply values, a set of initial supply values, which satisfies flow continuity conditions, must be assumed. (This problem is further discussed in the following section.)

5.9 Network flow controls

Flow controls in a water network may include non-return valves, pressure-reducing valves, pumps, and points of fixed pressures such as service reservoirs. The latter are always located at nodes while the remainder are non-nodal in location.

Non-return valves

If the computational procedure reveals a backward flow through a non-return valve, a correction can be made by introducing an equal and opposite loop flow or alternatively the network can be re-analysed with the pipe containing the non-return valve omitted.

Flow-regulating valves

Such devices cause a step-change in head h_v, which can usually be related to the flow Q by an equation of the form:

$$h_v = K_v Q^2 \tag{5.7}$$

where K_v is a variable coefficient whose value is a function of the extent of valve opening. (K_v-values may be calculated from the valve K-values given in Table 3.5.)

The flow correction ΔQ for a loop containing a valve is found by adding the appropriate valve terms to eqn (5.6):

$$\Delta Q = -\frac{\sum h_{ij} + h_v}{m \sum \left(\frac{h}{Q}\right)_{ij} + 2K_v Q}. \tag{5.8}$$

Pumps

Pumps cause a step-increase in head in the direction of flow, which can generally be determined from a pump characteristic equation of the form

$$h_p = A_0 + A_1 Q + A_2 Q^2 \tag{5.9}$$

where A_0 is the 'shut-off' head, and A_1 and A_2 are constant coefficients.

The flow correction ΔQ for a loop containing a pump is found by adding the appropriate pump terms to eqn (5.6):

$$\Delta Q = -\frac{\sum h_{ij} + h_p}{m \sum \left(\frac{h}{Q}\right)_{ij} + (2A_2 Q + A_1)}. \tag{5.10}$$

(Note: for the sign convention which associates positive h values with clockwise loop flow, h_p has a positive value if the pumping direction is anticlockwise.)

Points of fixed head

In network problems where supplies and demands are specified, the nodal pressure distribution can be found provided a single nodal pressure is known. Each additional specified node head constitutes an additional boundary condition. Suppose nodes x and y have specified node heads H_x and H_y, respectively. This boundary condition can be covered by an additional equation of the form

$$\sum h_{ij} = H_x - H_y$$

which, in terms of Q, becomes

$$\sum r_{ij} Q_{ij}^m - (H_x - H_y) = 0 \tag{5.11}$$

where the summation refers to all pipes in the connected path from node x to node y. If this equality is not fulfilled then a correction ΔQ can be applied to all pipes in that path, such that

$$\sum r_{ij}(Q_{ij} + \Delta Q)^m - (H_x - H_y) = 0.$$

This correction is found by the Newton–Raphson procedure to be

$$\Delta Q = -\frac{\sum h_{ij} - (H_x - H_y)}{m \sum \left(\frac{h}{Q}\right)_{ij}}. \tag{5.12}$$

The flow correction ΔQ is also applied to the supply and demand values at nodes x and y, resulting in variations from their assumed values. Thus, where more than one nodal head is fixed, the supplies and demands at points of fixed head must be treated as variables in network computation. If the number of nodes of fixed head (which must be supply or demand nodes) is N_f, the required number of additional boundary equations of the type (5.11) is $N_f - 1$.

Although eqn (5.12) is valid for any connected path between x and y, the most convenient path is that defined by the network tree.

5.10 Analysis of existing distribution systems

Existing water distribution systems may have a large number of interconnected pipes, varying in size from small-diameter service pipes to large-diameter trunk mains. In order to reduce the analytical task to manageable proportions, it is usually necessary to develop a simplified model of the network, which includes the main arteries of flow. This is conveniently carried out using a scale drawing of the network, on which the main pipe intersections (network nodes) are identified. It will usually be possible to replace minor pipes by equivalent supplies or demands at nodal points. A schematic outline of the simplified network can then be prepared.

The data required for network analysis include:

(1) pipe data: length, diameter, roughness, node elevations;

(2) supply and demand data;

(3) data on flow regulation: pumps, control valves, service reservoirs, and so on;

(4) field data from monitoring flows and heads at selected points in the network; this information is essential to verify that the simplified network is a satisfactory model of the actual system.

The model network is analysed and the results are compared with field data. Discrepancies are noted and the model network is appropriately modified and re-analysed. This procedure is repeated until the computed results are found to accord satisfactorily with the field observations. The schematic model can then be regarded as proven and can be reliably used as an analytical tool for the design of modifications and/or extensions.

5.11 Network analysis by computer: program PNA

The analysis of flow and pressure distribution in water networks is a tedious and time-consuming task, conveniently carried out by computer. As well as providing a means of analysis, the computer also allows the storage of network information on file, which is of particular benefit in network design where repeated modification and analysis are typically necessary.

Program PNA is a menu-driven program for the analysis of steady flow in pipe networks, catering for the range of flow controls already discussed, including multiple reservoir inputs, pumps, pressure-reducing valves, and non-return valves. It uses the Travers algorithm for loop selection and flow initialization, as previously described; analysis is based on the Hardy Cross iterative loop flow correction procedure.

The screen printout on the following pages illustrates the general mode of operation of the program, which is driven from a main menu containing eleven algorithms. These operations are selected by the analyst in the appropriate sequence for the entry of data and analysis of flow and pressure distribution.

Menu item 1, `NETWORK`, must be selected as the first operation when starting the analysis of a new network for which data are not already stored on disk. It offers the user a choice of flow units, l.p.s. or $m^3 s^{-1}$, and of flow formulae, Hazen–Williams or Darcy–Weisbach. It also requests the user to specify the required accuracy of flow computation, that is, the maximum residual integrated loop head loss (m).

Menu items 2–8, inclusive, are selected for entry or display of network data. These menu items are PIPE, PUMP, DEMAND/SUPPLY, VALVE N-R, VALVE P-R, HEAD, and ELEVATION.

PIPE is selected to enter or display pipe data. Each pipe is defined by head and tail node numbers, diameter, and *C*-value or wall roughness, as appropriate. The positive flow direction is from head to tail node. Thus, if the computed flow in a pipe is from its head node to its tail node, it is output as a positive value; if in the opposite direction, it is output as a negative value.

DEMAND/SUPPLY is selected to enter or display supply and demand values. The program requires that supplies are entered as negative values, demands as positive values.

PUMP is selected to enter or display pump data. The location of a pump is specified by the head and tail nodes of the pipe on which it is located. Pump characteristics are expressed in terms of the coefficients A_0, A_1, and A_2 in the pump equation (5.9). These coefficient values should be based on the computational flow units selected under menu item NETWORK.

Similarly, the data for non-return and pressure-reducing valves are entered by selecting VALVE N-R and VALVE P-R, respectively. HEAD and ELEVATION are selected to enter or display fixed heads and node elevations, respectively. If node elevations are not supplied as input data, a default value of 1000 m is assumed for all nodes. Fixed head values (service reservoirs) are entered as piezometric values relative to the same datum as the node elevations.

Menu item 9, FILE, writes the network data to a disk file on the current drive, thus providing a record for future analyses.

Menu item 10, *ANALYSIS*, activates the flow and pressure distribution analysis. The program is coded to run for a minimum of twenty iterations. If the specified accuracy is not reached after twenty iterations, computation continues until the required accuracy is achieved. As the computation proceeds, the iteration number and prevailing maximum error are printed on the screen. This enables the user to observe the rate of convergence.

The program output consists of two tables presenting flow and pressure distribution, respectively, as shown on the following illustrative example. The number of iterations executed to reach the specified error tolerance is also printed.

Sample computation

The network shown in Fig. 5.3 is used to illustrate the application of program PNA. This network has 17 pipes, 4 independent loops, 4 service reservoirs,

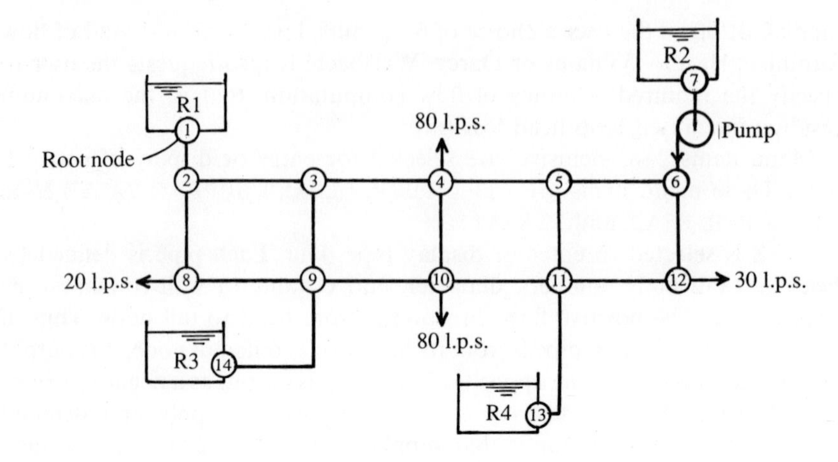

Fig. 5.3 Pipe network diagram.

and 1 pump. The input data in the form required for the program are given in Tables 5.7 and 5.8. In the absence of supplied node elevations (menu item 8, ELEVATION), the default value of 1000 m OD (over datum) is assumed for all nodes. It should be noted that head values (menu item 7, HEAD) are expressed in m relative to the same datum as the node elevations.

Table 5.7 Pipe data for the network shown in Fig. 5.3.

Node 1	Node 2	Length (m)	Diameter (mm)	Roughness (mm)
1	2	100	200	0.10
2	3	100	150	0.15
3	4	100	150	0.15
4	5	100	75	0.10
5	6	100	75	0.15
7	6	100	100	0.15
2	8	100	75	0.15
3	9	100	75	0.10
4	10	100	75	0.10
5	11	100	75	0.10
6	12	100	100	0.10
8	9	100	75	0.15
9	10	100	150	0.10
10	11	100	150	0.10
11	12	100	75	0.15
11	13	100	150	0.15
9	14	100	150	0.10

Table 5.8 Supply, demand, and head data for the network shown in Fig. 5.3. An arbitrary initial distribution of supplies between reservoirs is assumed, subject to the continuity requirement that the sum of supplies should equal the total demand.

Node	Supply (l.p.s.)	Demand (l.p.s.)	Head (m)
1	70	–	1100.0
4	–	80	–
7	30	–	1099.0
8	–	20	–
10	–	80	–
12	–	30	–
13	55	–	1101.0
14	55	–	1101.0

$$\text{Pump equation:} \quad h_p = 10 - 0.06Q - 0.0004Q^2$$

where the units are h_p (m) and Q (l.p.s.).

Program output is presented in Table 5.9 (flow distribution) and Table 5.10 (supplies, demands, and head distribution).

The pipe numbers, given in the first column of Table 5.9, are in order of increasing pipe resistance, that is, pipe number 1 has the lowest hydraulic resistance. Flow from node 1 (head node) to node 2 (tail node) is assigned a positive sign, while a negative sign indicates the reverse direction of flow.

In Table 5.10 supplies and demands are outputted in the 'Demand' column of the table, negative numbers being used to indicate supplies.

It will be noted that the computed supplies from the four reservoirs differ in magnitude from the assumed values given in Table 5.8. The entered values were arbitrarily distributed between the four reservoirs, subject to the continuity requirement that their sum equalled the total demand.

The validity of the computed flow and pressure distributions should be manually checked by verifying the consistency of the flow and node head differences for selected individual pipes.

```
RUN
          ****************************************
          *      PIPE NETWORK ANALYSIS PROGRAM    *
          *                                       *
          *               PNA.BAS                 *
          *                                       *
          ****************************************
          * DEPT. OF CIVIL ENGINEERING, UCD, 1991  *
          ****************************************

THIS PROGRAM ANALYSES FLOW IN WATER PIPE NETWORKS

DO YOU NEED INSTRUCTIONS ON HOW TO USE THIS PROGRAM? (Y/N)
? Y

                         NOTES ON PROGRAM USE

 This program calculates the flow distribution in pipe networks using the
 Hardy Cross loop flow correction procedure. The program selects
 a set of loops for the network and also an initial estimate of flow
 distribution. Thus, only that information which is required to define the
 problem, needs to be input by the user.
 First, make a sketch of your pipe network and assign a number to each node
 (pipe junction), numbering the nodes consecutively 1,2,3.....NN. Then draw
 up a pipe data table, which has the following data for each pipe: head node
 number, tail node number, length, diameter and C-value or roughness. When
 this table is completed, number the pipes consecutively 1,2,3...NP.
 Booster pump data is input by assigning values to the constants in the
 characteristic eqn.: H=A*Q**2+B*Q+C. Pressure-reducing valve data is input
 by assigning a value to the constant C in the relation H=C*Q**2.
 PROGRAM OPERATION: The program is operated through a main menu which offers
 the user a choice of operations. Each start-up of the program for a non-
 filed network is initiated by selecting operation 1 (NETWORK) from the
 main menu. The input data is filed on disk for future use by selecting
 operation 9 (FILE) from the main menu.

                      PRESS THE SPACE BAR TO CONTINUE
DO YOU WISH TO ANALYZE A NEW NETWORK (NN) OR A NETWORK FOR WHICH DATA
IS ON FILE (FN) ?; TYPE FN OR NN AS APPROPRIATE
? NN
ENTER TITLE OF NETWORK AND DATE FOR RECORDING ON OUTPUT
TITLE (up to 80 characters) ? BALLYVARRA  04-07-1991

BALLYVARRA  04-07-1991  12-21-1991

PROGRAM MENU

 1.    NETWORK              ENTER FOR EACH NEW NETWORK
 2.    PIPE                 ENTERS/DISPLAYS PIPE DATA
 3.    PUMP                 ENTERS/DISPLAYS PUMP DATA
 4.    DEMAND/SUPPLY        ENTERS/DISPLAYS SUPPLY/DEMAND DATA
 5.    VALVE N-R            ENTERS/DISPLAYS NR VALVE DATA
 6.    VALVE P-R            ENTERS/DISPLAYS PR VALVE DATA
 7.    HEAD                 ENTERS/DISPLAYS FIXED HEAD VALUES
 8.    ELEVATION            ENTERS/DISPLAYS NODE ELEVATION DATA
 9.    FILE                 WRITES NETWORK DATA TO A FILE
10.   *ANALYSIS*            RUNS ANALYSES ON DATA
11.    END PROGRAM


ENTER THE NUMBER OF YOUR CHOICE? 1

NETWORK IS BEING INITIALISED
```

```
INPUT UNITS
SPECIFY FLOW COMPUTATION UNITS:

              1. m**3/s - CUBIC METRES PER SECOND
              2. LPS    - LITRES PER SECOND

ENTER 1 OR 2, AS APPROPRIATE? 2

ENTER ACCURACY OF FLOW COMPUTATION
i.e. INTEGRATED LOOP HEAD LOSS TOLERANCE (m)? 0.05

SPECIFY PIPE FLOW FORMULA TO BE USED:

               1. HAZEN-WILLIAMS (C-VALUE) OR
               2. DARCY-WEISBACH (SURFACE ROUGHNESS)

ENTER 1 OR 2, AS APPROPRIATE? 2

PROGRAM MENU

 1.     NETWORK                 ENTER FOR EACH NEW NETWORK
 2.     PIPE                    ENTERS/DISPLAYS PIPE DATA
 3.     PUMP                    ENTERS/DISPLAYS PUMP DATA
 4.     DEMAND/SUPPLY           ENTERS/DISPLAYS SUPPLY/DEMAND DATA
 5.     VALVE N-R               ENTERS/DISPLAYS NR VALVE DATA
 6.     VALVE P-R               ENTERS/DISPLAYS PR VALVE DATA
 7.     HEAD                    ENTERS/DISPLAYS FIXED HEAD VALUES
 8.     ELEVATION               ENTERS/DISPLAYS NODE ELEVATION DATA
 9.     FILE                    WRITES NETWORK DATA TO A FILE
10.    *ANALYSIS*               RUNS ANALYSES ON DATA
11.     END PROGRAM

ENTER THE NUMBER OF YOUR CHOICE? 2

INPUT OF PIPE DATA

NUMBER OF PIPES IN NETWORK? 17

ENTER NUMERICAL VALUES, SEPARATED BY COMMAS

NODE1, NODE2, LENGTH(m), DIAMETER(mm), ROUGHNESS(mm)
? 1,2,100,200,0.10
? 2,3,100,150,0.15
? 3,4,100,150,0.15
? 4,5,100,75,0.10
? 5,6,100,75,0.15
?

      etc.
```

Table 5.9 Computed flows for the network shown in Fig. 5.3.

			Output of results				
Pipe no.	Node 1	Node 2	Flow l.p.s.	Length m	Diameter mm	Roughness mm	control
1	1	2	66.094	100	200.0	0.10	
2	3	4	60.516	100	150.0	0.15	
3	9	10	45.104	100	150.0	0.10	
4	2	3	54.880	100	150.0	0.15	
5	9	14	−59.526	100	150.0	0.10	
6	10	11	−43.985	100	150.0	0.10	
7	11	13	−57.948	100	150.0	0.15	
8	6	12	20.991	100	100.0	0.10	
9	5	6	−5.442	100	75.0	0.15	
10	3	9	−5.635	100	75.0	0.10	
11	5	11	−4.953	100	75.0	0.10	
12	11	12	9.009	100	75.0	0.15	
13	4	5	−10.395	100	75.0	0.10	
14	2	8	11.213	100	75.0	0.15	
15	4	10	−9.089	100	75.0	0.10	
16	8	9	−8.787	100	75.0	0.15	
17	7	6	26.432	100	100.0	0.15	Pump

Table 5.10 Computed heads for the network shown in Fig. 5.3.

Node no.	Demand l.p.s.	Gauge m	Elevation m	Total head m
1	−66.094	100.00	1000.00	1100.00
2	0.000	97.98	1000.00	1097.98
3	−0.000	91.32	1000.00	1091.32
4	80.000	83.17	1000.00	1083.17
5	0.000	91.60	1000.00	1091.60
6	0.000	94.20	1000.00	1094.20
7	−26.432	99.00	1000.00	1099.00
8	20.000	87.32	1000.00	1087.32
9	0.000	93.82	1000.00	1093.82
10	80.000	89.65	1000.00	1089.65
11	0.000	93.62	1000.00	1093.62
12	30.000	86.67	1000.00	1086.67
13	−57.948	101.00	1000.00	1101.00
14	−59.526	101.00	1000.00	1101.00

References

Cross, Hardy (1936). Analysis of flow in networks of pipe conduits or conductors. *Univ. of Illinois Bull.* 286.

Epp, R. and Fowler, A. G. (1970). Efficient code for steady state flows in networks. *J. Hyd. Div. ASCE*, **96**, No. HY1, 43–56.

Martin, D. W. and Peters, G. (1963). The application of Newton's method of network analysis by digital computer. *J. Inst. Wat. Eng.*, **17**, 115–29.

Travers, K. (1967). The mesh method in gas network analysis. *Gas Jour.*, **332**, 167–74.

Wood, D. J. and Charles, O. A. (1972). Hydraulic network analysis using linear theory. *J. Hyd. Div. ASCE*, No. HY7, 1157–70.

6
Unsteady flow in pipes

6.1 Introduction

Unsteady flow in pipes results primarily from the operation of flow regulation devices such as valves or pumps. Its practical significance is due to the fact that the associated pressure changes may exceed the permitted value or fluctuation range for the pipe material. Such transient pressures are dependent on a number of factors, including the rate of acceleration or deceleration of the fluid, the compressibility of the fluid, the elasticity of the pipe, and the overall geometry of the pipe system.

The audible noise sometimes associated with unsteady pipe flow is often described as 'waterhammer', due to the hammer-like sound sometimes emitted as a result of vapour pocket collapse or pipe vibration. The basic equations which describe unsteady flow in pipes are developed by applying the principles of continuity and momentum to a control volume, as illustrated in Fig. 6.1. These basic equations, together with appropriate boundary condition equations, define the flow regime and their solution allows the prediction of the variation of dependent variables, pressure (p) and flow velocity (v), with independent variables, time (t) and location (x).

6.2 The continuity equation

The continuity or mass balance equation is developed for the flow length ∂x:

Mass inflow rate − Mass outflow rate = Rate of change of contained mass

$$\rho A v - \left\{\rho A v + \frac{\partial}{\partial x}(\rho A v)\,\partial x\right\} = \frac{\partial}{\partial t}(\rho A\,\partial x) \tag{6.1}$$

where A is the pipe cross-sectional area and ρ is the fluid density. Simplifying (6.1), we get

$$\frac{\partial}{\partial x}(\rho A v)\,\partial x + \frac{\partial}{\partial t}(\rho A\,\partial x) = 0.$$

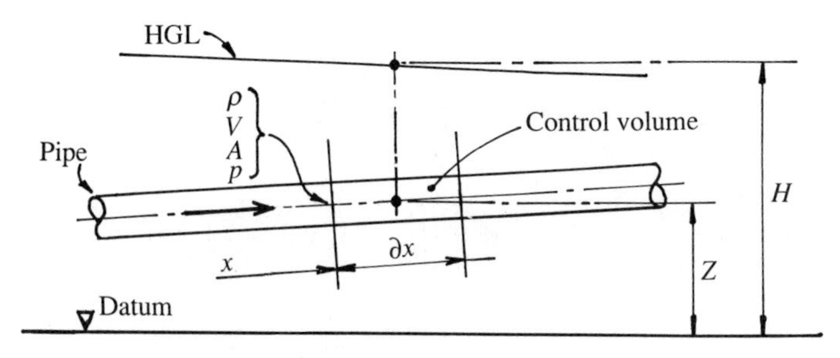

Fig. 6.1 Control volume definition.

Hence

$$v\frac{\partial}{\partial x}(\rho A)+\rho A\frac{\partial v}{\partial x}+\frac{\partial}{\partial t}(\rho A)=0$$

since

$$\frac{\mathrm{d}}{\mathrm{d}t}(\rho A)=v\frac{\partial}{\partial x}(\rho A)+\frac{\partial}{\partial t}(\rho A).$$

Hence

$$\rho A\frac{\partial v}{\partial x}+\frac{\mathrm{d}}{\mathrm{d}t}(\rho A)=0$$

$$\rho A\frac{\partial v}{\partial x}+A\frac{\mathrm{d}\rho}{\mathrm{d}t}+\rho\frac{\mathrm{d}A}{\mathrm{d}t}=0. \tag{6.2}$$

The change in fluid density is a function of the increase in pressure and the fluid bulk modulus. By definition, the bulk modulus $K=\mathrm{d}p/(-\mathrm{d}V/V)$, where V is the volume. Density and volume change are related thus: $-\mathrm{d}V/V=\mathrm{d}\rho/\rho$. Hence, we have the following correlation of density change with pressure change: $\mathrm{d}\rho=(\rho/K)\,\mathrm{d}p$. The change in pipe cross-sectional area is a function of the change in fluid pressure, the wall thickness of the pipe T, and the Young's modulus E of the pipe material. The change in area is $\mathrm{d}A=2\pi R\,\mathrm{d}R=(2A/R)\,\mathrm{d}R$, where R is the pipe radius. The increase in radius is $\mathrm{d}R=(\mathrm{d}p\,R^2)/TE$; hence $\mathrm{d}A=\mathrm{d}p(AD/TE)$, where D is the pipe diameter.

Equation (6.2) may therefore be written as follows:

$$\rho A\frac{\partial v}{\partial x}+\frac{A\rho}{K}\frac{\mathrm{d}p}{\mathrm{d}t}+\frac{\rho AD}{TE}\frac{\mathrm{d}p}{\mathrm{d}t}=0$$

$$\rho\frac{\partial v}{\partial x}+\frac{\mathrm{d}p}{\mathrm{d}t}\left(\frac{\rho}{K}+\frac{\rho D}{TE}\right)=0$$

$$\rho\frac{\partial v}{\partial x}+\frac{1}{\alpha^2}\frac{\mathrm{d}p}{\mathrm{d}t}=0$$

where

$$\alpha = \frac{1}{\sqrt{\rho/K + \rho D/TE}}.$$

Hence

$$\rho\frac{\partial v}{\partial x} + \frac{\rho g}{\alpha^2}\left(\frac{\partial H}{\partial x}v + \frac{\partial H}{\partial t} - \frac{\partial Z}{\partial x}v - \frac{\partial Z}{\partial t}\right) = 0 \tag{6.3}$$

where

$$p = \rho g(H - Z) \qquad \text{and} \qquad \frac{\mathrm{d}p}{\mathrm{d}t} = v\frac{\partial p}{\partial x} + \frac{\partial p}{\partial t}.$$

Equation (6.3) can be written in the form

$$\frac{\partial H}{\partial t} + v\frac{\partial H}{\partial x} - v\sin\theta + \frac{\alpha^2}{g}\frac{\partial v}{\partial x} = 0. \tag{6.4}$$

This is the desired form of the continuity equation. α is the speed of propagation of a pressure wave through the pipe; its magnitude is dependent on two factors: the bulk modulus of the fluid K and the rigidity of the pipe, as measured by thc ratio TE/D. The expression for α may be modified to take into account the Poisson effect on pipe expansion and the influence of pipe anchorage conditions (Wylie and Streeter 1978):

$$\alpha = \frac{1}{\sqrt{\rho/K + \rho CD/TE}} \tag{6.5}$$

where C is an anchorage coefficient with values as follows:

(1) pipe anchored at upstream end only: $C = 1 - \mu/2$;

(2) pipe anchored throughout against axial movement: $C = 1 - \mu^2$;

(3) pipe with expansion joints: $C = 1$.

where μ is Poisson's ratio for the pipe material.

The practical range of wavespeeds encountered in the water engineering field varies from about 1400 m s^{-1} for small-diameter steel pipes to about 280 m s^{-1} for low-pressure PVC pipes, with intermediate values for pipes in materials such as asbestos-cement and concrete (Creasey *et al.* 1977). A small amount of free gas (that is, undissolved gas) has a considerable influence on wavespeed, effecting a reduction in wavespeed as the pressure drops and the gas volume expands (Wylie and Streeter 1978). A free gas phase can arise from air intake through air valves or from air release from solution during negative gauge pressure or from the biological production of gases in wastewaters.

6.3 The momentum equation

The force–momentum relation is applied to the fluid contained in the control volume defined in Fig. 6.1:

$$pA - \left(pA + \frac{\partial}{\partial x}(pA)\,\partial x\right) - \tau_0 \pi D\,\partial x - \rho g A\,\partial x \sin\theta = \frac{\mathrm{d}v}{\mathrm{d}t}(\rho A\,\partial x). \quad (6.6)$$

Simplifying:

$$\frac{\partial}{\partial x}(pA) + \tau_0 \pi D + \rho g A \sin\theta + \rho A \frac{\mathrm{d}v}{\mathrm{d}t} = 0 \quad (6.7)$$

where τ_0 is the wall shear stress given by $\rho g R_\mathrm{h} S_\mathrm{f}$, R_h being the hydraulic radius, $D/4$, and S_f being the friction slope $fv|v|/2gD$. Expressing pressure in terms of H and Z, eqn (6.7) may be written in the form

$$\frac{\partial(H - Z)}{\partial x} + S_\mathrm{f} + \sin\theta + \frac{1}{g}\frac{\mathrm{d}v}{\mathrm{d}t} = 0$$

or

$$\frac{\partial H}{\partial x} - \frac{\partial Z}{\partial x} + \frac{fv|v|}{2gD} + \sin\theta + \frac{1}{g}\left(v\frac{\partial v}{\partial x} + \frac{\partial v}{\partial t}\right) = 0;$$

hence

$$g\frac{\partial H}{\partial x} + v\frac{\partial v}{\partial x} + \frac{\partial v}{\partial t} + \frac{fv|v|}{2D} = 0 \quad (6.8)$$

which is the desired form of the momentum equation. Note that $|v|$ means the absolute value of v. In expanding the term $\partial(pA)/\partial x$, it has been assumed that $\partial A/\partial x$ can be neglected; also $\partial Z/\partial x = \sin\theta$.

6.4 Solution by the method of characteristics

A general solution to the above pair of partial differential equations (variables v, H, x, t) is not available. They can, however, be transformed by the method of characteristics into a set of total differential equations which can be integrated to finite difference form for convenient solution by numerical methods. Note that v and H are the dependent variables, while x and t are the independent variables.

For ease of solution the two equations can be simplified by omitting the less important terms as follows:

the continuity equation $$\frac{\partial H}{\partial t} + v\frac{\partial H}{\partial x} - v\sin\theta + \frac{\alpha^2}{g}\frac{\partial v}{\partial x} = 0$$

becomes

$$\frac{\partial H}{\partial t}+\frac{\alpha^2}{g}\frac{\partial v}{\partial x}=0; \tag{6.9}$$

the momentum equation

$$g\frac{\partial H}{\partial x}+v\frac{\partial v}{\partial x}+\frac{\partial v}{\partial t}+\frac{fv|v|}{2D}=0$$

becomes

$$g\frac{\partial H}{\partial x}+\frac{\partial v}{\partial t}+\frac{fv|v|}{2D}=0. \tag{6.10}$$

Multiplying eqn (6.9) by a factor λ and adding to eqn (6.10):

$$\lambda\left(\frac{g}{\lambda}\frac{\partial H}{\partial x}+\frac{\partial H}{\partial t}\right)+\frac{\lambda\alpha^2}{g}\frac{\partial v}{\partial x}+\frac{\partial v}{\partial t}+\frac{fv|v|}{2D}=0$$

which can be written in total differential form as

$$\lambda\frac{\mathrm{d}H}{\mathrm{d}t}+\frac{\mathrm{d}v}{\mathrm{d}t}+\frac{fv|v|}{2D}=0 \tag{6.11}$$

provided that

$$\frac{\mathrm{d}x}{\mathrm{d}t}=\frac{g}{\lambda}=\frac{\lambda\alpha^2}{g}.$$

Hence $\lambda=\pm g/\alpha$, so

$$\frac{\mathrm{d}x}{\mathrm{d}t}=\pm\alpha. \tag{6.12}$$

Equations (6.11) and (6.12) are the equivalent total differential forms of the partial differential continuity and momentum equations. They can be written as two linked pairs of equations ('characteristic equations') as follows:

$$C^+\begin{cases}+\dfrac{g}{\alpha}\dfrac{\mathrm{d}H}{\mathrm{d}t}+\dfrac{\mathrm{d}v}{\mathrm{d}t}+\dfrac{fv|v|}{2D}=0 & (6.13)\\[2ex] \dfrac{\mathrm{d}x}{\mathrm{d}t}=+\alpha & (6.14)\end{cases}$$

$$C^-\begin{cases}-\dfrac{g}{\alpha}\dfrac{\mathrm{d}H}{\mathrm{d}t}+\dfrac{\mathrm{d}v}{\mathrm{d}t}+\dfrac{fv|v|}{2D}=0 & (6.15)\\[2ex] \dfrac{\mathrm{d}x}{\mathrm{d}t}=-\alpha. & (6.16)\end{cases}$$

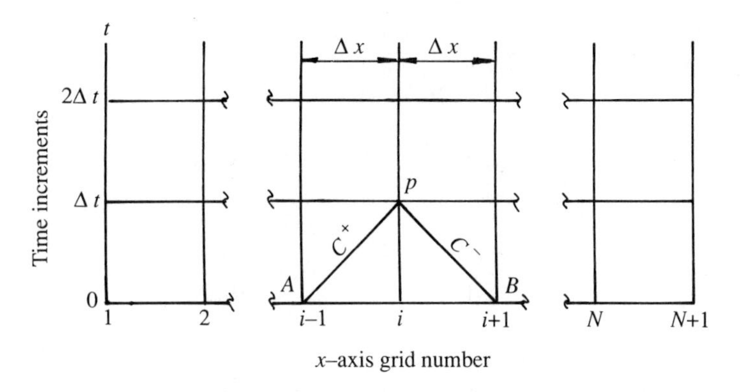

Fig. 6.2 The x–t finite difference grid.

Equations (6.14) and (6.16) are graphically represented as straight lines on the x–t plane, as illustrated in Fig. 6.2. Equations (6.13) and (6.15) define the variations of H and v with time subject to the x–t relationships of eqns (6.14) and (6.16), respectively.

6.4.1 Finite difference formulations

The pipeline is divided into N reaches, each of length Δx, from which the computational time step Δt is calculated:

$$\Delta t = \frac{\Delta x}{\alpha}. \tag{6.17}$$

Integrating eqn (6.13) along the C^+ characteristic:

$$\int \mathrm{d}H + \frac{\alpha}{g}\int \mathrm{d}v + \frac{\alpha f}{2gD}\int v|v|\,\mathrm{d}t = 0.$$

Replacing v by Q/A:

$$\int \mathrm{d}H + \frac{\alpha}{gA}\int \mathrm{d}Q + \frac{\alpha f}{2gDA^2}\int Q|Q|\,\mathrm{d}t = 0.$$

Integrating over the interval Δx:

$$H_P - H_A + \frac{\alpha}{gA}(Q_P - Q_A) + \frac{f\,\Delta x}{2gDA^2}Q_A|Q_A| = 0. \tag{6.18}$$

Similarly for the C^- characteristic equations:

$$H_P - H_B - \frac{\alpha}{gA}(Q_P - Q_B) - \frac{f\,\Delta x}{2gDA^2}Q_B|Q_B| = 0. \tag{6.19}$$

Equations (6.18) and (6.19) can be written as

$$C^+: \qquad H_P = H_A - B(Q_P - Q_A) - RQ_A|Q_A| \tag{6.20}$$

$$C^-: \qquad H_P = H_B + B(Q_P - Q_B) + RQ_B|Q_B| \tag{6.21}$$

where

$$B = \frac{\alpha}{gA} \quad \text{and} \quad R = \frac{f\,\Delta x}{2gDA^2}.$$

Thus if H_A, Q_A, H_B, and Q_B are known, the values of H_P and Q_P can be calculated by solution of eqns (6.20) and (6.21). Referring to the x–t plane, note the displacement in space and time of P from A and B.

Equations (6.20) and (6.21) can be written in grid reference form as follows:

$$C^+: \qquad H_{Pi} = H_{i-1} - B(Q_{Pi} - Q_{i-1}) - RQ_{i-1}|Q_{i-1}| \tag{6.22}$$

$$C^-: \qquad H_{Pi} = H_{i+1} + B(Q_{Pi} - Q_{i-1}) + RQ_{i+1}|Q_{i+1}|. \tag{6.23}$$

Assembling known values together:

$$H_{i-1} + BQ_{i-1} - RQ_{i-1}|Q_{i-1}| = CP$$

$$H_{i+1} - BQ_{i+1} + RQ_{i+1}|Q_{i+1}| = CM.$$

Equations (6.22) and (6.23) can thus be written as

$$H_{Pi} = CP - BQ_{Pi} \tag{6.24}$$

$$H_{Pi} = CM + BQ_{Pi}. \tag{6.25}$$

Solving for H_{Pi} and Q_{Pi}:

$$H_{Pi} = \frac{CP + CM}{2} \tag{6.26}$$

$$Q_{Pi} = \frac{CP - CM}{2B}. \tag{6.27}$$

Thus the computation procedure uses the current values of H and Q at points $i-1$ and $i+1$ to compute their values at point i at one time interval Δt later. Usually, the starting values are known from a prevailing prior steady flow condition.

6.5 Boundary conditions

In general, waterhammer results from a sudden change in the operational mode of a flow control device such as a pump, valve, or turbine. These devices may be located at either end of a pipeline or at some intermediate point. If the control is at the downstream end ($x = L$), the C^+ characteristic equation can be used, while if it is at the upstream end, the C^- characteristic equation can be applied. The second equation in each case is provided by the H/Q relation for the control device itself. The following are typical boundary condition equations.

6.5.1 Reservoir

1. At the upstream end of the line:

$$C^- \text{ characteristic:} \qquad H_{P1} = CM + BQ_{P1}$$

$$\text{Boundary condition:} \qquad H_{P1} = H_R$$

where H_R is the fixed reservoir head. Hence, compute Q_{P1}.

2. At the downstream end of the line:

$$C^+ \text{ characteristic:} \qquad H_{P(N+1)} = CP - BQ_{P(N+1)}$$

$$\text{Boundary condition:} \qquad H_{P(N+1)} = H_R.$$

Hence, compute $Q_{P(N+1)}$.

This is illustrated in Fig. 6.3.

6.5.2 Pump at the upstream end (running at fixed speed)

As already shown in Chapter 5, the characteristic head–discharge curve for a fixed-speed rotodynamic pump can be expressed as follows:

$$h_p = A_0 + A_1 Q + A_2 Q^2 \tag{5.9}$$

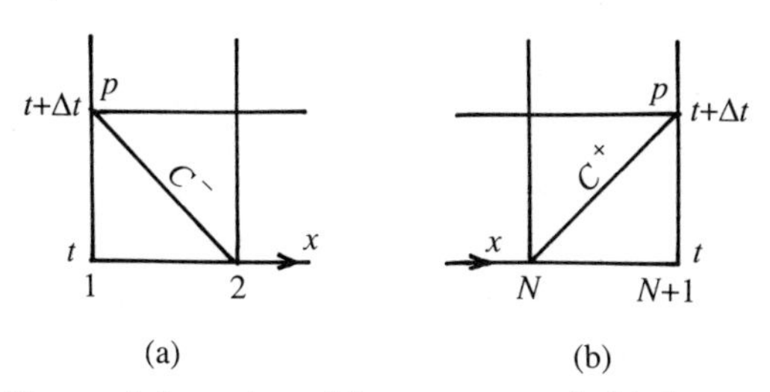

Fig. 6.3 Reservoir boundary: (a) upstream end; (b) downstream end.

where A_0 is the shut-off head and A_1 and A_2 are constant coefficients. Taking the water level in the pump sump as reference datum and neglecting losses in the suction line, the relevant equations may be written as

$$C^- \text{ characteristic:} \qquad H_{P1} = CM + BQ_{P1}$$

$$\text{Pump boundary condition:} \qquad H_{P1} = A_0 + A_1 Q_{P1} + A_2 Q_{P1}^2.$$

Solution of these equations yields the following:

$$Q_{P1} = \frac{1}{2A_2}\left[B - A_1 + \sqrt{(B - A_1)^2 + 4A_2(CM - A_0)}\right]$$

$$H_{P1} = A_0 + A_1 Q_{P1} + A_2 Q_{P1}^2.$$

6.5.3 Control valve at downstream end

As indicated in Chapter 5, the head loss h_v across a valve may be expressed in the form

$$h_v = K_v Q^2 \tag{5.7}$$

where K_v is a valve coefficient, the value of which may be computed from the valve K-values given in Table 3.5.

The downstream boundary condition equations thus become

$$C^+ \text{ characteristic:} \qquad H_{P(N+1)} = CP - BQ_{P(N+1)}$$

$$\text{Valve flow:} \qquad Q_{P(N+1)} = \sqrt{\frac{H_{P(N+1)}}{K_v}}$$

where the downstrean valve level is taken as datum, and there is a free discharge; where the discharge is to a fixed level reservoir, the reservoir level is taken as datum.

Solution of these two equations for Q yields the following:

$$Q_{P(N+1)} = -\frac{B}{2K_v} + \sqrt{\frac{1}{4}\left(\frac{B}{K_v}\right)^2 + \frac{CP}{K_v}}.$$

Knowing $Q_{P(N+1)}$, $H_{P(N+1)}$ is found from the C^+ characteristic equation.

The special case of a closed valve or 'dead end' at the downstream end of a pipeline is found from the foregoing equations by giving K_v an infinite value, resulting in the following expressions for H and Q:

$$H_{P(N+1)} = CP$$

$$Q_{P(N+1)} = 0.$$

6.5.4 Valve at an intermediate location

The upstream side of the valve is denoted by the subscript i and the downstream side by the subscript $i + 1$, as shown in Fig. 6.4.

The defining equations are:

$$C^+ \text{ characteristic:} \qquad H_{Pi} = CP - BQ_{Pi}$$

$$C^- \text{ characteristic:} \qquad H_{P(i+1)} = CM + BQ_{P(i+1)}$$

$$\text{Valve flow:} \qquad Q_{Pi} = \sqrt{\frac{H_{Pi} - H_{P(i+1)}}{K_v}}$$

$$\text{Continuity:} \qquad Q_{Pi} = Q_{P(i+1)}.$$

Solution of these four equations for Q yields

$$Q_{Pi} = -\frac{B}{K_v} + \sqrt{\left(\frac{B}{K_v}\right)^2 - \frac{CM - CP}{K_v}}.$$

Hence H_{Pi} and $H_{P(i+1)}$ can be found from the C^+ and C^- characteristic equations, respectively.

6.5.5 Change in pipe size

Assuming the junction is at node i, as illustrated in Fig. 6.5, the following equations apply:

$$C^+ \text{ characteristic:} \qquad H_{Pi} = CP_1 - B_1 Q_{Pi}$$

$$C^- \text{ characteristic:} \qquad H_{Pi} = CM_2 + B_2 Q_{Pi}.$$

Hence

$$Q_{Pi} = \frac{CP_1 - CM_2}{B_1 + B_2}$$

where the subscript 1 refers to pipe 1 and the subscript 2 refers to pipe 2, respectively. It should be noted that the pipe length between node points

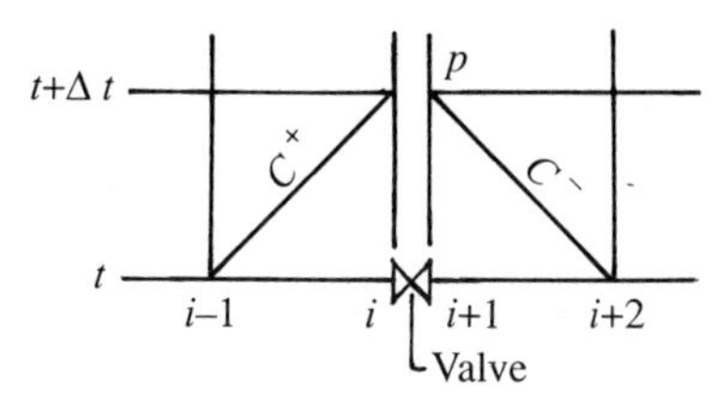

Fig. 6.4 Internal valve.

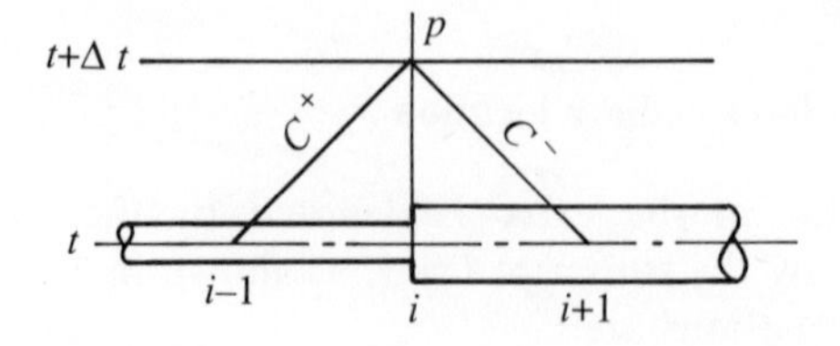

Fig. 6.5 Change in pipe size.

will not be the same in the two pipes if their wavespeeds differ, since this length is $\Delta x = \Delta t \cdot \alpha$, where α is the wavespeed.

The same approach can be applied to any other change in pipe properties and can also be extended to pipe junctions.

6.6 Pressure transients due to pump starting and stopping

The start-up and stopping of pumps give rise to rapid change in pipeline velocity and hence waterhammer effects. This is particularly so in pump cut-out due to power failure. This latter condition must be evaluated in all pumping installations to ensure that pressures are contained within permissible value ranges.

6.6.1 Pump characteristics

This discussion is confined to pumps of the rotodynamic type. The performance of rotodynamic pumps is defined for a normal or rated speed in terms of the parameters h_R, Q_R, N_R, and T_R, referring to head, discharge, speed, and torque, respectively. The values of these parameters at any other speed N can be related to the rated values as follows:

$$\frac{Q_N}{Q_R} = \frac{N}{N_R}; \qquad \frac{h_N}{h_R} = \frac{N^2}{N_R^2}; \qquad \frac{T_N}{T_R} = \frac{N^2}{N_R^2}. \tag{6.28}$$

The torque–discharge pump characteristic may be expressed in a quadratic form similar to the head–discharge equation:

$$T = B_0 + B_1 Q + B_2 Q^2. \tag{6.29}$$

By applying the homologous relationships of (6.28) to the head–discharge and torque–discharge equations, the following expressions for pump head and pump torque, at any speed N, are found:

$$h_N = A_0\left(\frac{N}{N_R}\right)^2 + A_1\left(\frac{N}{N_R}\right)Q + A_2Q^2 \tag{6.30}$$

$$T_N = B_0\left(\frac{N}{N_R}\right)^2 + B_1\left(\frac{N}{N_R}\right)Q + B_2Q^2 \tag{6.31}$$

When a pump motor cuts out, the inertia of the rotating parts maintains a decreasing pump output in accordance with the deceleration relationship:

$$T = -I\frac{\mathrm{d}\omega}{\mathrm{d}t} \tag{6.32}$$

where T is the reactive torque of the fluid, I is the moment of inertia of the rotating elements of the pump set, and $-\mathrm{d}\omega/\mathrm{d}t$ is the angular deceleration. Thus

$$\mathrm{d}\omega = -\frac{T}{I}\,\mathrm{d}t$$

or

$$\mathrm{d}N = -\frac{60}{2\pi}\frac{T_N}{I}\,\mathrm{d}t \tag{6.33}$$

where N is the rotational speed in r.p.m. The pump characteristics at the reduced speed can be determined from the rated values using the relationships (6.30) and (6.31).

6.6.2 Pump cut-out: governing equations for pump node

The pump node is illustrated in Fig. 6.6. The pump operational condition being considered is deceleration under zero external power input. The governing equations are as follows:

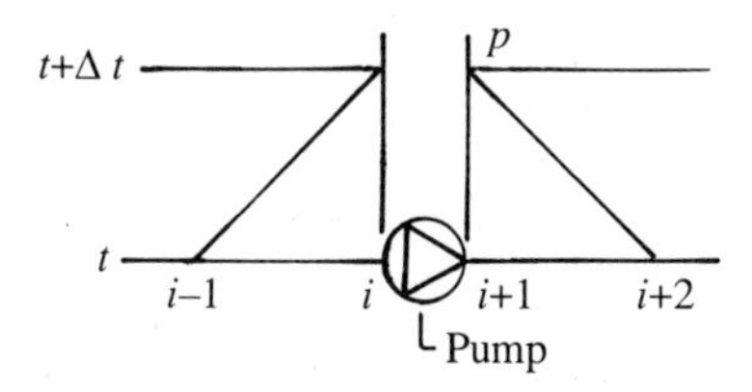

Fig. 6.6 Pump boundaries.

C^+ characteristic: $H_{Pi} = CP - BQ_{Pi}$

C^- characteristic: $H_{P(i+1)} = CM + BQ_{P(i+1)}$

Pump discharge: $H_{P(i+1)} - H_{Pi} = h_N$

Continuity: $Q_{Pi} = Q_{P(i+1)}$

where h_N is the pump manometric head at pump speed N, as expressed by eqn (6.30). Simultaneous solution of the above equations gives the following value for Q_{Pi}:

$$Q_{Pi} = \frac{-(A_{1N} - 2B) + \sqrt{(A_{1N} - 2B)^2 - 4A_2(A_{0N} - CM + CP)}}{2A_2}$$

where $A_{1N} = A_1(N/N_R)$ and $A_{0N} = A_0(N/N_R)^2$. Knowing Q_{Pi}, H_{Pi} and $H_{P(i+1)}$ can be determined from the foregoing C^+ and C^- characteristic equations. The step change in N for each time interval Δt is found from eqn (6.33). The value of T_N for any Q and N is found from eqn (6.31).

6.7 Waterhammer control

The most frequently encountered waterhammer problems in water engineering relate to (a) abrupt pump stopping, as in power failure, and (b) rapid valve closure. Practical control devices, which can be used to limit the waterhammer effects due to pump cut-out, include:

(1) use of a flywheel to increase the pumpset inertia;

(2) installation of an air vessel or accumulator;

(3) use of a surge tank;

(4) use of air valves.

6.7.1 Flywheel

The increased inertia provided by a flywheel extends the stopping time of a pump (eqn (6.32)) and hence reduces waterhammer. Its effect is computed by simply adding the flywheel inertia to that of the pumpset.

6.7.2 Air vessel

Air vessels are frequently used on pump rising mains for the control of transient pressures. They are typically located close to the pump, downstream of the non-return valve, as shown in Fig. 6.7.

When the pump stops, the delivery pressure immediately drops, causing

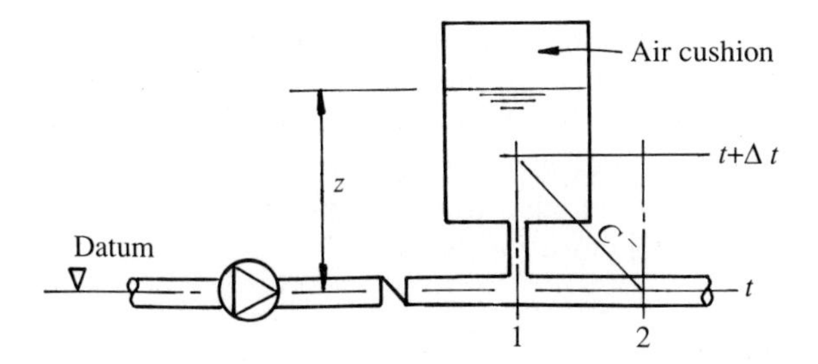

Fig. 6.7 Air vessel boundary.

a rapid discharge from the air vessel and immediate closure of the non-return valve. In the associated waterhammer analysis it is usually assumed that closure of the non-return valve is instantaneous and coincident with pump stopping; hence the air vessel becomes the effective upstream boundary control.

The governing equations are as follows:

C^- characteristic:	$H_{P1} = CM + BQ_{P1}$
Air volume/head:	$h_a V_a^\gamma = \text{constant}$
Throttle head loss:	$h_L = CQ_{P1}^2$
h_a/H_{P_1} relation:	$h_a = H_{P1} + H_{atm} + h_L - Z$
Continuity:	$\Delta V_a = \frac{1}{2}\Delta t(Q_{P1} + Q_1)$

where h_a is the absolute pressure head, H_{atm} is the atmospheric pressure head, and C is a head loss coefficient for flow between the air vessel and the rising main. Generally the throttle is designed to have a lesser head loss during outflow from the air vessel than during flow into the vessel, resulting in different C-values for inflow and outflow. The air volume change occurs rapidly and hence closely approximates an adiabatic process ($\gamma = 1.4$ for air). A γ-value of 1.35 is recommended for practical flow computations.

The above equations are solved for H_{P1} and Q_{P1}; initially, Q_1 may be assumed to be zero. The values of Z and V_a are modified for the next time step as follows:

$$Z_{t+\Delta t} = Z_t - \frac{\Delta t(Q_{P1} + Q_1)}{2A_v}$$

$$V_{a(t+\Delta t)} = V_{a(t)} + \tfrac{1}{2}\Delta t(Q_{P1} + Q_1)$$

where A_v is the cross-sectional area of the air vessel.

The air cushion functions rather like a spring, expanding during periods of water outflow from the vessel and thus exerting an increasing restraining force on the water outflow, eventually causing flow reversal. The resulting inflow to the vessel causes a compression of the enclosed gas volume, which in turn exerts an increasing flow-resisting force. The natural frequency of the corresponding water mass oscillation is much lower than that of the waterhammer wave, which also travels back and forth along the pipe and is superimposed on the mass oscillation.

6.7.3 Surge tank

Surge tanks are open-top vessels connected to the pipe system in which pressure transients are to be controlled. They are similar to air vessels, differing in the respect that the overlying air pressure remains constant. They give rise to a similar mass oscillation under transient flow conditions and are often used in hydropower plants, as illustrated in Fig. 6.8.

When flow to the turbine is throttled back, the penstock is subjected to waterhammer transient pressure. The surge tank prevents these transients from reaching the supply main connecting the reservoir to the surge tank. This system may be analysed using the waterhammer equations already presented. Alternatively, the transient behaviour of the reservoir–pipe–surge tank part of the system can be modelled as a simple mass oscillation:

Momentum equation: $$\rho g(H_R - H_{ST})A - \rho g h_f A = \rho A L \frac{\Delta v}{\Delta t}$$

Continuity: $$\Delta H_R = -\frac{Q\Delta t}{A_R} \quad \text{and} \quad \Delta H_{ST} = \frac{Q\Delta t}{A_{ST}}$$

where h_f is the flow head loss between reservoir and surge tank, A_R and A_{ST} are the reservoir and surge tank plan areas, respectively, and A is the pipe cross-sectional area.

This set of equations can be solved numerically to determine the variation of H_R and H_{ST} with time, resulting from an abrupt change in flow to the turbine.

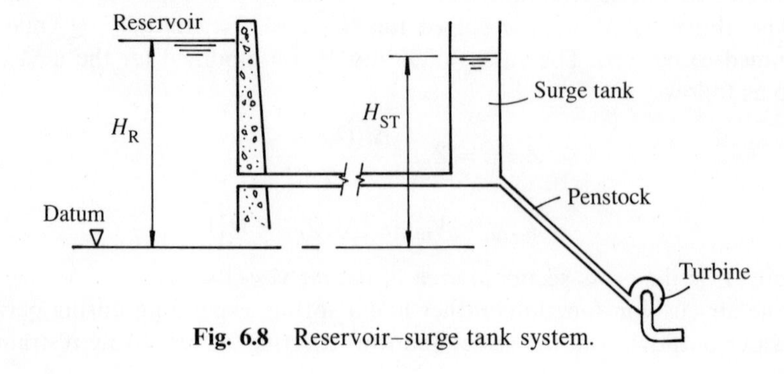

Fig. 6.8 Reservoir–surge tank system.

6.7.4 Air valves

Air valves are installed at high points on rising mains to allow escape of air. They also admit air when the gauge pressure drops below atmospheric pressure and hence can be used to limit the pressure downsurge under waterhammer conditions.

6.8 Column separation, entrained gas

Column separation occurs when the pressure drops to vapour level resulting in the formation of a vapour cavity which grows in size as long as the pressure remains below vapour level. When the pressure rises the vapour cavity collapses, resulting in a high pressure at the cavity location due to the collision of two water masses moving in opposite directions. The resulting collision pressure head rise is

$$\Delta H = \frac{\alpha}{2gA}(Q_u - Q)$$

where Q_u is the inflow to the section and Q is the outflow. While analytical procedures, which take vapour formation into account, are available (Wylie and Streeter 1978), computed and observed pressure levels do not always show good agreement. From a design viewpoint, as later discussed, it is good practice to control pressure transients so as to avoid vapour pocket formation.

Dissolved gases are released from solution when the water pressure drops below their solution pressure. The rate of release of dissolved air depends on the saturation excess and the degree of turbulence in the flow. The primary effect of a dispersed gas phase is to reduce the wavespeed in accordance with the following correlation (Wylie and Streeter 1978)

$$\alpha = \sqrt{\frac{K/\rho}{1 + KD/ET + mKR\Theta/p^2}} \tag{6.34}$$

where m is the mass of gas per unit volume of fluid, Θ is the absolute temperature (K), and p is the absolute pressure. Figure 6.9 illustrates the influences of entrained air and pressure on wavespeed.

It is worth noting that wastewaters such as raw sewage may have sufficient entrapped gases of biological origin to significantly reduce the waterhammer wavespeed. Conditions favourable to the production of biogases exist in sewage rising mains of low gradient and low pumping velocity, resulting in permanent organic solids deposition.

The expulsion of entrapped air, following pump start-up or filling of a

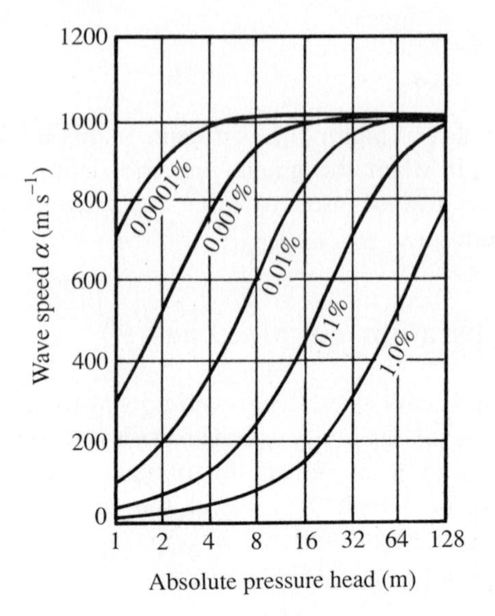

Fig. 6.9 The influence of the dispersed gas phase on wavespeed. The wavespeed is for air–water mixtures; air volumes are at standard pressure:

$$\alpha = \sqrt{\frac{(K/\rho)_L}{1 + K_L D/ET + mK_L R\Theta/p^2}}$$

where m = mass of gas per volume of liquid; $K_L = 2.05 \times 10^9$ N m^{-2}; $\rho_L = 1000$ kg m^{-3}; $TE/D = 2.00 \times 10^9$ N m^{-2}; $\Theta = 293$ K.

pressure main, may sometimes give rise to severe pressure transients. This would happen, for example, where a main is charged rapidly with water at one end while the contained air is displaced through an air valve at the other end. The rapidly moving water column will be halted abruptly as the air valve closes once the pipe is completely filled with water. This problem can be avoided by a gradual charging of empty mains.

In general, however, the presence of a dispersed gas phase tends to suppress waterhammer effects.

6.9 Transient pressure limits

As a general rule, specifications permit a maximum transient pressure during waterhammer conditions, which is in excess of the permissible maximum sustained working pressure. In addition, pipes which are vulnerable to fatigue failure, notably prestressed concrete and plastics, have a limiting value for the maximum pressure fluctuation (max. – min.) which is permitted under waterhammer conditions.

For cast iron, ductile iron, steel, and asbestos-cement pressure pipes it is generally recommended (BS 8010, 1987) that the maximum pressure due to waterhammer should not exceed the maximum permissible sustained working pressure by more than 10 per cent.

For prestressed concrete it is recommended (Creasey and Sanderson 1977) that (a) the maximum pressure under waterhammer conditions should not exceed the maximum permissible sustained working pressure by more than 20 per cent, and (b) the maximum pressure fluctuation amplitude should not exceed 40 per cent of the maximum permissible sustained working pressure.

For uPVC pipe it is recommended (BSI 1973) that (a) the maximum pressure under waterhammer conditions should not exceed the maximum permissible sustained working pressure, (b) the maximum pressure fluctuation amplitude should not exceed 50 per cent of the maximum permissible sustained working pressure, and (c) Class B pipe should not be used where the pressure, under waterhammer conditions, is likely to fall below atmospheric pressure.

Where analysis indicates that the pressure is likely to drop to vapour level and thus cause some column separation, it is generally advisable to install some form of waterhammer control to either prevent such cavitation occurring or to curtail the resulting pressure transients to predictable limits.

6.10 Computer program: WATHAM

The program WATHAM analyses unsteady flow in a pump and rising main system due to sudden pump cut-out, using the analytical procedures set out in this chapter. The program is coded to deal with a system bounded by fixed-level reservoirs at its upstream and downstream ends. The analysis presumes a non-return valve on the pump delivery. Provision is also made for the optional inclusion of an air vessel in the system. It is presumed to be connected to the rising main by a throttle pipe on the downstream side of the non-return valve.

The user may therefore choose to analyse the system with the air vessel included or omitted. In the absence of an air vessel, the analysis assumes that the non-return valve closes simultaneously with the occurrence of zero forward flow in the rising main. As discussed later, this may not always happen in practice. Valve closure may not occur until there is a significant reverse velocity through the valve, resulting in more severe transient pressures than predicted by program WATHAM (see Fig. 6.14).

Where the system includes an air vessel, the non-return valve is assumed to close simultaneously with pump cut-out and hence the air vessel becomes the effective upstream boundary of the system.

In most practical cases, the suction main is very much shorter than the rising main. The rising main is divided into a number of segments (ten or a multiple thereof). The corresponding segmental length for the suction main is calculated and, if found to be greater than the actual suction main length, the latter is defined as 'short'. In this circumstance the suction main pressure is assumed to remain constant and its piezometric head value is taken to correspond with the suction reservoir level.

The required input data include the following:

(1) PIPES: length, diameter, surface roughness, wall thickness, Young's modulus;
(2) PUMPS: constant coefficients for the head–discharge and torque–discharge equations, rated speed, moment of inertia of the rotating mass;
(3) AIR VESSEL: volume of the vessel, air volume in the vessel under steady flow conditions, throttle pipe discharge coefficients (inflow and outflow), air vessel cross-sectional area, height of water in the vessel above the connection point to the rising main;
(4) RESERVOIRS: water surface elevation relative to the piezometric head datum.

The program requests the user to input the necessary data for the analysis of the system. It computes the initial steady state flow and, following pump cut-out, the variations in pressure and flow at the specified system node points. The maximum and minimum pressure envelopes are also computed. The latter enable the user to check whether gauge pressures remain within permissible design limits under the prevailing transient conditions.

Sample program run

```
RUN
        ********************************
        * WATERHAMMER ANALYSIS PROGRAM *
        *                              *
        *         WATHAM.BAS           *
        ********************************

    This program computes the transient flow and pressure
    conditions in a suction/rising main system due to pump
    cut-out. It relates to a system bounded by reservoirs or
    points of fixed head at its upstream and downstream ends
    and which includes a non-return valve on the pump delivery.

    It also offers the option of analysis of the above system
    protected by an air vessel connected to the rising main on
    the downstream side of the non-return valve.

    The program computes the initial steady flow in the system
    from the given pump and system data.

    Pump performance is defined by quadratic expressions
    for head H and torque T, as functions of discharge Q:

        HEAD        H = A0 + A1*Q + A2*Q*Q
        TORQUE      T = B0 + B1*Q + B2*Q*Q
```

```
          PRESS THE SPACE BAR TO CONTINUE

    It should be noted that program computations are based
    on piezometric head values and, hence, locations at
    which the gauge pressure may have dropped to vapour
    level are not sign-posted by the program.

    The piezometric head datum is that to which the specified
    upstream and downstream reservoir levels relate.

         PRESS THE SPACE BAR TO CONTINUE

WHICH SYSTEM DO YOU WISH TO ANALYSE ?

         1.  PUMP/RISING MAIN SYSTEM
         2.  PUMP/AIR VESSEL/RISING MAIN SYSTEM

ENTER 1 OR 2, AS APPROPRIATE? 1

INPUT OF PUMP CHARACTERISTICS
THE PUMP HEAD/DISCHARGE CORRELATION IS AS FOLLOWS:

    H = A0 + A1*Q + A2*Q*Q (Units: H(m), Q(m**3/s))

ENTER VALUES OF A0, A1 AND A2, SEPARATED BY COMMAS? 33,-214,-4515

THE PUMP TORQUE/DISCHARGE CORRELATION IS AS FOLLOWS:

    T = B0 + B1*Q + B2*Q*Q  (Units: T (Nm), Q (m**3/s))

ENTER VALUES OF B0,B1 AND B2, SEPARATED BY COMMAS? 50,207,0

ENTER DUTY POINT HEAD (m)? 22.5
ENTER DUTY POINT DISCHARGE (m**3/s)? 0.03
ENTER DUTY POINT TORQUE (Nm)? 56.2
ENTER PUMP SPEED (rpm)? 1500
ENTER MOMENT OF INERTIA OF PUMP SET (kg*m*m)? 1.5

PIPE DATA INPUT
ENTER SUCTION PIPE LENGTH (m),DIAM(m),WALL THICKNESS (m),WALL ROUGHNESS (m),
 YOUNG'S MOD (N/m**2):? 5,0.2,0.006,0.0001,15E+10
ENTER DELIVERY PIPE LENGTH (m), DIAM (m), WALL THICKNESS (m), WALL ROUGHNESS(m),
  YOUNG'S MOD (N/m**2):? 500,0.2,0.006,0.0001,15E+10

ENTER  SUCTION RESERVOIR LEVEL (mOD)? 100
DELIVERY RESERVOIR LEVEL (mOD)? 120

ENTER NO OF SEGMENTS INTO WHICH DELIVERY MAIN IS DIVIDED
THIS SHOULD BE 10 OR A MULTIPLE THEREOF? 10

ENTER NO OF COMPUTATION ITERATIONS:? 300

DATA ENTRY NOW COMPLETE

         ....computation in progress, please wait....

PUMP STEADY DISCHARGE (m**3/s) IS  3.038112E-02
PUMP STEADY HEAD (m) IS  22.33104

SUCTION MAIN WAVE SPEED (m/s)=  1215.571
DELIVERY MAIN WAVE SPEED (m/s)=  1215.571
```

PUMP DISCHARGE (m**3/s)	PUMP SPEED (rpm)	HEAD U/S PUMP (m)	HEAD D/S PUMP (m)	TIME (s)
+0.0304	+1500.0	+100.0	+122.3	0.00
+0.0303	+1485.3	+100.0	+121.8	0.04
+0.0301	+1470.8	+100.0	+121.3	0.08
+0.0300	+1456.6	+100.0	+120.8	0.12
+0.0299	+1442.7	+100.0	+120.3	0.16
+0.0298	+1429.0	+100.0	+119.9	0.21
+0.0297	+1415.6	+100.0	+119.4	0.25

etc. up to 300 iterations

NODE	RISING MAIN		SUCTION MAIN	
	MAX PRESSURE (m)	MIN PRESSURE (m)	MAX PRESSURE (m)	MIN PRESSURE (m)
1	133.5	106.5	100.0	100.0
2	132.4	107.7		
3	131.2	108.8		
4	130.0	110.1		
5	128.7	111.3		
6	127.4	112.6		
7	126.1	113.9		
8	124.7	115.3		
9	123.3	116.7		
10	121.8	118.2		
11	120.0	120.0		

```
DO YOU WISH TO PLOT THE PRESSURE TRANSIENT
AT THE PUMP DELIVERY ON THE SCREEN (Y/N)? N

DO YOU WISH TO RE-RUN THE PROGRAM (Y/N)? N
Ok
```

6.11 Examples of waterhammer computation

The computed results in the examples which follow are presented in graphical form as (a) envelopes of the maximum and minimum pressures experienced over the pipe length, and (b) the variation of discharge and pressure with time at the point of origin of the waterhammer effect.

Examples 1 and 2 illustrate computed waterhammer flow transients due to pump trip-out on short and long rising mains, respectively. Example 3 illustrates the computation of flow transients in a rising main system protected by an air vessel. Example 4 illustrates the waterhammer effects of rapid valve closure.

Example 1

This is a pump/rising main system where the static head is significantly larger than the friction head. The waterhammer condition analysed is that caused by pump trip-out. The required data input for the system is summarized in Fig. 6.10(a), which shows the normal hydraulic gradient and also the

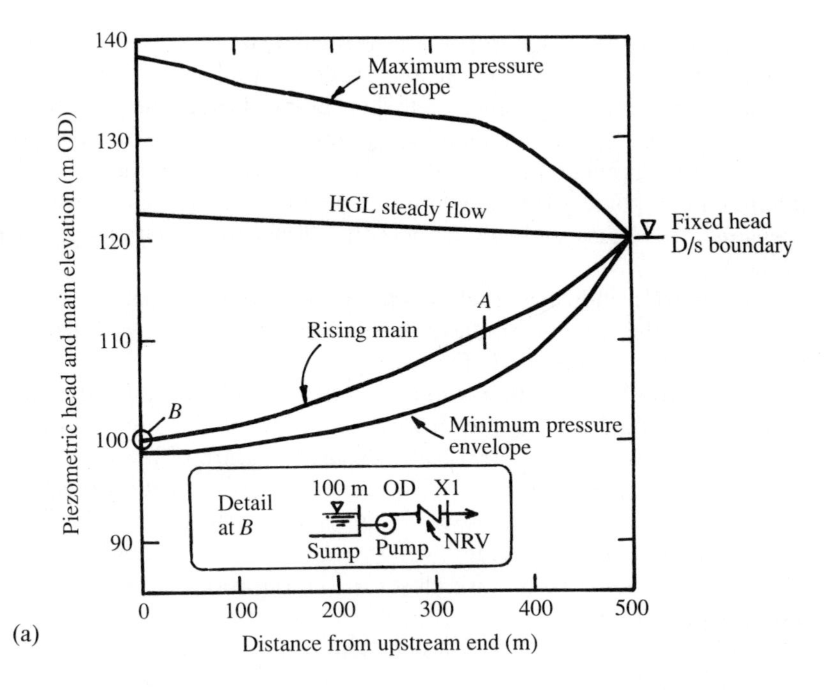

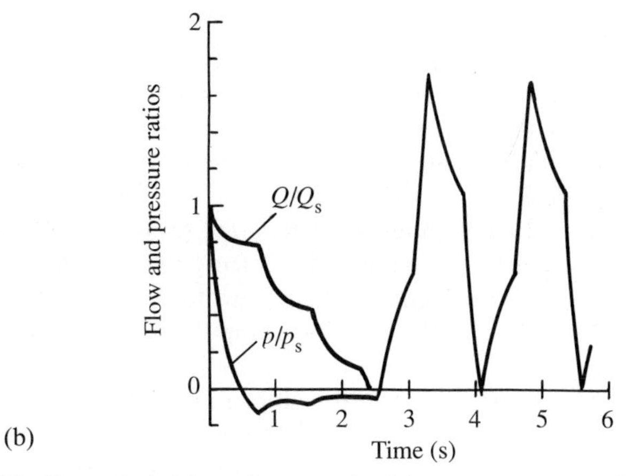

Fig. 6.10 Example 1: (a) maximum and minimum pressure envelopes; (b) computed flow and pressure transients at the upstream end of a rising main (point X1) where Q = transient flow, Q_s = initial steady flow, p = transient gauge pressure, and p_s = initial steady gauge pressure.

waterhammer pressure envelopes. The vertical distance between an envelope line and the pipe profile line at any section represents the maximum or minimum pressure experienced at that point during the waterhammer occurrence. The critical maximum gauge pressure is clearly at the upstream end of the main, while the critical minimum value is at point A, where the minimum computed gauge pressure is -5.5 m, which is well above vapour pressure for ordinary ambient temperature. (Refer to the data list at the end of the chapter.)

The analysis indicates that waterhammer control measures are not required as the computed maximum gauge pressure is well below the permitted maximum pressure for the pipe.

Figure 6.10(b) shows the computed variation of flow and pressure with time at the point X1 on the downstream side of the non-return valve. These plots illustrate the intrinsically unsteady nature of the waterhammer phenomenon. The decrease in discharge exhibits a periodicity corresponding to the return travel time for the waterhammer wave (*c.* 0.76 s). The pressure drops rapidly to a low level, following pump trip-out and remains at a low level until the non-return valve closes at the onset of reversed flow in the rising main. From this point onwards in time the upstream boundary is effectively a closed valve or dead end. The periodic time of the subsequent pressure transient is about 1.5 s or twice the wave travel time.

Example 2

Example 2 also relates to a pump/rising main system. The input data differ from those in Example 1 in two respects. The rising main length is 2500 m (500 m in Example 1) and secondly the static and friction heads have approximately equal values; their sum, which is the pump operating head, is the same as in Example 1.

Figure 6.11(a) shows the computed waterhammer pressure envelopes for pump trip-out, while Fig. 6.11(b) shows the corresponding discharge and pressure transients at the pump end of the main (downstream side of the non-return valve). It will be noted that the computed time for the pump discharge to reduce to zero is about 14 s, considerably longer than in Example 1, where, as may be observed from Fig. 6.10(b), the corresponding time was 2.4 s. Example 2 may be regarded as typical of a long main/low static head system where the critical waterhammer effect is likely to be the downsurge which, depending on the vertical profile of the rising main, may cause the pressure to drop to vapour level with resulting column separation. This does not occur in this case because of the very flat gradient of the main over most of its length. The positive or upsurge pressure transient, which is caused by the stopping of the reverse flow by the non-return valve, is seen not to be critical in the long rising main/low static head system.

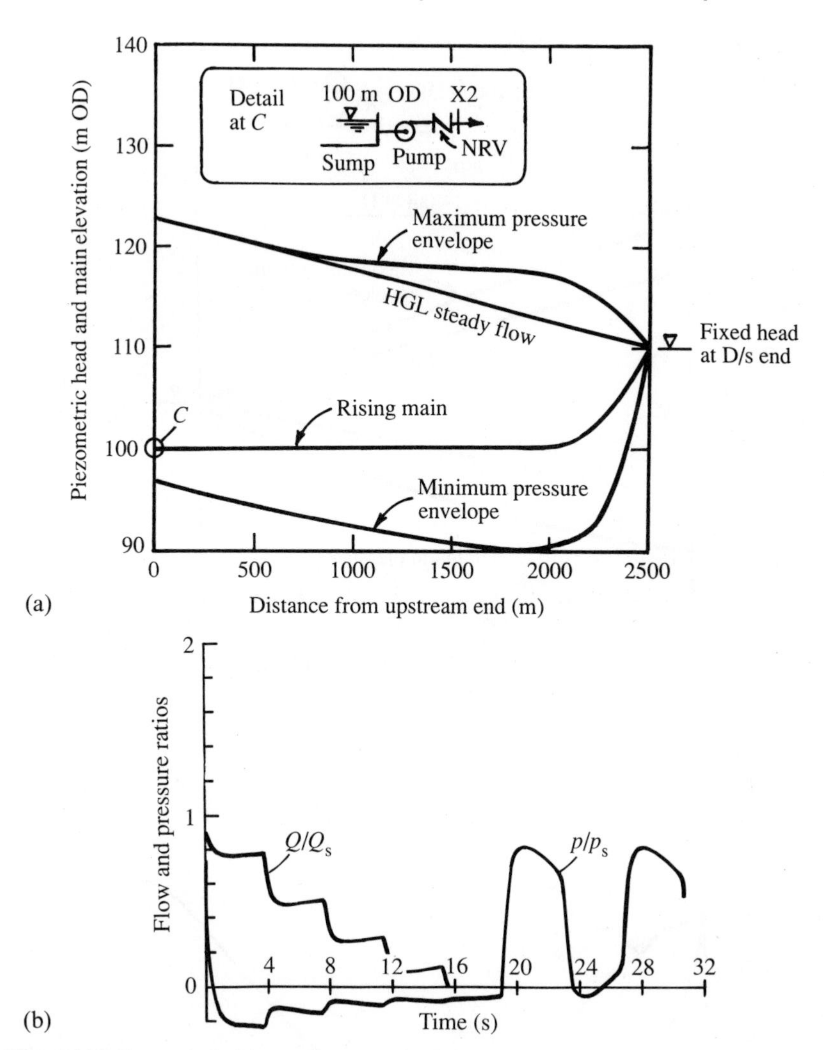

Fig. 6.11 Example 2: (a) maximum and minimum pressure envelopes; (b) computed flow and pressure transients at the upstream end of a rising main (point X2) where Q = transient flow, Q_s = initial steady flow, p = transient gauge pressure, and p_s = initial steady gauge pressure.

Example 3

The hydraulic system for Example 3 is the same as in Example 2 except that the rising main vertical profile has been altered to a uniform gradient, as shown in Fig. 6.12(a). It is clear from the results presented in Fig. 6.11(a) that pump trip-out would result in cavitating flow in the rising main in its

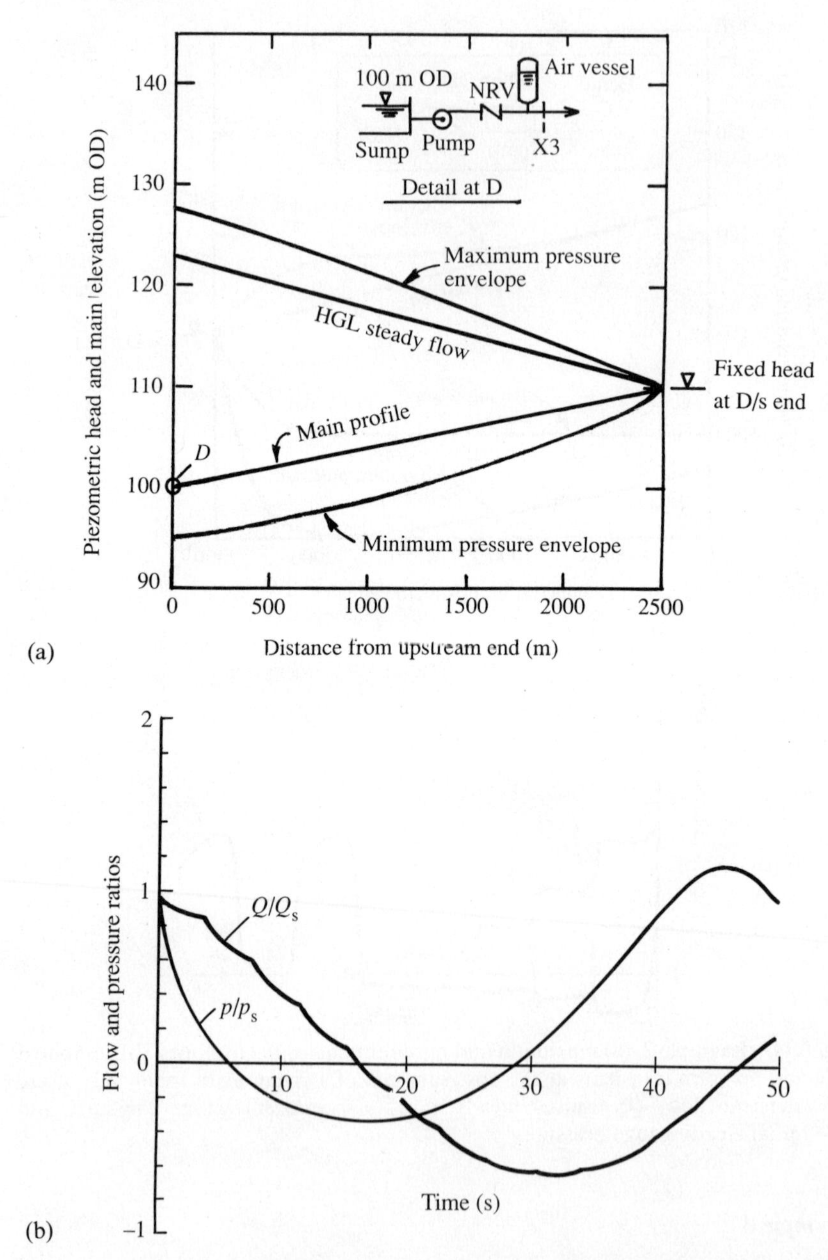

Fig. 6.12 Example 3: (a) maximum and minimum pressure envelopes; (b) computed flow and pressure transients at an air vessel (point X3) where p and p_s are gauge pressures.

revised vertical profile. To avoid this condition, an air vessel is used as a waterhammer control device. It is connected to the rising main immediately downstream of the non-return valve.

On pump trip-out, the pressure simultaneously drops, resulting in outflow from the air vessel and almost simultaneous closure of the non-return valve. The compressed air cushion acts in a manner similar to a mechanical spring, causing a mass oscillation of the water in the rising main and at the same time suppressing the waterhammer transient. The periodicity of the latter can just about be observed on the graph of air vessel discharge shown on Fig. 6.12(a), superimposed on the much slower mass oscillation of the system.

The computed results presented in Fig. 6.12(a) indicate that the maximum and minimum gauge pressures are within acceptable design limits, the maximum being within +5 m of the steady flow maximum pressure and the minimum being approximately −5 m gauge.

Example 4

The transient flow conditions caused by valve closure are illustrated in this example. When a valve is closed, it causes a progressive retardation of flow. If closure is rapid, it causes a waterhammer effect which results in an upsurge or positive pressure front being transmitted upstream from the valve and a negative pressure wave being transmitted in the downstream direction.

The hydraulic system is shown in schematic outline in Fig. 6.13(a). It consists of a 2500 m long gravity main connecting two reservoirs which have a 10 m difference in water level. The valve is located midway between the reservoirs.

Two closure rates are examined: the first is a linear closure rate of 1 deg/s (the valve is a butterfly valve which is closed by rotation of the valve disc through 90°) and the second is a two-step closure routine, the first 50° at a rate of 2 deg/s and the remaining 40° at a rate of 0.615 deg/s, giving an overall closure time of 90 s, as in the first case.

Examination of the results presented in Fig. 6.13(a) and (b) shows that the first 40° of closure reduced the flow by less than 10 per cent while the following 40° of closure reduced the flow by over 80 per cent, under both closure routines. While the differences in flow values for any particular valve angle do not appear to be significant for the two closure rates, the computed pressure/time profiles for the upstream and downstream sides of the valves are, as shown in Fig. 6.13(a) and (b), significantly different. The two-step closure, in which the valve was closed more rapidly in the first half of closure and at a reduced rate in the latter half of closure, is seen to have resulted in a smaller deviation from the final static pressure condition than that resulting from the uniform closure rate.

Thus it is feasible to design a multi-step automatic valve closure routine

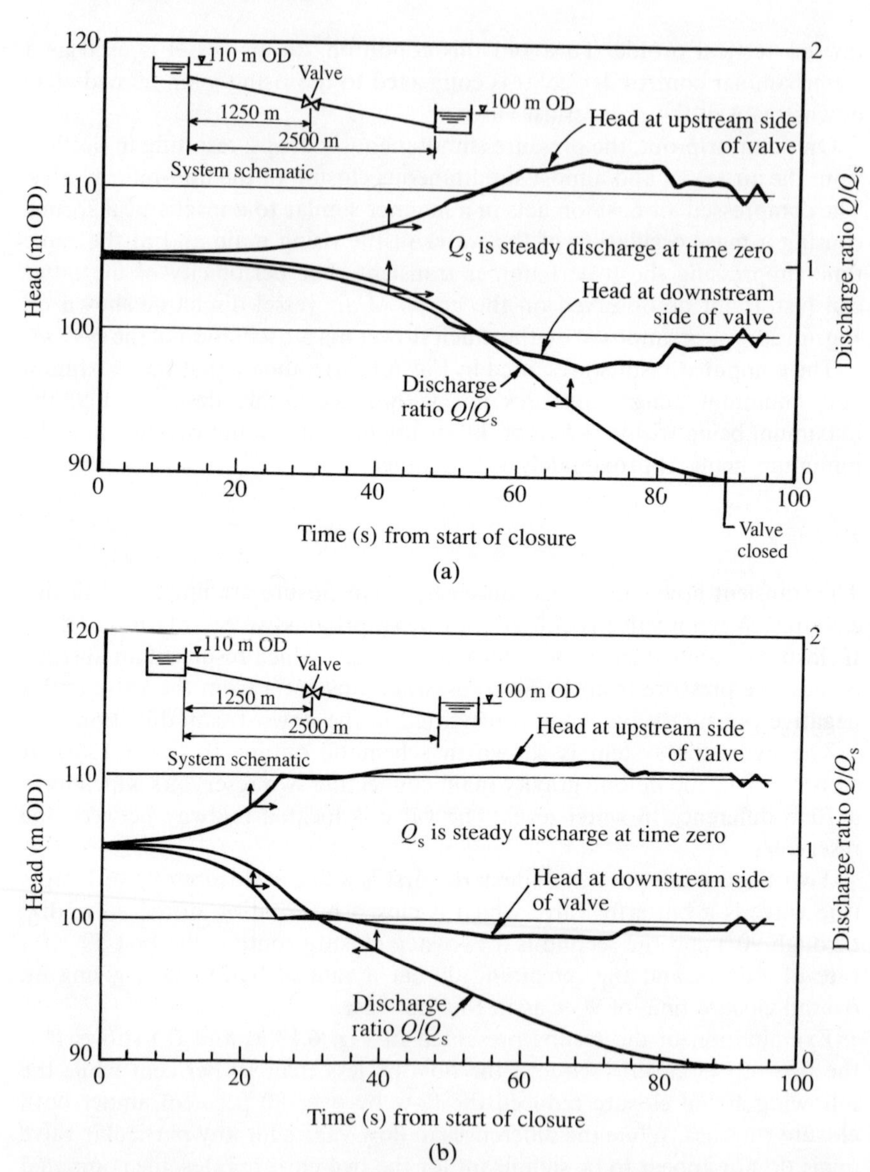

Fig. 6.13 Example 4: (a) computed transient flow and head values due to valve closure at a uniform rate of 1 degree/s from a fully open position; (b) computed transient flow and head values due to two-speed valve closure: 2 degrees/s for the first 50 degrees and 0.615 degree/s for the remaining 40 degrees to fully closed position.

which ensures closure in a minimum time while not exceeding the permissible pressures for the pipe system. This procedure is known as valve stroking.

6.12 Some practical design considerations

In practical pipeline design in respect of waterhammer effects it is the obvious goal of the designer to ensure that pressure extremes remain within permissible limits, as previously outlined. Where the analysis indicates that cavitation is likely to occur, appropriate design steps should, as a general rule, be taken to eliminate the possibility of vapour pocket formation because of the risks associated with vapour pocket collapse. This will usually be achieved by the use of air valves or an air vessel or other control device appropriate to the prevailing circumstances.

It is also good practice to verify, where possible, analytical predictions by post-installation field measurements, particularly where there is a degree of uncertainty in the predicted response of the system under investigation. The measurement of waterhammer pressure transients requires a fast-response pressure-sensing device such as a piezoresistive transducer, linked to a compatible recording system.

Figures 6.14 and 6.15 illustrate recorded pressure transients in which the system response was found to differ significantly from that which would be predicted from conventional analysis of the respective systems.

Figure 6.14 shows measured transient pressure traces following pump trip-out on a laboratory test rig rising main. The relevant rising main data were as follows: low-density polyethylene pipe, 150 m long, 50 mm diameter. The steady flow forward velocity was 1.25 m s^{-1} and the static lift was 6.0 m. The rising main was fitted with a standard hinged flap non-return valve (NRV) on the delivery side of the pump.

Pressure traces 1 and 2 show the recorded pressure transients, following pump trip-out, at a point just downstream of the NRV. Prior to the recording of pressure trace 2 the NRV flap was spring-loaded so as to make its closure coincide with the reduction of the forward velocity to zero value. The delayed closure of the NRV (the time duration from pump cut-out to the first rising leg of pressure trace 1 was about 3 s longer than the corresponding time for pressure trace 2) is clearly seen to cause a dramatic jump in the maximum positive waterhammer pressure, raising the piezometric value from about 9.0 m to about 25.0 m. This increase in pressure is due to the fact that the delayed closure of the NRV flap allowed a significant reverse velocity to develop prior to closure. The instantaneous stoppage of this reverse flow, as the flap closed, gave rise to the observed increase in peak transient pressure. The sharp impact of closure under such circumstances typically produces an audible valve-slam sound.

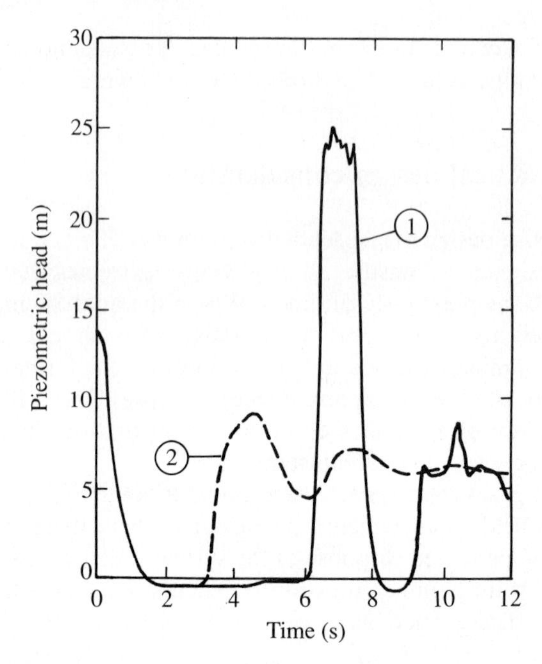

Fig. 6.14 Measured pressure traces following pump trip-out; ① is unmodified NRV and ② is spring-loaded NRV (Purcell 1991, personal communication).

Figure 6.15 shows a measured waterhammer pressure trace, recorded on the downstream side of the non-return valve on a rising main. As the system data given in Fig. 6.15 indicates, the pump/rising main system belongs to the same category as the system analysed in Example 1, that is, it is a short rising main in which the friction head is small relative to the static lift. The measured pressure variation with time is generally similar to that computed for Example 1, as presented in Fig. 6.10(b). There are, however, a number of significant differences. The computed wavespeed for this rising main was $1153\ \mathrm{m\ s^{-1}}$, resulting in an estimated wave return travel time of 0.32 s; the measured return travel time was 0.75 s, indicating an actual average wavespeed of $493\ \mathrm{m\ s^{-1}}$. This low wavespeed is indicative of the presence of dispersed gas bubbles in the pumped liquid, which was municipal sewage. As there was no possibility of air entry to the system or air release from solution, it was concluded that the gas phase was of biological origin.

A second notable feature of the measured pressure trace is the pronounced damping in the system as is evident from the decrease in magnitude of successive pressure peaks. This decrease is much higher than that predicted by the usual computational assumption that flow friction under waterhammer conditions is the same as in steady flow at the same velocity. This is clearly

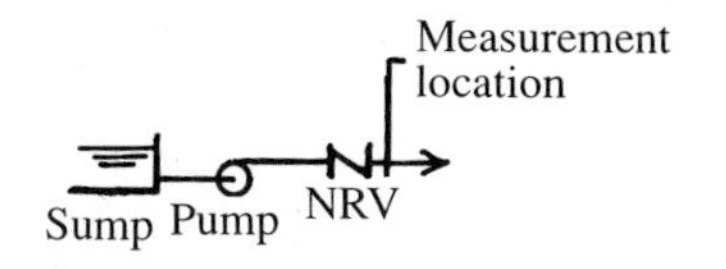

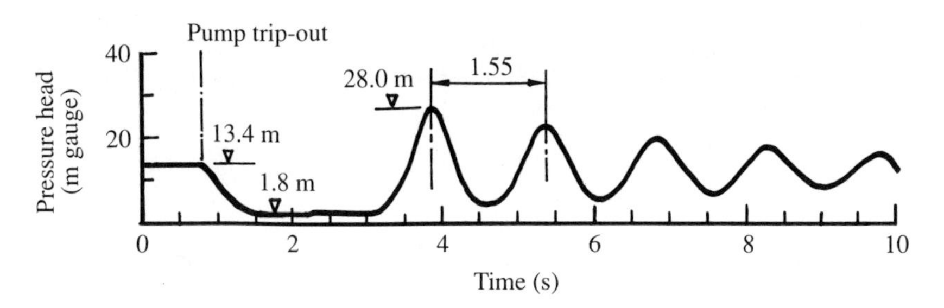

Fig. 6.15 Measured pressure trace following pump trip-out. Experimental results: maximum head rise = 14.6 m; maximum head drop = 11.6 m. The graph is for a cast iron rising main 185 m long with an internal diameter of 389 mm; discharge rate = 0.100 $m^3 s^{-1}$; static lift = 12.7 m; and friction head = 0.67 m.

not the case; it would appear that the effective energy dissipation under conditions of rapid velocity change is significantly higher than that which obtains under steady flow conditions.

6.13 Some relevant material properties

Bulk modulus of water (K) = 2.05×10^9 N m^{-2}

Young's modulus (E) for pipe materials

Material	Cast iron	Ductile iron	Steel	Asbestos-cement	Prestressed concrete	uPVC
E (10^{10} N m^{-2})	11.2	15.0	20.0	2.5	3.7	0.3

Vapour pressure of water

Temperature	5	10	15	20	25	°C
Vapour pressure	89	125	174	239	323	mm water

References

British Standards Institution (1973). CP312 Part 2.
British Standards Institution (1987). BS 8010, Part 2.
Creasey, J. D. and Sanderson, P. R. (1977). Surge in water and sewage pipelines, TR 51, Water Research Centre, Medmenham, UK.
Wylie, E. B. and Streeter, V. L. (1978). *Fluid Transients*, McGraw-Hill, New York.

7
Steady flow in open channels

7.1 Introduction

The flow of water in open channels is characterized by the existence of a 'free' surface, that is, an upper boundary in contact with air at atmospheric pressure. Gravity is the motive force. The term 'steady' implies that the velocity vector at a particular location does not change with time. We can distinguish three categories of steady flow in open channels:

(1) uniform flow;

(2) gradually varied flow;

(3) rapidly varied flow.

As defined in Chapter 2, steady uniform flow exists when the velocity vector is constant with respect to both time and space variables. In steady varied flow the magnitude of the velocity varies along the flow path. This variation may be gradual, for example, upstream of a flow-measuring device, or rapid, such as in a spillway discharge.

7.2 Hydraulic resistance to flow

The fundamental nature of the hydraulic resistance to flow in open channels is the same as that outlined in Chapter 3 for pipes flowing full. The flow Reynolds number for open channels is defined in terms of the channel hydraulic radius R_h:

$$R_e = \frac{vR_h\rho}{\mu}. \tag{7.1}$$

When the Reynolds number exceeds about 1100, as is invariably the case for water flow in open channels, the flow is turbulent. The flow energy

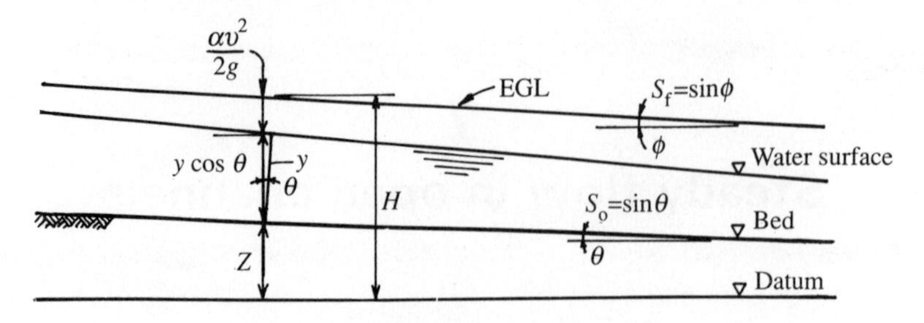

Fig. 7.1 Components of total head H.

parameters are defined in Fig. 7.1:

$$H = z + y\cos\theta + \alpha\frac{v^2}{2g} \tag{7.2}$$

where H is the total head relative to a horizontal datum. The term specific head or specific energy E_s is also an important parameter in open channel flow. It is defined as

$$E_s = y\cos\theta + \alpha\frac{v^2}{2g}, \tag{7.3}$$

that is, it is the total head relative to the channel bed as datum.

The EGL or friction slope S_f reflects the hydraulic resistance to flow. The expressions relating friction slope to pipe flow, presented in Chapter 3, can be adapted to open channel flow geometry by replacing the pipe diameter D by the channel hydraulic radius R_h, using the relationship $D = 4R_h$.

Using this relationship, the open channel form of the Darcy–Weisbach equation becomes:

$$S_f = \frac{fv^2}{8gR_h} \tag{7.4}$$

where the friction factor f is given by the appropriately adapted form of the Colebrook–White equation (3.25):

$$\frac{1}{\sqrt{f}} = -0.88\ln\left[\frac{k}{14.8R_h} + \frac{0.625}{R\sqrt{f}}\right]. \tag{7.5}$$

Equations (7.4) and (7.5) may be combined to give the following explicit expression for velocity:

$$v = -\sqrt{6.2gR_hS_f}\,\ln\left[\frac{k}{14.8R_h} + \frac{0.625}{R_e\sqrt{f}}\right]. \tag{7.6}$$

Where flow is in the rough turbulent category, as in most natural channels, eqn (7.6) may be simplified by the omission of the Reynolds number term, giving

$$v = 7.8\sqrt{R_h S_f}\,\ln\left[\frac{14.8R_h}{k}\right]. \tag{7.7}$$

As already noted in Chapter 3, the Manning equation was developed for open channel flow computation. Its generally used form is

$$v = \frac{1}{n}R_h^{0.67}S_f^{0.5}. \tag{3.32}$$

The following correlation of Manning's n and equivalent sand roughness k is derived from eqns (7.7) and (3.32):

$$\frac{1}{n} = 7.8R_h^{-0.167}\ln\left[\frac{14.8R_h}{k}\right] \tag{7.8}$$

which may be simplified to the form

$$\frac{1}{n} = \frac{C_n}{k^{0.167}} \tag{7.9}$$

where C_n is a function of the ratio R_h/k, as defined by eqns (7.8) and (7.9):

$$C_n = 7.8\left(\frac{R_h}{k}\right)^{-0.167}\ln\left[\frac{14.8R_h}{k}\right]$$

resulting in the following numerical value range:

R_h/k	10	100	1000	10 000
C_n	26.48	26.39	23.63	19.94

(Note that k is in *m* in the foregoing correlations with the Manning n-value.)

Recommended surface roughness values for use in design are given in Table 7.1 (refer also to Table 3.1 for additional surface roughness data relating to pipes).

7.2.1 Influence of channel shape on flow resistance

The most commonly used channel cross-sections are rectangular, trapezoidal, or circular. Expressions for the section flow parameters for these sections are given in Fig. 7.2.

Table 7.1 Recommended values for the surface roughness parameter k (Hydraulics Research 1990).

	k-value (mm)		
	Good	Normal	Poor
Slimed sewers*			
(a) Half-full velocity about $0.75\ \text{m s}^{-1}$			
Concrete, spun or vertically cast	–	3.0	6.0
Asbestos-cement	–	3.0	6.0
Clayware	–	1.5	3.0
uPVC	–	0.6	1.5
(b) Half-full velocity about $1.2\ \text{m s}^{-1}$			
Concrete, spun or vertically cast	–	1.5	3.0
Asbestos-cement	–	0.6	1.5
Clayware	–	0.3	0.6
uPVC	–	0.15	0.3
Unlined rock tunnels			
Granite and other homogeneous rocks	60	150	300
Diagonally bedded slates	–	300	600
Earth channels			
Straight uniform artificial channels	15	60	150
Straight natural channels, free from shoals, boulders, and weeds	150	300	600

* The roughness of a slimed sewer varies considerably during any year. The normal value is that roughness which is exceeded for approximately half the time. The poor value is that which is exceeded, generally on a continuous basis, for one month of the year. The value of k should be interpolated for velocities between 0.75 and $1.2\ \text{m s}^{-1}$.

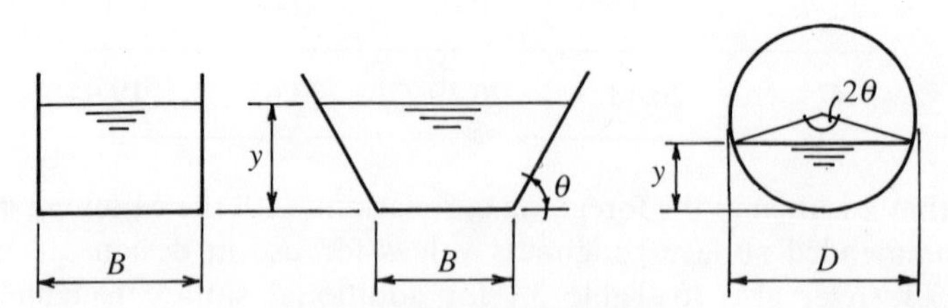

Fig. 7.2 Section parameters.

	Rectangular	Trapezoidal	Circular
A	By	$y(B + y/\tan\theta)$	$\frac{D^2}{4}(\theta - 0.5\sin 2\theta)$
R_h	$\frac{By}{B + 2y}$	$\frac{y(B + y/\tan\theta)}{B + 2y/\sin\theta}$	$\frac{D}{4}\left(\frac{\theta - 0.5\sin 2\theta}{\theta}\right)$

The hydraulic resistance to flow in an open channel of a given cross-sectional area is minimized by minimizing its wetted perimeter length. For a rectangular section of given cross-sectional area A, the channel proportions, which minimize the perimeter length P are found as follows:

$$P = B + 2y = \frac{A}{y} + 2y$$

$$\frac{\mathrm{d}P}{\mathrm{d}y} = -\frac{A}{y^2} + 2 = 0 \qquad \text{for } P_{\text{min}}.$$

Hence at P_{min}, $B = 2y$, that is, the most efficient shape of rectangular section from a discharge viewpoint is one in which the flow depth is half the width. It may also be noted that an inscribed semicircle can be fitted to this section profile.

In a trapezoidal section of given area A, the wetted perimeter length can be expressed as a function of the flow depth, y, and the angle of inclination, θ, of the sidewall to the horizontal:

$$P = B + \frac{2y}{\sin\theta} = \frac{A}{y} - \frac{y}{\tan\theta} + \frac{2y}{\sin\theta}.$$

The values of y and θ, which minimize P for a given value of A, are found from the relationships:

$$\frac{\mathrm{d}P}{\mathrm{d}y} = \frac{\partial P}{\partial y} + \frac{\partial P}{\partial \theta}\frac{\mathrm{d}\theta}{\mathrm{d}y} = 0$$

$$\frac{\mathrm{d}P}{\mathrm{d}\theta} = \frac{\partial P}{\partial \theta} + \frac{\partial P}{\partial y}\frac{\mathrm{d}y}{\mathrm{d}\theta} = 0$$

which simplify to the following conditions:

$$\frac{\partial P}{\partial y} = 0 \tag{7.10}$$

$$\frac{\delta P}{\delta \theta} = 0. \tag{7.11}$$

Differentiation of P with respect to y in accordance with eqn (7.10) results in the relationships

$$B + \frac{2y}{\tan\theta} = \frac{2y}{\sin\theta}. \tag{7.12}$$

Differentiation of P with respect to θ in accordance with eqn (7.11) results

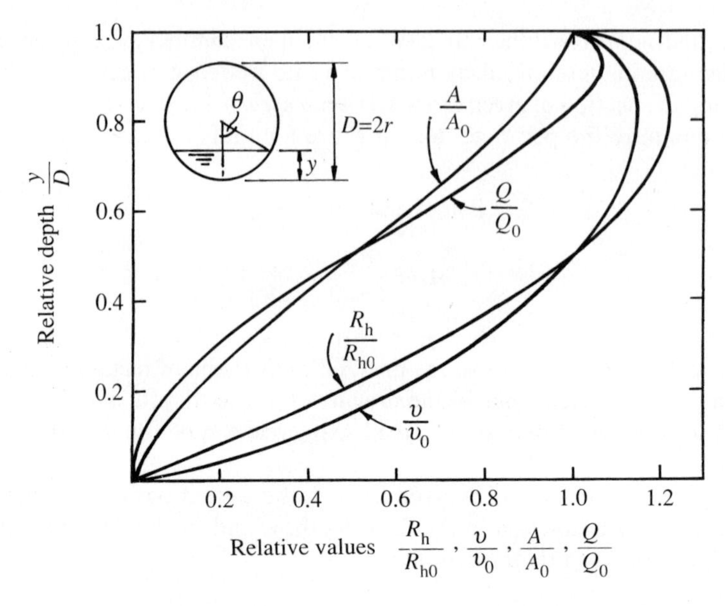

Fig. 7.3 Flow in part-full circular pipes. The subscript 0 refers to the flowing full value of the parameter; v and Q relations are based on the Manning equations.

$$\frac{R_h}{R_{h0}} = \left(1 - \frac{\sin 2\theta}{2\theta}\right) \qquad \frac{v}{v_0} = \left(1 - \frac{\sin 2\theta}{2\theta}\right)^{2/3}$$

$$\frac{A}{A_0} = \frac{1}{\pi}\left(\theta - \frac{\sin 2\theta}{2}\right) \qquad \frac{Q}{Q_0} = \left(1 - \frac{\sin 2\theta}{2\theta}\right)^{2/3}\left(\theta - \frac{\sin 2\theta}{2}\right)\frac{1}{\pi}$$

in the relationship

$$\cos \theta = 0.5. \tag{7.13}$$

Equation (7.12) infers a surface width equal to twice the sidewall length, while eqn (7.13) indicates a 60° sidewall inclination to the horizontal. These requirements are satisfied by an equilateral trapezoidal section which can be circumscribed by a semicircle having the surface width as diameter.

Circular section conduits (pipes) are widely used in open channel flow mode, particularly in sewerage systems. The variations with depth of flow of the flow parameters of principal interest are shown in Fig. 7.3. It is of design interest to note that the velocity at half depth is the same as the flowing-full value. The theoretical maximum discharge occurs at a flow depth ratio y/D equal to 0.87. Attention is drawn to the computational instabilities which may be encountered at flow depths in the vicinity of the maximum flow depth ratio. For practical design purposes it is recommended that the flowing-full value be used as the effective maximum discharge capacity for circular section conduits used as open channels.

7.3 Computation of uniform flow

The Manning and Darcy–Weisbach equations for open channel flow incorporate the three flow variables, velocity v, friction slope S_f, and hydraulic radius R_h. Thus, given any two of these variables the third may be computed. In many cases, designers may prefer to replace the variables v and R_h by the related variables, discharge Q and flow depth y, respectively.

Assembling the known terms on the right-hand side:

Manning:

$$Q = \frac{A}{n} R_h^{0.67} S_f^{0.5} \tag{7.14}$$

$$A^{1.5} R_h = (nQ)^{1.5} S_f^{-0.75} \tag{7.15}$$

$$S_f = \left(\frac{nQ}{A}\right)^2 R_h^{-1.33} \tag{7.16}$$

Darcy–Weisbach:

$$Q = A\left(\frac{8gR_h S_f}{f}\right)^{0.5} \tag{7.17}$$

$$A^2 R_h = \frac{fQ^2}{8gS_f} \tag{7.18}$$

$$S_f = \frac{f}{8gR_h}\left(\frac{Q}{A}\right)^{0.5}. \tag{7.19}$$

The parameters A and R_h are functions of the flow depth y.

7.3.1 Computer program: OCSF

Program OCSF computes the unknown flow parameter in steady uniform open channel flow, given the remaining two parameters, using the Manning equations (7.14), (7.15), and (7.16), or the Darcy–Weisbach equations (7.17), (7.18), and (7.19), as selected by the user. The program caters for rectangular, trapezoidal, and circular flow sections. The relevant geometric properties of these sections are summarized in Fig. 7.2.

Program use is illustrated by the following sample program run. A program listing is included in Appendix A.

Sample run: program OCSF

```
RUN
Program OCSF

THIS PROGRAM COMPUTES DISCHARGE, FLOW DEPTH AND FRICTION
SLOPE IN OPEN CHANNEL STEADY UNIFORM FLOW USING EITHER
THE MANNING OR DARCY-WEISBACH EQUATIONS.


DO YOU WISH TO USE:  1 MANNING  OR   2 DARCY-WEISBACH ?
ENTER 1 OR 2, AS APPROPRIATE? 2
ENTER WALL ROUGHNESS (mm)? 1.5

ENTER CHANNEL DATA:
IS SECTION   1 CIRCULAR  2 RECTANGULAR  3 TRAPEZOIDAL ?
ENTER 1, 2 OR 3, AS APPROPRIATE? 1
ENTER DIAMETER (mm)? 1000

DO YOU WISH TO COMPUTE:  1 DISCHARGE
                         2 FRICTION SLOPE
                         3 FLOW DEPTH ?
ENTER 1, 2  OR  3, AS APPROPRIATE? 3
ENTER DISCHARGE (m**3/s)? 0.5
ENTER FRICTION SLOPE? 0.001

.....Computation in progress, please wait.....

FLOW DEPTH (mm)  =  598.5136

DO YOU WISH TO MAKE ANOTHER COMPUTATION (Y/N)? N
Ok
```

7.4 Specific energy

The specific energy E_s is defined as the total head relative to the channel bed:

$$E_s = y \cos \theta + \alpha \frac{v^2}{2g}. \tag{7.20}$$

When $\cos \theta$ is taken as unity and v is replaced by Q/A, we get

$$E_s = y + \frac{\alpha Q^2}{2gA^2}. \tag{7.21}$$

For a given discharge Q, the value of the flow depth y and, hence E_s, changes as the channel slope is changed. The flow depth at which E_s has a minimum value is known as the critical depth y_c. Its value is found by differentiating E_s with respect to y:

$$\frac{dE_s}{dy} = 1 - \frac{\alpha Q^2}{gA^3}\left(\frac{dA}{dy}\right).$$

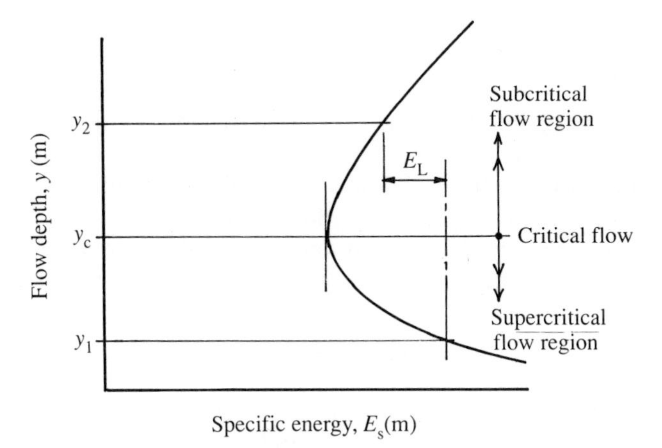

Fig. 7.4 Energy-defined flow regimes.

For minimum E_s, $dE_s/dy = 0$, hence

$$\frac{\alpha Q^2}{gA^3}\left(\frac{dA}{dy}\right) = 1. \tag{7.22}$$

Since $dA = W\,dy$, where W is the channel width at the water surface, eqn (7.22) may be written in the form

$$\frac{\alpha W Q^2}{gA^3} = 1. \tag{7.23}$$

Equation (7.23) defines the condition of minimum specific energy and hence the critical depth y_c.

For a rectangular section and taking $\alpha = 1$, eqn (7.23) simplifies to

$$\frac{v^2}{gy_c} = 1. \tag{7.24}$$

The ratio $v/\sqrt{gy}$ is known as the Froude number F_r. It is essentially a measure of the ratio of inertial to gravitational forces in the flow regime. When the depth of flow exceeds y_c ($F_r < 1$), the flow is described as subcritical or tranquil; when the depth of flow is less than y_c ($F_r > 1$), the flow is described as supercritical. The variation of E_s as a function of flow depth for a rectangular channel is plotted in Fig. 7.4.

An important attribute of critical flow is that it defines a unique relationship between mean velocity and flow depth and hence the creation of critical flow provides the basis of many open channel flow measurement structures.

7.5 Rapidly varied steady flow: the hydraulic jump

It is clear from Fig. 7.4 that for a given specific energy value, two flow depths are feasible, one subcritical and the other supercritical. A smooth transition from supercritical to subcritical flow is not feasible (unconfined deceleration); instead we get a hydraulic jump, as illustrated in Fig. 7.5.

The relation between the incident and sequent depth in a hydraulic jump is found by applying the momentum principle to the control volume between sections 1 and 2:

$$\rho g A_1 \bar{y}_1 - \rho g A_2 \bar{y}_2 + \rho g \bar{A}\,\mathrm{d}x\,S_0 - \rho g \bar{A}\,\mathrm{d}x\,S_\mathrm{f} = \rho Q(v_2 - v_1) \quad (7.25)$$

where $\bar{y}_1$ and $\bar{y}_2$ represent the depth of the centroid at sections 1 and 2, respectively, and $\bar{A}$ is the mean of A_1 and A_2. Neglecting the weight and friction terms, this simplifies to

$$A_1\bar{y}_1 - A_2\bar{y}_2 - \frac{Q}{g}(v_2 - v_1) = 0. \quad (7.26)$$

Applied to a rectangular channel of width B:

$$y_1^2 - y_2^2 - \frac{2Q}{Bg}(v_2 - v_1) = 0$$

which simplifies to

$$y_1^2 y_2 + y_1 y_2^2 - \frac{2q^2}{g} = 0 \quad (7.27)$$

where $q = Q/B$ is the discharge per unit width.

Solving for the sequent depth y_2:

$$y_2 = -\frac{y_1}{2} + \left(\frac{y_1^2}{4} + \frac{2q^2}{gy_1}\right)^{1/2} \quad (7.28)$$

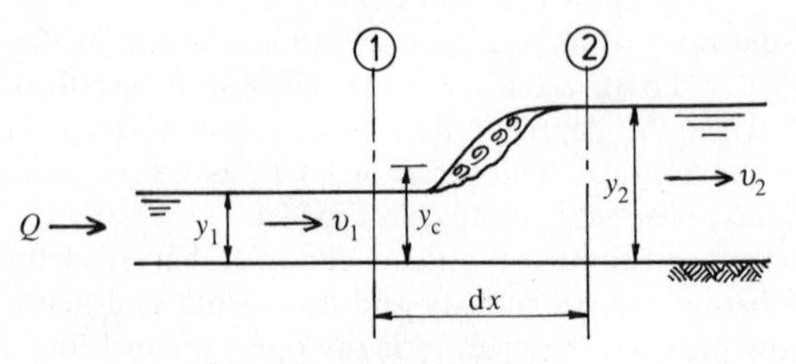

Fig. 7.5 Hydraulic jump.

or, expressed in terms of the Froude number,

$$y_2 = \frac{y_1}{2}\{(1 + 8F_{r1}^2)^{1/2} - 1\} \tag{7.29}$$

where $F_{r1} = v_1/\sqrt{gy_1}$. Similarly, eqn (7.27) may be solved to find the incident depth y_1, when the sequent depth y_2 is known:

$$y_1 = \frac{y_2}{2}\{(1 + 8F_{r2}^2)^{1/2} - 1\}. \tag{7.30}$$

The character of the hydraulic jump can be qualitatively classified by its incident Froude number value (Chow 1959), with best performance in the F_{r1} value range 4.5–9.0. In this range the channel length over which the jump takes place is about six times the downstream depth.

There is a loss of energy in the hydraulic jump, as illustrated in Fig. 7.4. This loss E_L is

$$E_L = E_1 - E_2 = \left(y_1 + \frac{v_1^2}{2g}\right) - \left(y_2 + \frac{v_2^2}{2g}\right) \tag{7.31}$$

which simplifies to

$$E_L = \frac{(y_2 - y_1)^3}{4y_1y_2}. \tag{7.32}$$

7.5.1 Computer program: HJUMP

Program HJUMP computes the incident and sequent depths in hydraulic jumps in rectangular, trapezoidal, and circular section open channels. The computation is based on the general correlation of incident and sequent depths as defined by eqn (7.26).

Sample run: Program HJUMP

```
RUN

Program HJUMP

THIS PROGRAM COMPUTES THE INCIDENT AND
SEQUENT DEPTHS FOR AN HYDRAULIC JUMP
IN RECTANGULAR, TRAPEZOIDAL AND CIRCULAR CHANNELS

DO YOU WISH TO COMPUTE THE
 1  INCIDENT DEPTH   OR   2  SEQUENT DEPTH ?
ENTER 1 OR 2, AS APPROPRIATE? 2
ENTER INCIDENT DEPTH (mm)? 200
```

```
ENTER CHANNEL DATA:
IS SECTION   1 CIRCULAR   2 RECTANGULAR   3 TRAPEZOIDAL ?
ENTER 1, 2 OR 3, AS APPROPRIATE? 1
DIAMETER (mm)? 1000

DISCHARGE (m**3/s)? 0.5

       Computation in progress; please wait ....

CRITICAL DEPTH (mm) =  398.7832
SEQUENT DEPTH (mm)  =  722.5831

DO YOU WISH TO MAKE ANOTHER COMPUTATION (Y/N)? N
Ok
```

7.6 Gradually varied flow

In gradually varied steady open channel flow, velocity and depth vary along the channel length but are invariant with time at any particular location. Such flow occurs in the vicinity of control sections, in channel transitions where there is a change of channel slope or cross-section, in collector channels, as used in sedimentation tanks and sand filters, and in channels with side overflow weirs, as used for storm overflow purposes. Some typical gradually varied water surface profiles are illustrated in Fig. 7.6. Referring to Fig. 7.1, the total head H, as expressed by eqn (7.2), may be written in the form

$$H = Z + y\cos\theta + \frac{\alpha Q^2}{2gA^2}. \tag{7.33}$$

Differentiating with respect to x:

$$\frac{dH}{dx} = \frac{dZ}{dx} + \frac{dy}{dx}\cos\theta + \alpha\left(\frac{Q}{gA^2}\frac{dQ}{dx} - \frac{Q^2}{gA^3}\frac{dA}{dx}\right). \tag{7.34}$$

The term dA/dx can be expressed as the product

$$\frac{dA}{dy}\frac{dy}{dx} = W\frac{dy}{dx},$$

where W is the channel width at water surface level. Assuming $\cos\theta$ equal to unity, eqn (7.34) can be written as

$$-S_f = -S_0 + \frac{\alpha Q}{gA^2}\frac{dQ}{dx} + \frac{dy}{dx}\left(1 - \frac{\alpha W Q^2}{gA^3}\right)$$

which results in the following expression for the water surface slope $\mathrm{d}y/\mathrm{d}x$:

$$\frac{\mathrm{d}y}{\mathrm{d}x} = \frac{S_0 - S_\mathrm{f} - \dfrac{\alpha Q}{gA^2}\dfrac{\mathrm{d}Q}{\mathrm{d}x}}{1 - \alpha\dfrac{WQ^2}{gA^3}}. \qquad (7.35)$$

Where the lateral inflow/outflow is zero ($\mathrm{d}Q/\mathrm{d}x = 0$), the foregoing expression becomes

$$\frac{\mathrm{d}y}{\mathrm{d}x} = \frac{S_0 - S_\mathrm{f}}{1 - \dfrac{\alpha WQ^2}{gA^3}}. \qquad (7.36)$$

The water surface in gradually varied flow is found by integration of eqn (7.35) or (7.36), as appropriate, subject to the particular prevailing upstream and/or downstream boundary values for the flow depth y.

7.7 Computation of gradually varied flow

Equation (7.35) or (7.36) can be numerically integrated using a fourth-order Runge–Kutta numerical computational scheme (Chapra and Canale 1985) in which the flow depth change $y_{i+1} - y_i$, over a channel length Δx, is calculated as follows:

$$y_{i+1} = y_i + \frac{\Delta x}{6}(k_1 + 2k_2 + 2k_3 + k_4) \qquad (7.37)$$

where

$$k_1 = f(x_i, y_i), \qquad k_2 = f(x_i + 0.5\Delta x,\ y_i + 0.5\Delta x\, k_1),$$

$$k_3 = f(x_i + 0.5\Delta x,\ y_i + 0.5\Delta x\, k_2), \qquad k_4 = f(x_i + \Delta x,\ y_i + \Delta x\, k_3).$$

In this case $f(y) = \mathrm{d}y/\mathrm{d}x$, as given by (7.35) or (7.36).

The k-terms may be regarded as measures of $\mathrm{d}y/\mathrm{d}x$ within the channel interval Δx under consideration, $\frac{1}{6}(k_1 + 2k_2 + 2k_3 + k_4)$ being the estimate of its mean value. The cumulative discretization error associated with the fourth-order Runge–Kutta method is proportional to Δx^4. Computational accuracy is thus greatly increased by reducing the channel step length Δx. In this regard, it should be noted that the surface water slope changes rapidly when the flow depth is close to the critical depth and hence, to achieve accuracy in this region, Δx should be assigned a small value. Computation starts from a control section at which the flow depth y is known. Proceeding along the channel in steps of Δx, successive flow depths are calculated using eqn (7.37). The channel reach in which gradually varied flow prevails may be upstream or downstream of the control section. It should be noted that Δx is positive in the downstream direction and negative in the upstream direction.

7.7.1 Computer program: GVF

Program GVF computes the water surface profile from a specified starting depth using a fourth-order Runge–Kutta numerical integration scheme. It caters for rectangular, trapezoidal, and circular section channels and for the following three categories of gradually varied flow:

(1) computation of water surface profiles in channels without lateral inflow;

(2) computation of water surface profiles in collector channels with a uniform lateral inflow rate;

(3) computation of water surface profile and overflow discharge in channels with side weirs.

Figure 7.6(a) and (b) illustrate water surface profiles, typical of gradually varied flow belonging in flow category (1). The water surface slope for this category is defined by eqn (7.36). The program user must input a starting flow depth, which may be at a downstream control, as in Fig. 7.6(a), or at an upstream control, as in Fig. 7.6(b). In addition to computing the flow depth at specified Δx intervals in the channel reach of interest, the program also computes the normal and critical depths for the given discharge, as these frequently represent boundary or limiting values for the computation. The program outputs the flow depth at channel intervals corresponding to the selected computational step length Δx.

Figure 7.6(c) illustrates a gradually varied flow profile characteristic of flow category (2), that is, flow in a collector/decanting channel of the type widely used in water processing units such as sand filters and sedimentation tanks. The illustrated channel has a uniform lateral inflow over its length and a free overfall at its outlet end. Equation (7.35) defines the water surface slope for this category of gradually varied flow. The program user must specify the total lateral inflow which is assumed to be uniformly distributed over the channel length. Computation starts from the outlet end where the flow depth is assumed to be the critical depth. As a practical means of avoiding the computational difficulties associated with the critical depth, referred to above, the program uses a value of $1.02y_c$ as its starting depth at the outlet end of the channel. The program outputs the flow depth at intervals equal to the computational step Δx along the channel, starting from the discharge end.

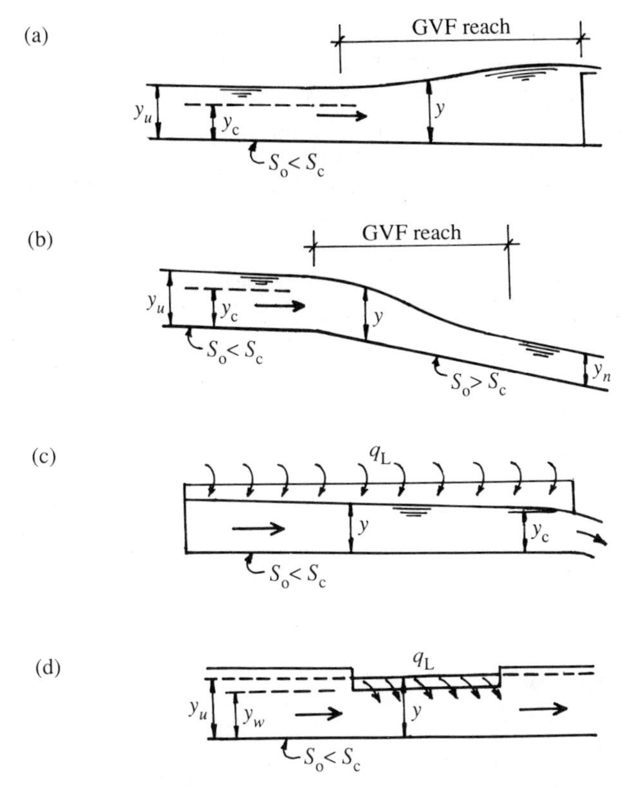

Fig. 7.6 Typical gradually varied flow examples (S_0 = bottom slope, S_c = critical slope): (a) GVF profile in channel of subcritical slope; (b) GVF profile due to bed slope change; (c) GVF profile in collector channel; (d) GVF profile in channel with side weir overflow.

Figure 7.6(d) illustrates a gradually varied flow profile characteristic of flow category (3), that is, flow in a channel reach in which there is a lateral discharge over side overflow weirs. The water surface slope for this flow type is described by eqn (7.35). In this instance the lateral discharge $q_L = -dQ/dx$ is not constant but varies with the weir head H_w in accordance with the weir equation:

$$q_L = C_w H_w^{1.5} \tag{7.38}$$

where C_w is a variable weir coefficient. Program GVF uses the following empirical expression for the weir coefficient C_w, based on experimental results reported by Frazer (1957):

$$C_w = 4.15 - \frac{1.81 y_c}{y} - \frac{0.14 y_c}{L_w} \tag{7.39}$$

where L_w is the side weir length. The computational scheme used in program GVF assumes constant overflow discharge over the computational step length Δx, based on the value of the weir head H_w at the upstream end of the Δx reach. Computation starts at the upstream end of the weir where the flow depth is assumed to be the normal depth and proceeds downstream in steps of Δx. It should be noted that this computation procedure ignores downstream influences on water level which may have a critical effect on side weir discharge in many practical applications. The program outputs the overflow weir head, the overflow discharge, and the residual channel discharge rate, for each computational step, over the side weir length.

Three sample program runs are presented, illustrating the use of program GVF for the analysis of flow categories (1), (2), and (3). Program listings are presented in Appendix A.

Sample run: program GVF, flow category (1)

```
RUN

Program GVF

THIS PROGRAM ANALYSES THREE CATEGORIES OF STEADY GRADUALLY
VARIED OPEN CHANNEL FLOW; IT CATERS FOR RECTANGULAR, TRAPEZOIDAL
AND CIRCULAR CHANNEL SECTIONS AND OFFERS A CHOICE BETWEEN THE
MANNING AND DARCY-WEISBACH FLOW EQUATIONS. THE ANALYSIS USES A
FOURTH ORDER RUNGE-KUTTA NUMERICAL COMPUTATIONAL SCHEME IN THE
THE SOLUTION OF THE RELEVANT WATER SURFACE SLOPE EQUATION.

THE THREE FLOW CATEGORIES ARE:

        (1) GVF WIHOUT LATERAL INFLOW OR OUTFLOW
        (2) GVF WITH LATERAL INFLOW (COLLECTOR CHANNEL)
        (3) GVF WITH LATERAL OUTFLOW (SIDE WEIR CHANNEL)

ENTER NUMBER OF YOUR CHOICE? 1
ANALYSIS OF GVF IN CHANNELS WITHOUT LATERAL INFLOW/OUTFLOW
The program computes the flow depth at specified intervals along
the channel, starting from a control point at which the depth
is specified. The programs outputs distance from the control
point and corresponding flow depth. Note that distances measured
upstream from the control point are printed as negative values.

DATA ENTRY:

DO YOU WISH TO USE   1 MANNING OR   2 DARCY-WEISBACH ?
ENTER 1 OR 2, AS APPROPRIATE? 2
WALL ROUGHNESS (mm)? 1.5

ENTER CHANNEL DATA:
IS SECTION   1 CIRCULAR  2 RECTANGULAR  3 TRAPEZOIDAL ?
ENTER 1, 2 OR 3, AS APPROPRIATE? 1
DIAMETER (mm)? 1000
```

```
DEPTH AT CONTROL SECTION (mm)? 990

ENTER CHANNEL BED SLOPE? 0.001

ENTER DISCHARGE (m**3/s)? 0.5

IS COMPUTATION PROCEEDING UPSTREAM (UP) OR
DOWNSTREAM (DN) FROM CONTROL SECTION ?
ENTER UP OR DN, AS APPROPRIATE? UP

ENTER CHANNEL STEP COMPUTATION LENGTH (m)? 10

ENTER NUMBER OF COMPUTATION STEPS? 8

.....data input complete; computation now in progress.....

CRITICAL DEPTH (mm) =  398.7832

NORMAL DEPTH (mm) =  598.5136

 DISTANCE (m)  DEPTH (mm)
       0.0      990.0
     -10.0      984.0
     -20.0      978.0
     -30.0      971.9
     -40.0      965.7
     -50.0      959.5
     -60.0      953.3
     -70.0      947.1
     -80.0      940.8

DO YOU WISH TO MAKE ANOTHER COMPUTATION (Y/N)?
```

Sample run: program GVF, flow category (2)

```
RUN

Program GVF

THIS PROGRAM ANALYSES THREE CATEGORIES OF STEADY GRADUALLY
VARIED OPEN CHANNEL FLOW; IT CATERS FOR RECTANGULAR, TRAPEZOIDAL
AND CIRCULAR CHANNEL SECTIONS AND OFFERS A CHOICE BETWEEN THE
MANNING AND DARCY-WEISBACH FLOW EQUATIONS. THE ANALYSIS USES A
FOURTH ORDER RUNGE-KUTTA NUMERICAL COMPUTATIONAL SCHEME IN THE
THE SOLUTION OF THE RELEVANT WATER SURFACE SLOPE EQUATION.

THE THREE FLOW CATEGORIES ARE:

      (1) GVF WIHOUT LATERAL INFLOW OR OUTFLOW
      (2) GVF WITH LATERAL INFLOW (COLLECTOR CHANNEL)
      (3) GVF WITH LATERAL OUTFLOW (SIDE WEIR CHANNEL)

ENTER NUMBER OF YOUR CHOICE? 2
ANALYSIS OF COLLECTOR CHANNEL FLOW

The analysis relates to collector channels having a uniform
lateral inflow over the channel length and a free overfall at
the outlet end. The analysis starts from the outlet end where
the flow depth is taken as the critical depth. For practical
computational reasons, as explained in Chapter 7, the starting
depth value is taken as 1.02 times the critical depth.
```

```
DATA ENTRY:

DO YOU WISH TO USE   1 MANNING OR   2 DARCY-WEISBACH ?
ENTER 1 OR 2, AS APPROPRIATE? 1
MANNING N-VALUE? 0.01

ENTER CHANNEL DATA:
IS SECTION   1 CIRCULAR  2 RECTANGULAR  3 TRAPEZOIDAL ?
ENTER 1, 2 OR 3, AS APPROPRIATE? 2
CHANNEL WIDTH (mm)? 500

ENTER CHANNEL BED SLOPE? 0

ENTER DISCHARGE (m**3/s)? 0.25

ENTER CHANNEL LENGTH (m)? 8

ENTER CHANNEL STEP COMPUTATION LENGTH (m)? 1

.....data input complete; computation now in progress.....

CRITICAL DEPTH (mm) =  294.3134

 DISTANCE (m)  DEPTH (mm)  Q (m**3/s)
      0.00      300.20        0.25
     -1.00      475.95        0.22
     -2.00      489.93        0.19
     -3.00      500.45        0.16
     -4.00      508.33        0.13
     -5.00      514.07        0.09
     -6.00      517.99        0.06
     -7.00      520.25        0.03
     -8.00      521.00        0.00

DO YOU WISH TO MAKE ANOTHER COMPUTATION (Y/N)? N
Ok
```

Sample run: program GVF, flow category (3)

```
RUN

Program GVF

THIS PROGRAM ANALYSES THREE CATEGORIES OF STEADY GRADUALLY
VARIED OPEN CHANNEL FLOW; IT CATERS FOR RECTANGULAR, TRAPEZOIDAL
AND CIRCULAR CHANNEL SECTIONS AND OFFERS A CHOICE BETWEEN THE
MANNING AND DARCY-WEISBACH FLOW EQUATIONS. THE ANALYSIS USES A
FOURTH ORDER RUNGE-KUTTA NUMERICAL COMPUTATIONAL SCHEME IN THE
THE SOLUTION OF THE RELEVANT WATER SURFACE SLOPE EQUATION.

THE THREE FLOW CATEGORIES ARE:

      (1) GVF WIHOUT LATERAL INFLOW OR OUTFLOW
      (2) GVF WITH LATERAL INFLOW (COLLECTOR CHANNEL)
      (3) GVF WITH LATERAL OUTFLOW (SIDE WEIR CHANNEL)

ENTER NUMBER OF YOUR CHOICE? 3
ANALYSIS OF GVF IN CHANNEL WITH SIDE OVERFLOW WEIRS
```

```
The analysis relates to flow in a channel in which there is
a lateral outflow over sharp-edged side weirs of specified crest
level and crest length. The program computes the normal flow
depth and the variation in depth over the weir length.It outputs
the weir head and weir overflow rate, the channel flow rate and
flow depth at the specified computational step intervals over
the weir length.

DATA ENTRY:

DO YOU WISH TO USE   1 MANNING OR    2 DARCY-WEISBACH ?
ENTER 1 OR 2, AS APPROPRIATE? 2
WALL ROUGHNESS (mm)? 1.5

ENTER CHANNEL DATA:
IS SECTION   1 CIRCULAR  2 RECTANGULAR  3 TRAPEZOIDAL ?
ENTER 1, 2 OR 3, AS APPROPRIATE? 3
BOTTOM WIDTH (m)? 0.5
ANGLE OF SIDE TO HORL (deg)? 45

ENTER SINGLE SIDE WEIR LENGTH (m)? 1.6
ENTER NUMBER OF SIDE WEIRS (1 OR 2)? 2

ENTER HT OF WEIR CREST ABOVE CHANNEL BED AT U/S END (mm)? 480

ENTER CHANNEL BED SLOPE? 0.001

ENTER DISCHARGE (m**3/s)? 0.5

ENTER COMPUTATIONAL STEP LENGTH (m)? 0.2

.....data input complete; computation now in progress.....

CRITICAL DEPTH (mm) =  364.9902

NORMAL DEPTH (mm) =  504.2223
  DIST ALONG  WEIR HEAD  OVERFLOW     CHANNEL FLOW)
  WEIR (m)      (mm)    (m**3/s.m)    (m**3/s)
     0.00      24.22       0.02         0.50
     0.20      25.01       0.02         0.50
     0.40      25.83       0.02         0.49
     0.60      26.68       0.02         0.49
     0.80      27.57       0.03         0.48
     1.00      28.48       0.03         0.48
     1.20      29.42       0.03         0.47
     1.40      30.40       0.03         0.47

DO YOU WISH TO MAKE ANOTHER COMPUTATION (Y/N)? N
Ok
```

7.8 Channel transitions

A channel transition is typically a short length of channel over which there is a change in cross-section and/or slope and hence a change in velocity under steady flow conditions. Transitions are used to connect channels of different sizes and are often associated with flow-measurement structures.

Where the flow regime throughout a transition channel is entirely subcritical or entirely supercritical, eqns (7.35) or (7.36), as appropriate, may be used to compute the variation in flow depth. In expanding or contracting channels the channel width W at water surface level may be expressed as a function of the flow depth y and space variable x.

Where the flow regime changes from subcritical to supercritical within a transition section, passing through critical depth in the process, the transition acts as a 'control section', that is, a section which effectively controls the upstream flow depth. This type of transition forms the basis of many open channel flow measurement structures, as discussed in Chapter 8.

Where the flow changes from supercritical to subcritical in a transition via a hydraulic jump, the location of the latter may be determined using the computational procedures outlined in Section 7.5.

Abrupt changes in channel cross-section give rise to local energy losses which are analogous to the form losses associated with pipe fittings. The energy loss ΔH due to an abrupt expansion may be shown from momentum considerations to be

$$\Delta H = \frac{(v_U - v_D)^2}{2g} \tag{7.40}$$

where v_U and v_D are the mean upstream and downstream velocities, respectively. The head loss can be reduced to about one-third of this value by using a side wall taper of 1:4 in the channel expansion (Henderson 1966).

The energy losses associated with channel contractions are less than those for channel expansions. They are conveniently expressed in terms of the downstream velocity head:

$$\Delta H = K \frac{v_D^2}{2g}$$

where K has a value of about 0.23 for sharp-edged contractions in rectangular channels and about 0.11 when the edges are rounded.

7.8.1 Entry flow to closed conduits

The transition from open to closed conduit flow is a frequently encountered flow regime in water engineering practice. It merits particular attention because entry conditions may sometimes exert a controlling influence on the discharge capacity of the closed conduit.

Where the closed conduit entry is fully submerged and flowing full throughout its length, the computation of its steady discharge rate is straightforward, being a function of the difference between the upstream and downstream water levels and the integrated hydraulic resistance of the closed conduit.

Where the conduit entry is partly submerged or operating at a low

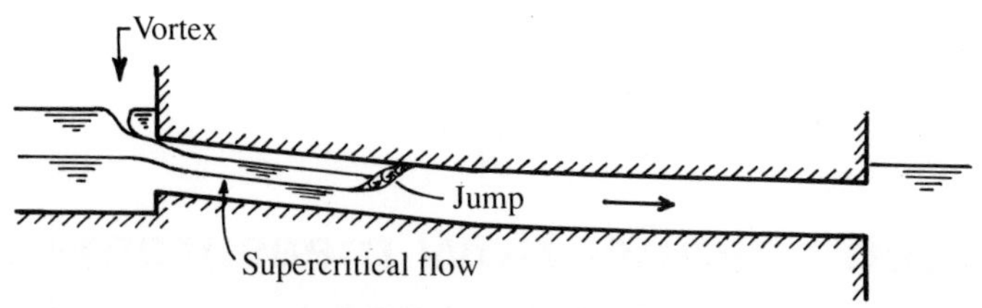

Fig. 7.7 Flow profiles at the entrance to a closed conduit of steep slope at inlet end.

submergence, entry conditions may effectively control the discharge rate if the immediate closed conduit has a supercritical slope, as illustrated in Fig. 7.7. Under these circumstances the flow passes through the critical depth at the conduit entrance, which acts as a control section. This condition may prevail even when the conduit entrance is submerged, being maintained by air entrainment due to vortex formation associated with flow acceleration towards the entrance. The transition from open channel to fully closed conduit flow is brought about by a hydraulic jump at some distance downstream of the entrance, as shown in Fig. 7.7.

In general, the entrainment of air at closed conduit entrances is undesirable and may significantly reduce the discharge capacity. Where flow acceleration is unavoidable at closed conduit entrances, air intake at low submergence levels can be prevented by the use of anti-vortex baffles (Blaisdell 1960).

References

Blaisdell, F. W. (1960). Hood inlet for closed conduit spillways. *Proc. ASCE*, **86**, No. HY5, 7.

Chapra, S. C. and Canale, R. P. (1985). *Numerical Methods for Engineers*, McGraw-Hill, New York.

Chow, V. T. (1959). *Open-channel Hydraulics*, McGraw-Hill, New York.

Frazer, W. (1957). The behaviour of side weirs in prismatic rectangular channels. *Proc. Inst. Civ. Eng.*, **6**, 305–27.

Henderson, F. M. (1966). *Open Channel Flow*, Macmillan, New York.

Hydraulics Research, Wallingford (1990). *Charts for the Hydraulic Design of Channels and Pipes*, (6th edn). Thomas Telford Ltd., London.

8
Open channel flow measurement structures

8.1 Introduction

Open channel flow measurement structures are so designed that the flow rate can be reliably determined from measurement of the upstream head relative to a reference level in the 'control section' of the structure. The control section may incorporate a weir, orifice plate, or critical depth flume.

To insure that there is a unique relationship between the upstream head and flow rate it is essential that the upstream head is not influenced by variations in the downstream or 'tailwater' level. When flow is not influenced by the tailwater level, conditions are said to be 'modular' and the upstream head is entirely determined by the control section of the measuring structure.

The following measuring structures are discussed in this chapter:

(1) broad-crested weirs;

(2) sharp-crested weirs;

(3) long-throated flumes;

(4) sharp-edged orifices.

8.2 The broad-crested weir

The broad-crested weir has a raised horizontal cill of sufficient length in the flow direction to effect a horizontal surface and hydrostatic pressure distribution for at least a short distance, as shown in Fig. 8.1. Neglecting any energy losses between the upstream and control sections:

$$H_1 = 1.5h_c = 1.5\left(\frac{Q^2}{gb^2}\right)^{0.33}. \qquad (8.1)$$

Hence

$$Q = \tfrac{2}{3}b\sqrt{\tfrac{2}{3}g}\,H_1^{1.5}. \qquad (8.2)$$

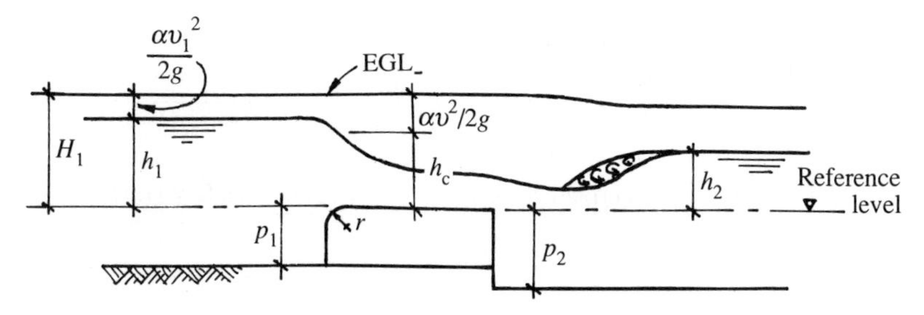

Fig. 8.1 Broad-crested weir.

For application to practical flow measurement, this equation is written in the form

$$Q = C_d C_v \tfrac{2}{3} b \sqrt{\tfrac{2}{3} g}\, h_1^{1.5} \tag{8.3}$$

where C_d is an empirically determined discharge coefficient and C_v is an approach velocity coefficient that allows for the replacement of H_1 by h_1 in the discharge equation.

The discharge coefficient C_d is a function of the upstream head over the cill h_1, the cill length L in the flow direction, the crest width b, and the roughness of the flow surface. The following expression for C_d has been proposed by Bos (1976):

$$C_d = \left[1 - \frac{2x(L-r)}{b}\right]\left[1 - \frac{x(L-r)}{h_1}\right]^{1.5} \tag{8.4}$$

where x is a surface roughness parameter which for well-finished concrete may be taken as 0.005 and for smooth surfaces as 0.003.

The approach velocity coefficient for discharge measurement structures in general is given by the relation

$$C_v = \left(\frac{H_1}{h_1}\right)^n \tag{8.5}$$

where n is the head parameter exponent in the discharge equation, in this instance having the value 1.5.

The following practical design recommendations have been proposed by Bos (1976):

(1) $h_1 \geq 0.06$ m or $\geq 0.05L$, whichever is greater;

(2) radius of cill nose $r = 0.2H_{1(\max)}$;

(3) $p_1 \geq 0.15$ m or $\geq 0.67H_1$, whichever is greater;

Table 8.1 Modular limit values for broad-crested weirs (Harrison 1967).

	H_2/H_1	
H_1/p_2	Vertical back face	Sloping back face (1:4)
0.1	0.71	0.74
0.2	0.74	0.79
0.4	0.78	0.85
0.6	0.82	0.88
0.8	0.84	0.91
1.0	0.86	0.92
2.0	0.90	0.96
4.0	0.94	0.97
7.0	0.96	0.98
10.0	0.98	0.99

(4) $20H_1 \geq L \geq 2H_1$, to ensure parallel flow while avoiding undulations over the cill;

(5) $b \geq 0.3$ m or $\geq H_1$ or $\geq L/5$, whichever is greater;

(6) to ensure modular flow conditions, the downstream depth and step height p_2 should comply with the modular limit values set out in Table 8.1.

8.3 The sharp-crested weir

Sharp-crested or thin plate weirs are widely used for the measurement of small to medium discharges. The control section opening may be of rectangular, triangular (V-notch), or exponential shape (Sutro). The thickness of the plate crest in the direction of flow is generally less than 2 mm. If the plate thickness exceeds 2 mm, a bevelled edge is formed, as illustrated on Fig. 8.2. Sharp-crested weirs are placed vertically, the weir plate being normal to the direction of flow.

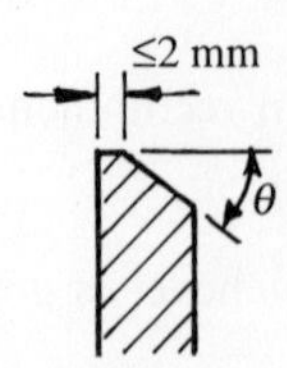

Fig. 8.2 Profile of sharp-crested weir edge; $\theta = 45$ degrees for rectangular weirs and $\theta = 60$ degrees for non-rectangular weirs.

8.3.1 Rectangular sharp-edge weirs

Figure 8.3 defines the dimensional parameters for rectangular sharp-edged weirs. Such weirs can be treated as simple orifices to give the theoretical discharge equation:

$$Q = \tfrac{2}{3}\sqrt{2g}\, b h_1^{1.5} \tag{8.6}$$

where the velocity of approach is considered negligible. For practical use in flow measurement this equation can be written in modified form as proposed by Kindsvater and Carter (1957):

$$Q = C_e \tfrac{2}{3}\sqrt{2g}\, b_e h_e^{1.5} \tag{8.7}$$

where the discharge coefficient $C_e = K_1 + K_2(h_1/p_1)$; the effective weir width $b_e = b + K_b$; and the effective weir head $h_e = h_1 + 0.001$ m. Numerical values for the empirical coefficients K_1, K_2, and K_b are given in Table 8.2.

The following design limits for practical application are suggested by Bos (1976).

(1) $h_1 \geq 0.03$ m;

(2) $h_1/p_1 \leq 2$; $p_1 \geq 0.10$ m;

(3) $b \geq 0.15$ m;

(4) to allow unobstructed weir overspill (aerated nappe) the tailwater level should be at least 0.05 m below the weir crest level.

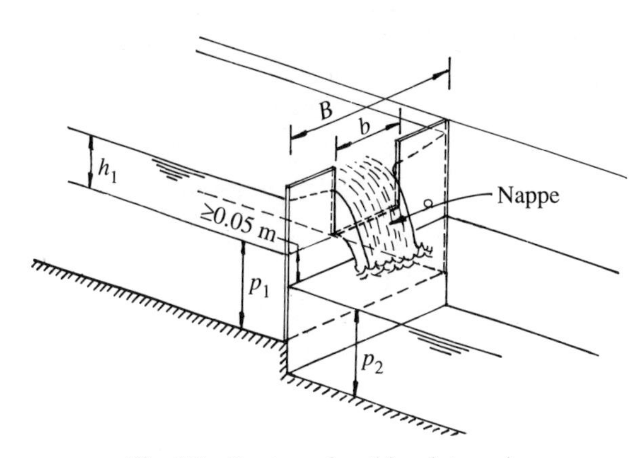

Fig. 8.3 Rectangular thin plate weir.

Table 8.2 Coefficient values for rectangular sharp-crested weirs (Kindsvater and Carter 1957).

b/B	K_1	K_2	K_b
1.0	0.602	0.075	−0.0009
0.9	0.599	0.064	0.0037
0.8	0.597	0.045	0.0043
0.7	0.595	0.030	0.0041
0.6	0.593	0.018	0.0037
0.5	0.592	0.011	0.0030
0.4	0.591	0.0058	0.0027
0.3	0.590	0.0020	0.0025
0.2	0.589	−0.0018	0.0024
0.1	0.588	−0.0021	0.0024
0	0.587	−0.0023	0.0024

8.3.2 V-notch weirs

Figure 8.4 defines the dimensional parameters for thin plate V-notch weirs. The basic theoretical discharge equation for such weirs is

$$Q = \tfrac{8}{15}\sqrt{2g}\,\tan(\theta/2)h_e^{2.5}. \tag{8.8}$$

Kindsvater and Carter (1957) proposed the following practical form of this equation:

$$Q = C_e\tfrac{8}{15}\sqrt{2g}\,\tan(\theta/2)h_e^{2.5} \tag{8.9}$$

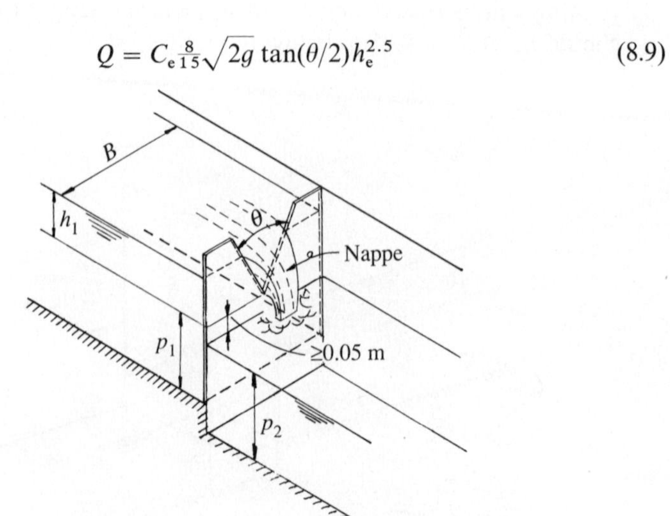

Fig. 8.4 Thin plate V-notch weir.

Table 8.3 V-notch sharp-crested weir coefficients (Kindsvater and Carter 1957).

Notch angle θ (degrees)	20	40	60	80	100
C_e	0.595	0.581	0.577	0.577	0.580
K_h (mm)	2.8	1.8	1.2	0.85	0.80

where the discharge coefficient C_e is a function of the notch angle θ, as given in Table 8.3. The effective head $h_e = h_1 + K_h$, where K_h is an empirical head correction factor, being a function of the notch angle θ, as given in Table 8.3.

The following design limits for practical application of sharp-crested V-notch weirs are recommended by Bos (1976):

(1) $h_1/p_1 \leq 1.2$;

(2) $h_1/B \leq 0.4$; $B \geq 0.60$ m;

(3) $0.60 \geq h_1 \geq 0.05$ m;

(4) $p_1 \geq 0.10$ m;

(5) $100° \geq \theta \geq 25°$;

(6) tailwater level ≥ 0.05 m below the vertex of the V-notch.

8.3.3 The proportional-flow (Sutro) weir

When installed in a rectangular channel, the proportional-flow weir regulates flow such that the discharge is linearly related to the upstream depth and in consequence the mean upstream velocity remains constant. Hence, it is sometimes used as an outlet control device on grit-separation channels in sewage treatment plants. The geometric outline of the weir profile is given in Fig. 8.5, which also defines its dimensional parameters.

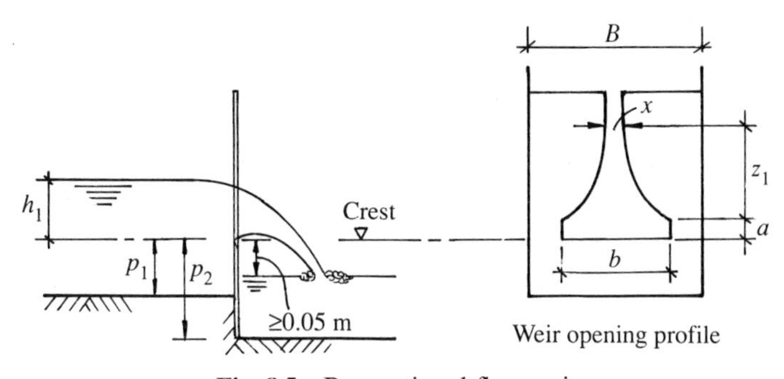

Fig. 8.5 Proportional-flow weir.

Table 8.4 Proportional-flow weir discharge coefficient C_d.

a (m)	b (m) 0.15	0.23	0.30	0.38	0.46
0.006	0.608	0.613	0.617	0.618	0.619
0.015	0.606	0.611	0.615	0.617	0.617
0.030	0.603	0.608	0.612	0.613	0.614
0.046	0.601	0.606	0.610	0.612	0.612
0.061	0.599	0.604	0.608	0.610	0.610
0.076	0.598	0.603	0.607	0.608	0.609
0.091	0.597	0.602	0.606	0.608	0.608

The weir opening has a lower rectangular portion connected to an upper curved portion, the width of which reduces according to the relation

$$\frac{x}{b} = 1 - \frac{2}{\pi} \tan^{-1} \sqrt{\frac{z_1}{a}}. \qquad (8.10)$$

The discharge equation for this weir type may be written as follows (Bos 1976):

$$Q = C_d b \sqrt{2ga}\,(h_1 - \tfrac{1}{3}a). \qquad (8.11)$$

Recommended values for the discharge coefficient C_d, as a function of a and b, are given in Table 8.4.

The values given in Table 8.4 may also be used for crestless weirs, provided the weir width b is not less than 0.15 m (Singer and Lewis 1966).

The following design limits for practical application are recommended by Bos (1976):

(1) $h_1 \geq 2a$ or ≥ 0.03 m, whichever is greater;

(2) $a \geq 0.005$ m;

(3) $b \geq 0.15$ m;

(4) $b/p_1 \geq 1$;

(5) $B/b \geq 3$;

(6) tailwater level ≥ 0.05 m below the weir crest.

8.4 The critical depth flume

The critical depth flume is created by reducing the cross-sectional area of a channel sufficiently for critical flow to occur at the constricted section or throat of the flume. In 'long-throated' flumes the prismatic throat section has a sufficient length in the flow direction to achieve parallel flow and associated hydrostatic pressure distribution, at least over a short length. The flume has a horizontal invert which constitutes the reference level for flow measurement.

The following analysis relates to the general case of a trapezoidal long-throated flume in a trapezoidal channel, as illustrated in Fig. 8.6. Rectangular long-throated flumes in rectangular or trapezoidal channels may be regarded as particular examples of this category of flow-measuring structure.

Critical flow conditions are created in the throat (control) section. Under such conditions, the discharge can be expressed, as shown in Chapter 7, as follows:

$$Q = \left(\frac{gA_c^3}{\alpha W_c}\right)^{1/2} \tag{8.12}$$

$$H_1 = y_c + \frac{A_c}{2W_c} \tag{8.13}$$

where y_c, A_c, and W_c represent the depth, flow cross-sectional area, and water surface width at the critical section, respectively. Combining these two

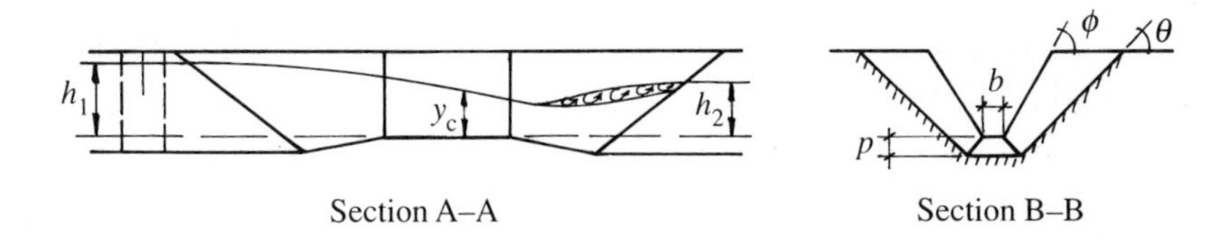

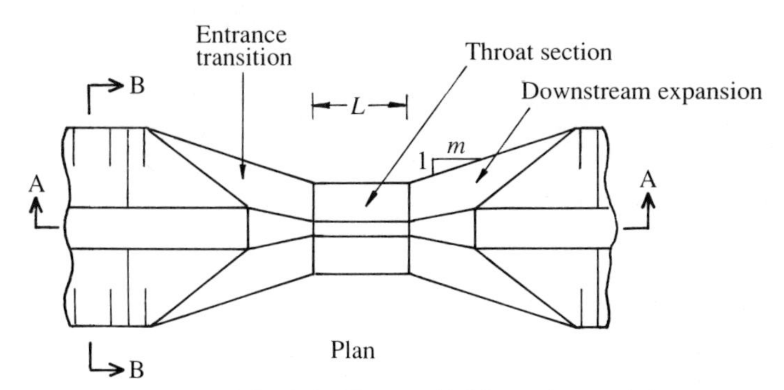

Fig. 8.6 Trapezoidal long-throated flume.

relations, assuming $\alpha = 1$, we get

$$Q = A_c\sqrt{2g(H_1 - y_c)}. \tag{8.14}$$

For practical computational purposes, this relation is written in the form:

$$Q = C_d C_v A_c\sqrt{2g(h_1 - y_c)}. \tag{8.15}$$

The discharge coefficient C_d is a measure of the variation of the discharge from its theoretical value as expressed by eqn (8.14). This variation is due to friction head loss between the point at which the upstream head is measured and the throat, and is also influenced by boundary layer separation in the throat (Ackers *et al.* 1979). Based on the experimental data plotted in Fig. 8.7 (Bos 1976) C_d can be empirically correlated with H_1/L using the following linear approximations:

$$0.1 < H_1/L < 0.2 \qquad C_d = 0.89 + 0.20(H_1/L)$$

$$0.2 < H_1/L < 1.0 \qquad C_d = 0.95 + 0.05(H_1/L).$$

The velocity coefficient $C_v = (H_1/h_1)^{1.5}$ allows the substitution of the measured head h_1 for the total head H_1.

The modular limit requirement is satisfied if the available head difference between the upstream and downstream water levels can accommodate the head losses through the structure. A major component of this head loss is that due to flow expansion from the throat cross-section to the downstream channel section. The influence of the expansion rate on the modular limit is indicated in Table 8.5.

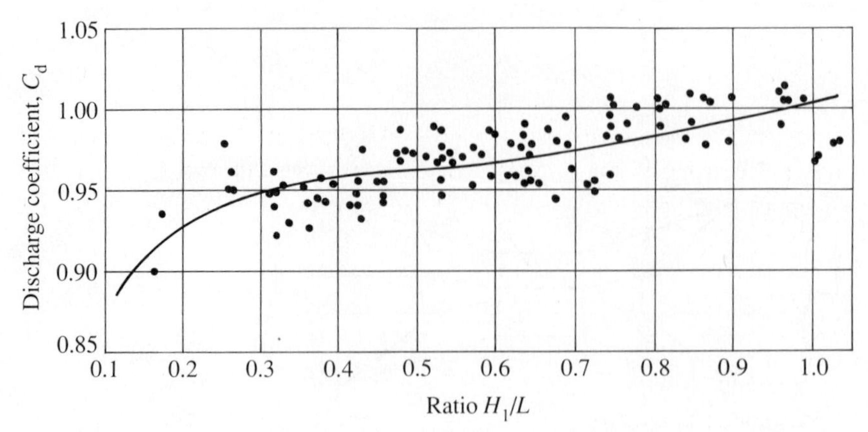

Fig. 8.7 Discharge coefficient for long-throated flumes (source: Bos 1976).

Table 8.5 Effect of the downstream expansion rate on the modular limit (Ackers *et al.* 1979).

Expansion rate (1: m, Fig. 8.6)	Modular limit (H_2/H_1)
1:20	0.91
1:10	0.83
1:6	0.80
1:3	0.74

Where the head loss is not too critical, an expansion rate in the range 1:3 to 1:6 is recommended. Where the head loss is critical, the designer may select a more gradual downstream expansion to reduce the head loss through the structure.

On the inflow side, a convergence rate of about 1:3 is typically recommended for the transition from the upstream channel section to the throat section.

For accurate flow measurement application, Bos (1976) recommends the following design limits:

(1) $h_1 \geq 0.06$ m or $\geq 0.1L$, whichever is greater;

(2) Froude number $F_r = v_1/(gA_1/W_1)^{1/2}$ in approach channel to be ≤ 0.5;

(3) $1.0 \geq H_1/L \geq 0.1$;

(4) $W_t \geq 0.3$ m or $\geq H_{1(\max)}$ or $\geq L/5$, whichever is greater, where W_t is the width of the water surface in the throat at maximum discharge.

8.5 Sharp-edged orifices

Figure 8.8 shows the typical installation configuration for an orifice plate in an open channel. Such an orifice may have a free discharge to air or may submerged on the downstream side. The basic discharge equation for the

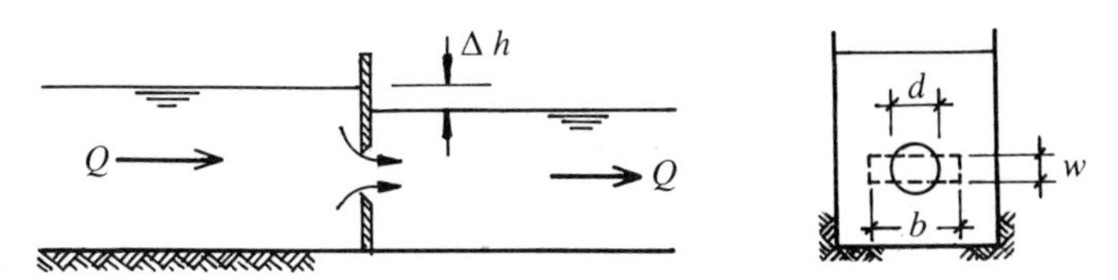

Fig. 8.8 Sharp-edged orifice with a section showing circular and rectangular orifices.

orifices is

$$Q = C_d C_v A \sqrt{2g\, \Delta h} \tag{8.16}$$

where Δh is the differential head for submerged orifices and is the upstream head relative to the centre of the orifice for freely discharging orifices; A is the orifice area. Orifices are preferably installed where the velocity of approach is negligible. The edge profile for orifice plates should comply with the specification recommended for thin plate weirs (Fig. 8.2).

8.5.1 The circular sharp-edged orifice

Values of C_d for free flow and submerged sharp-edged circular orifices are given in Table 8.6. The following practical design limits are recommended:

(1) edge distance $\geq d/2$;

(2) upstream channel cross-sectional area $\geq 10 \times$ orifice area;

(3) upstream submergence of top of orifice $\geq d$;

(4) $\Delta h \geq 0.03$ m.

8.5.2 The rectangular sharp-edged orifice

The rectangular orifice has a width b and depth w; $A = bw$. Under fully contracted, submerged conditions, where the velocity of approach is negligible, the discharge coefficient C_d may be taken as 0.61. If the contraction is suppressed along part of the orifice perimeter the coefficient of discharge is

Table 8.6 Discharge coefficient C_d for sharp-edged circular orifices (negligible approach velocity).

Orifice diameter d (m)	C_d	
	free flow	submerged flow
0.020	0.61	0.57
0.025	0.62	0.58
0.035	0.64	0.61
0.045	0.63	0.61
0.050	0.62	0.61
0.065	0.61	0.60
≥0.075	0.60	0.60

modified (Bos 1976) as follows:

$$C_d = 0.61(1 + 0.15r) \tag{8.17}$$

where r is the ratio of suppressed length to the total perimeter length.

Where both side and bottom contractions are suppressed, for example, as in flow under a sluice gate, the discharge equation may be written in the form

$$Q = C_d C_v bw\sqrt{2g(y_1 - y)} \tag{8.18}$$

where y_1 is the depth of flow upstream of the sluice gate and y is the contracted downstream depth. Introducing the parameters $n = y_1/w$ and $\delta = y/w$ (δ is the coefficient of contraction), eqn (8.18) becomes

$$Q = C_d C_v\, bw^{1.5}[2g(n - \delta)]^{0.5} \tag{8.19}$$

or

$$Q = KA\sqrt{2gw} \tag{8.20}$$

where K is a function of n, δ, C_d, and C_v; K-values are given in Table 8.7.

To ensure free flow below a sluice gate (modular flow conditions), the downstream water depth should not exceed the sequent hydraulic jump depth, as computed from the known supercritical depth y on the discharge side of the sluice gate.

For accurate flow measurement, the following design limits are recommended by Bos (1976) for sharp-edged rectangular orifices:

(1) to ensure fully contracted flow, the edge distance to the channel boundaries should be at least twice the least dimension of the orifice;

(2) $\Delta h \geq 0.03$ m;

(3) submergence of the top edge $\geq w$;

(4) $w \geq 0.02$ m.

Table 8.7 Sluice gate discharge coefficients (Bos 1976).

y_1/w	δ	C_d	K
1.5	0.648	0.600	0.614
1.7	0.637	0.598	0.665
1.9	0.632	0.597	0.713
2.2	0.628	0.596	0.780
2.6	0.y26	0.597	0.865
3.0	0.625	0.599	0.944
4.0	0.624	0.604	1.124
5.0	0.624	0.607	1.279

8.6 Selection and design of flow measurement structures

The installation of a flow measurement structure will in most instances cause an increase in upstream water levels throughout the flow range. This characteristic is often critically significant in the selection and design of such structures and is an aspect of selection and design that must always be carefully considered. It is also important to check that modular flow conditions prevail over the full flow range.

It is desirable to have an unobstructed approach channel of reasonably uniform cross-section and straight for a distance of 10–20 times the channel width. The selected location for head measurement should be at a distance of $3h_1$ to $4h_1$ upstream of the control section.

Manual head-measuring devices include the staff gauge, the point gauge, and the hook gauge, while a variety of water level-sensing devices is available for automatic recording of head. Stilling chambers are normally used in conjunction with automatic head-measuring systems. They offer the advantage of eliminating water surface ripples and similar transient level variations.

8.7 Computer programs

The following programs relate to the material presented in this chapter:

Title	Function
BROAD	Design of round-nosed broad-crested weirs
SHARP1	Design of rectangular sharp-edged weirs
SHARP2	Design of V-notch weirs
SUTRO	Design of proportional-flow weirs
FLUME	Design of long-throated flumes

The programs are written in an interactive design-oriented style, in which the user is requested to input appropriate design data and to select structure dimensions. These are then checked for compliance with the recommended practical design limits and verified as conforming or not conforming to these limits.

The program output provides either tabulated head and discharge values or a plotted head–discharge relationship for the structure in question.

Program listings are given in Appendix A. The following is a sample program run, illustrating the use of program FLUME for the design of a long-throated flume.

Sample program run: program FLUME

```
RUN
Program FLUME

****DESIGN OF TRAPEZOIDAL LONG-THROATED FLUMES****

Design parameters:
                   throat bottom width
                   throat side wall inclination to horizontal
                   throat length
                   throat step height

(above parameters are illustrated on Fig 8.6, Chap. 8)


Press the space bar to continue
INPUT OF FLOW AND CHANNEL DATA

Enter max expected flow (m**3/s)? 0.2
Enter min expected flow (m**3/s)? 0.02
Enter channel slope (sin theta)? 0.001
Enter channel bed width (m)? 0.5
Enter channel side slope angle to horl. (deg)? 60
Enter channel Manning n-value? 0.01

INPUT OF CONTROL SECTION DATA

Enter value for b (throat bottom width,m)? 0.25
Enter value for angle of inclination to horl. of throat wall (deg)? 60
Enter value for L (throat length, m)? 0.6
Enter value for P, the throat step height (m)
Note:P value is not limited by any specifications.? 0.1

INPUT OF DOWNSTREAM TRANSITION DATA

The expansion from the throat section to the downstream channel
should be gradual; a sidewall and bottom expansion in the range
of 1:4 to 1:6 is recommended but is not obligatory.
Enter value of m, where expansion is expressed as 1:m? 4

DATA INPUT IS NOW COMPLETE

Press the space bar to continue
DESIGN CHECK FOR MAX FLOW CONDITION
===================================
Specifications for hl
---------------------
Lower limit for hl is 0.06m or 0.1*L= .06  whichever is greater
Actual value of hl= .4164726
You are within specifications
```

```
Specifications for Froude number Fr.
------------------------------------
Fr in the approach channel shouldn't exceed 0.5
Fr = V1/(g*A1/BC)©0.5 =  .2526087
V1,A1,BC are velocity, x-sectional area and water surface width,
respectively, in the approach channel.
You are within specifications

Press the space bar to continue

Specifications for H1/L
--------------------
H1/L should be between 0.1 and 1.0
Actual H1/L= .7141153
You are within specifications

Specifications for width of the water surface in the throat
-----------------------------------------------------------
Width of the water surface in the throat at the maximum stage should  not be
less than 0.3m,nor less than H1max= .4284692  nor less than L/5= .12
Actual width =  .6122514
You are within specifications

Press the space bar to continue

Specifications for the modular limit
------------------------------------
The modular ratio limit H2/H1 should not be exceeded
Modular limit H2/H1= .75
TAILWATER DEPTH (m) =  .3052526
Actual H2/H1= .5907041
You are within specifications

Press the space bar to continue

DESIGN CHECK AT MINIMUM FLOW
============================

Specifications for h1
---------------------
Lower limit for h1 is 0.06m or 0.1*L= .06  whichever is greater
Actual value of h1= .1189988
You are within specifications

Specifications for H1/L
--------------------
H1/L should be between 0.1 and 1.0
Actual H1/L= .2001367
You are within specifications

Check for modular flow condition at min flow
--------------------------------------------
Modular limit H2/H1= .75
TAILWATER DEPTH (m) =  .0757662
Actual H2/H1=-.1017773
You are within specifications

Press the space bar to continue

SUMMARY OF DIMENSIONAL DATA

throat width b =  .25 (m)
throat sidewall angle to horl. =  59.99236 (deg)
throat length L =  .6 (m)
throat step height P =  .1 (m)
downstream divergence of sidewalls and bottom 1: 4
```

```
The upstream convergence of side walls and bottom should be about 1:3
The floor of the entrance transition and of the approach channel should be
flat and level ,and at no point higher than the invert of the throat, up to
a distance 1.0*Hlmax= .4284692 (m) upstream of the head measurement station.
This head measurement station should be located upstream of the flume at a
distance equal to between 2 & 3 Hlmax ie.   .8569384  and  1.285408


Enter required form of head/discharge relationship
1-Tabular form.
2-Graph form.
? 1
Enter flow increment (m**3/s) for tabulation? 0.02


Q(m3/s)        yc(m)        hl(m)         Hl(m)
=========================================
 0.020         0.082        0.119         0.120
 0.040         0.123        0.177         0.180
 0.060         0.157        0.222         0.226
 0.080         0.187        0.258         0.264
 0.100         0.214        0.291         0.298
 0.120         0.238        0.320         0.328
 0.140         0.258        0.347         0.356
 0.160         0.277        0.373         0.382
 0.180         0.297        0.395         0.405
 0.200         0.314        0.416         0.428
 0.220         0.331        0.437         0.450

yc is the critical depth at the control section


Enter (1) to run again;    enter (2) to quit
? 2
Ok
```

References

Ackers, P., White, W. R., Perkins, J. A., and Harrison, A. J. M. (1979). *Weirs and Flumes for Flow Measurement*. Wiley, New York.

Bos, M. G. (ed.) (1976). *Discharge Measurement Structures*. International Institute for Land Reclamation and Improvement, Wageningen, The Netherlands.

Harrison, A. J. M. (1967). The streamlined broad-crested weir. *Proc. ICE*, **38**, 657–78.

Kindsvater, C. E. and Carter, R. W. C. (1957). Discharge characteristics of rectangular thin-plate weirs. *J. Hyd. Div. ASCE*, HY 6, 83.

Singer, J. and Lewis, D. C. G. (1966). Proportional-flow weirs for automatic sampling or dosing. *Water and Water Engineering*, **70**, 105–11.

Related reading

Bos, M. G., Replogle, J. A., and Clemmens, A. S. (1984). *Flow Measuring Flumes for Open Channel Systems*. Wiley, New York.

Cheremisinoff, N. P. (1979). *Applied Fluid Flow Measurements*. Marcel Dekker, Basel.

Scott, R. W. W. (1982). *Developments in Flow Measurement*. Applied Science Publishers, London.

9
Dimensional analysis, similitude, and hydraulic models

9.1 Introduction

Exact theoretical solutions to fluid flow problems are generally only available for laminar flow conditions and simple boundary conditions, circumstances rarely found in civil engineering. Recourse to experiment may be necessary, especially where the physical boundaries are complex. One of the difficulties which faces the analyst is the large number of variables which may influence a particular flow phenomenon. However, by judicious grouping of the variables involved into composite variable groups, it is possible to reduce the number of variables used to define a particular flow problem. This can be accomplished by application of the principle that, in a physically correct equation, all terms must have the same dimensions. The primary dimensions which characterize fluid flow systems are mass M, length L, and time T. All system parameters, such as force, power, velocity, and so on, can be expressed in MLT terms:

$$\begin{aligned} \text{force} &= \text{mass} \times \text{acceleration} = MLT^{-2} \\ \text{power} &= \text{force} \times \text{velocity} \qquad = ML^2T^{-3} \\ \text{dynamic viscosity} &= \text{force} \times \text{time/area} \quad = ML^{-1}T^{-1} \end{aligned}$$

9.2 Dimensionless quantities

Each fluid flow characteristic is dependent on a number of variables. For example, the force F in a particular flow environment can be expressed in the form

$$F = \phi(v, d, \rho, \mu) \tag{9.1}$$

where ϕ means 'function of'. Any such functions can be represented as a power series sum:

$$F = v^{a1}d^{b1}\rho^{c1}\mu^{d1} + v^{a2}d^{b2}\rho^{c2}\mu^{d2} + \cdots$$

where $a1$, $b1$, $a2$, $b2, \ldots$ are numerical indices. Dividing across by the first term on the right-hand side:

$$\frac{F}{v^{a1} d^{b1} \rho^{c1} \mu^{d1}} = 1 - v^{a2-a1} d^{b2-b1} \rho^{c2-c1} \mu^{d2-d1} + \cdots.$$

Since the first term on the right-hand side is dimensionless, all terms in the equation must be non-dimensional, that is,

$$\left[\frac{F}{v^a d^b \rho^c \mu^d} \right] = 0 \tag{9.2}$$

where [ ] indicates 'dimensions of'.

9.3 The Buckingham π theorem

A phenomenon which is a function of n variables can be modelled as follows:

$$\phi(x_1, x_2, x_3, \ldots, x_n) = 0.$$

Such a phenomenon can also be described as a function of $n - m$ non-dimensional group variables, where m is the number of basic component dimensions of the variables $x_1, \ldots, x_n$. In fluid flow these basic dimensions are M, L, and T, so that $m = 3$. The corresponding non-dimensional functional relationship is

$$F(\pi_1, \pi_2, \pi_3, \ldots, \pi_{n-m}) = 0.$$

Each π term is a non-dimensional grouping of $m + 1$ variables, m of which are repeated in all terms. For example, the pressure drop in pipe flow can be expressed as a function of six variables:

$$\phi(\Delta p, L, \rho, v, D, \mu, k) = 0 \tag{9.3}$$

where Δp is the pressure drop over a pipe length L and k is the pipe wall roughness. In this case $n = 7$ and $m = 3$. Taking v, ρ, and D as the three repeated variables, the alternative non-dimensional functional relationship is:

$$F(\pi_1, \pi_2, \pi_3, \pi_4) = 0.$$

Each π term is a grouping of four (that is, $m + 1$) variables, three (that is, m) of which are repeated in all π terms. Taking the first term π_1:

$$\pi_1 = v^\alpha \rho^\beta D^\gamma \Delta p$$

and is non-dimensional. Hence

$$[(LT^{-1})^\alpha (ML^{-3})^\beta L^\gamma ML^{-1}T^{-2}] = 0$$

from which it follows that $\alpha = -2$, $\beta = -1$, and $\gamma = 0$, giving π_1 the following value:

$$\pi_1 = \frac{\Delta p}{\rho v^2}.$$

Similarly

$$\pi_2 = \frac{L}{D}, \qquad \pi_3 = \frac{\mu}{\rho v D} \qquad \text{and} \qquad \pi_4 = \frac{k}{D}.$$

The resulting non-dimensional functional relationship for pressure drop in pipe flow is

$$F\left(\frac{\Delta p}{\rho v^2}, \frac{L}{D}, \frac{\mu}{\rho v D}, \frac{k}{D}\right) = 0. \tag{9.4}$$

The number of variables has been reduced from seven to four. The non-dimensional group variables can be combined by multiplication or division:

$$F\left(\frac{\Delta p D}{L \rho v^2}, \frac{\mu}{\rho v D}, \frac{k}{D}\right) = 0.$$

Hence

$$\frac{\Delta p}{L} = \frac{\rho v^2}{D} \phi\left(\frac{k}{D}, R_e\right). \tag{9.5}$$

In the Darcy–Weisbach equation for pipe flow the friction factor f is related to pressure drop as follows:

$$\frac{\Delta p}{L} = \frac{\rho v^2}{D} f.$$

Hence $f = \phi(k/D, R_e)$, as in the Colebrook–White equation. In this case, however, the derivation has been based on dimensional reasoning and a judicious selection of the three repeated variables v, D, and ρ.

9.4 Physical significance of non-dimensional groups

The force components in fluid systems arise from gravity, viscosity, elasticity, surface tension, and pressure influences. The resultant force is called the inertial force and the ratio of each of the above force components to the resultant force indicates the relative significance of each on overall system behaviour.

1. Gravity

$$\frac{F_i}{F_g} = \frac{MLT^{-2}}{Mg} = \frac{v^2}{gL} = F_r^2$$

where F_r is the Froude number.

2. Viscosity

$$\frac{F_i}{F_\mu} = \frac{MLT^{-2}}{\mu L^2 T^{-1}} = \frac{\rho L^4 T^{-2}}{\mu L^2 T^{-1}} = \frac{\rho L v}{\mu} = R_e$$

where R_e is the Reynolds number.

3. Surface tension

$$\frac{F_i}{F_\sigma} = \frac{MLT^{-2}}{\sigma L} = \frac{\rho L^4 T^{-2}}{\sigma L} = \frac{\rho L v^2}{\sigma} = W_e^2$$

where W_e is the Weber number.

9.5 Similarity requirements in model studies

Dynamic similarity between model and prototype requires that the ratios of the inertial force to its individual force components are the same in model and prototype. This implies that the Reynolds, Froude and Weber numbers have the same values in model and prototype.

Geometric similarity is assured by adopting a fixed scale ratio for all dimensions.

If model and prototype are dynamically and kinematically similar then the flow patterns will be the same at both scales, resulting in kinematic similarity.

1. R_e similarity

$$\left(\frac{vL\rho}{\mu}\right)_m = \left(\frac{vL\rho}{\mu}\right)_p$$

where the subscripts m and p relate to model and prototype, respectively. If $\rho_m = \rho_p$ and $\mu_m = \mu_p$ then

$$\frac{v_m}{v_p} = \frac{L_p}{L_m}.$$

2. F_r similarity

$$\left(\frac{v}{\sqrt{gL}}\right)_m = \left(\frac{v}{\sqrt{gL}}\right)_p.$$

If $g_m = g_p$ then

$$\frac{v_m}{v_p} = \left(\frac{L_m}{L_p}\right)^{0.5}.$$

Thus if the same fluid is used in model and prototype (that is, ρ and μ are the same), it is not possible to achieve complete similarity because of the conflicting operational requirements for R_e and F_r similarity. In practice, a compromise is reached by basing scaling relationships on the predominant force component. For flows without a free surface, for example, pipe flow and flow around submerged bodies such as submarines, aircraft, motor vehicles, and buildings, R_e is taken as the scaling criterion. For free surface phenomena such as hydraulic structures, ships, and so on, F_r scaling is used. An illustrative example of F_r is shown in Fig. 9.1.

Where ρ, μ, and g are assumed to be the same in model and prototype, the scale ratios for the various flow parameters, as determined by R_e and F_r scaling, can be expressed in terms of the length scale ratio λ, where $\lambda = L_m/L_p$. Table 9.1 shows scale ratios for flow variables.

The influence of surface tension (Weber number) is generally not significant in hydraulic model studies—refer to related comments in Section 9.5.2.

Fig. 9.1 1:50 scale model of the spillway on an impounding reservoir on the River Dodder at Bohernabreena, County Dublin. (Hydraulics Laboratory, University College, Dublin. Courtesy A. L. Dowley, by permission of K. O'Donnell, Chief Engineer, Dublin Corporation.).

Table 9.1 Scale ratios for flow variables.

Variable	Dimensions	R_e scaling	F_r scaling
Time	T	λ^2	$\lambda^{0.5}$
Velocity	LT^{-1}	λ^{-1}	$\lambda^{0.5}$
Acceleration	LT^{-2}	λ^{-3}	λ^0
Discharge	L^3T^{-1}	λ	$\lambda^{2.5}$
Force	ρL^4T^{-2}	λ^0	λ^3
Pressure	ρL^2T^{-2}	λ^{-2}	λ
Power	ρL^5T^{-3}	λ^{-1}	$\lambda^{3.5}$

9.5.1 Pumps and turbines

Discharge

The discharge rate Q through a pump or turbine can be expressed in terms of the geometric characteristics of the device, the operating head, and the fluid properties, in the general functional form

$$f(Q, N, D, B, gH, \rho, \mu) = 0 \tag{9.6}$$

where N is the rotational speed, D and B are the impeller or runner diameter and width, respectively, and gH is a measure of the operating pressure rise/drop across the device. Replacing this functional relationship with its non-dimensional equivalent by taking ρ, μ, and D as the repeated variables in the transforming procedure, yields the following:

$$\phi\left(\frac{B}{D}, \frac{Q}{ND^3}, \frac{gH}{N^2D^2}, \frac{\mu}{\rho ND^2}\right) = 0. \tag{9.7}$$

If the numerical values of the non-dimensional groups in model and prototype are equal, then complete similarity is achieved. The ratio B/D infers geometric similarity; the non-dimensional groups gH/N^2D^2 and $\mu/\rho ND^2$ can be recognized as F_r^{-2} and R_e^{-1}, respectively.

Power

Similarly, pump or turbine power P can be expressed in a form similar to eqn (9.6):

$$f(P, N, D, B, gH, \rho, \mu) = 0. \tag{9.8}$$

Using ρ, N, and D as the repeated variables, the corresponding non-dimensional relation is found:

$$\phi\left(\frac{P}{\rho N^3 D^5}, \frac{B}{D}, \frac{gH}{N^2 D^2}, \frac{\mu}{\rho N D^2}\right) = 0 \tag{9.9}$$

Specific speed

It follows from eqn (9.7) that, if the viscosity influence is neglected, dynamic similarity is achieved in geometrically similar pumps, if the following relationships are satisfied:

$$\left(\frac{Q}{ND^3}\right)_{\mathrm{m}} = \left(\frac{Q}{ND^3}\right)_{\mathrm{p}}, \qquad \text{hence} \qquad \frac{N_{\mathrm{m}}}{N_{\mathrm{p}}} = \frac{Q_{\mathrm{m}}}{Q_{\mathrm{p}}}\left(\frac{D_{\mathrm{p}}}{D_{\mathrm{m}}}\right)^3$$

and

$$\left(\frac{N^2 D^2}{gH}\right)_{\mathrm{m}} = \left(\frac{N^2 D^2}{gH}\right)_{\mathrm{p}}, \qquad \text{hence} \qquad \left(\frac{D_{\mathrm{p}}}{D_{\mathrm{m}}}\right)^3 = \left(\frac{(gH)_{\mathrm{p}}^{1.5}}{(gH)_{\mathrm{m}}^{1.5}}\right)\left(\frac{N_{\mathrm{m}}}{N_{\mathrm{p}}}\right)^3.$$

From these relations it follows that

$$\frac{N_{\mathrm{m}}}{N_{\mathrm{p}}} = \left(\frac{Q_{\mathrm{p}}}{Q_{\mathrm{m}}}\right)^{0.5}\left(\frac{(gH)_{\mathrm{m}}}{(gH)_{\mathrm{p}}}\right)^{0.75}. \tag{9.10}$$

If the model is defined as having unit values of Q and H, the rotational speed of the model can be expressed, using eqn (9.10), as

$$N_{\mathrm{m}} = \frac{N_{\mathrm{p}}(Q_{\mathrm{p}})^{0.5}}{(gH)_{\mathrm{p}}^{0.75}}. \tag{9.11}$$

The model speed, thus defined, is known as the 'specific speed' N_{s}:

$$N_{\mathrm{s}} = \frac{NQ^{0.5}}{(gH)^{0.75}}. \tag{9.12}$$

In the form presented in eqn (9.12), the specific speed is a non-dimensional index, which can be used to categorize pump types, as discussed in Chapter 11. In pump technology literature, the gravity constant is often omitted from the specific speed expression, resulting in the following dimensional form of the specific speed characteristic:

$$N_{\mathrm{s}} = \frac{NQ^{0.5}}{H^{0.75}}. \tag{9.13}$$

9.5.2 The use of distorted scales

For practical reasons it may be desirable to use different vertical and horizontal scales. In rivers and estuaries the horizontal dimensions of the reach to be modelled may be very large relative to the water depth. In order to accommodate a model within a reasonable plan area it is often necessary to select a horizontal scale that is smaller than the vertical scale. The vertical scale should, as a general rule, be not less than 1:100 and should not lead to water depths that are likely to be significantly influenced by surface tension effects.

Distorted scales influence scale relationships. Since the circumstances in which they are necessary invariably involve free surface flow, scale relationships are governed by Froude law scaling. Representing the horizontal scale as λ_x and the vertical as λ_y, the scale relations for velocity v and discharge Q, as dictated by the Froude number, are as follows:

$$\text{Velocity} = f(y), \qquad \text{hence} \qquad \frac{v_m}{v_p} = \lambda_y^{0.5}$$

$$\text{Discharge} = f(v, x, y), \qquad \text{hence} \qquad \frac{Q_m}{Q_p} = \lambda_y^{0.5}\lambda_x\lambda_y.$$

9.6 Concluding comments

Dimensional analysis is a valuable aid to modelling of flow phenomena. It enables the effects of a number of variables to be considered together. However, it does not yield any information on whether particular variables are important or not; this knowledge must be obtained from a physical examination of the problem.

Related reading

Allen, J. (1952). *Scale Models in Hydraulic Engineering*. Longmans, London.
Francis, J. R. D. and Minton, P. (1984). *Civil Engineering Hydraulics*. Edward Arnold, London.
Novak, P. and Cabelka, J. (1981). *Models in Hydraulic Engineering*. Spon, London.
Featherstone, R. E. and Nalluri, C. (1982). *Civil Engineering Hydraulics*. Collins, London.

10
Unsteady flow in open channels

10.1 Introduction

Unsteady flow in open channels differs from that in closed conduits in that the existence of a free surface allows the flow cross-section to change freely, a factor which has an important influence on the rate of transient change propagation. Unsteady open channel flow is encountered in flood flow in rivers, in headrace canals supplying hydropower stations, in river estuaries, and so on.

10.2 Basic equations

As in the case of closed conduits the basic equations are derived from continuity and momentum considerations. In deriving these equations the following assumptions are made.

1. Hydrostatic pressure prevails at every point in the channel.
2. Velocity is uniformly distributed over each cross-section.
3. The slope of the channel bed is small and uniform.
4. The frictional resistance is the same as for steady flow.

Figure 10.1 defines a control volume and the dimensional parameters used to develop the continuity equation.

The continuity equation balances mass inflow and mass outflow with the rate of change of the contained mass within the control volume:

In time $\mathrm{d}t$: Inflow − Outflow = Change in mass of fluid in CV

$$\rho v A\,\mathrm{d}t + \rho q_1\,\mathrm{d}t\,\mathrm{d}x - \rho\left(v + \frac{\partial v}{\partial x}\mathrm{d}x\right)\left(A + \frac{\partial A}{\partial x}\mathrm{d}x\right)\mathrm{d}t = \rho\frac{\partial A}{\partial t}\mathrm{d}t\,\mathrm{d}x.$$

Dividing across by $\rho\,\mathrm{d}t\,\mathrm{d}x$ and neglecting higher-order terms:

$$A\frac{\partial v}{\partial x} + v\frac{\partial A}{\partial x} + \frac{\partial A}{\partial t} = q_1. \tag{10.1}$$

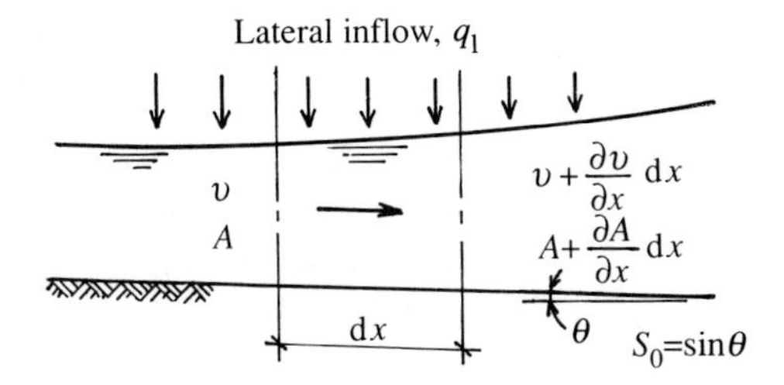

Fig. 10.1 Reference diagram for the continuity equation.

Equation (10.1) can also be written in the form

$$\frac{\partial}{\partial x}(vA) + \frac{\partial A}{\partial t} = q_1$$

or

$$\frac{\partial Q}{\partial x} + \frac{\partial A}{\partial t} = q_1. \tag{10.2}$$

For a rectangular channel with zero lateral inflow, eqn (10.1) simplifies to

$$y\frac{\partial v}{\partial x} + v\frac{\partial y}{\partial x} + \frac{\partial y}{\partial t} = 0. \tag{10.3}$$

Hence

$$\frac{\partial}{\partial x}(vy) + \frac{\partial y}{\partial t} = 0$$

or

$$\frac{\partial q}{\partial x} + \frac{\partial y}{\partial t} = 0 \tag{10.4}$$

where q is the discharge per unit width of channel.

Figure 10.2, which shows the forces acting on a fluid control volume, defines the parameters required for the development of the momentum equation. The momentum equation relates net force to momentum

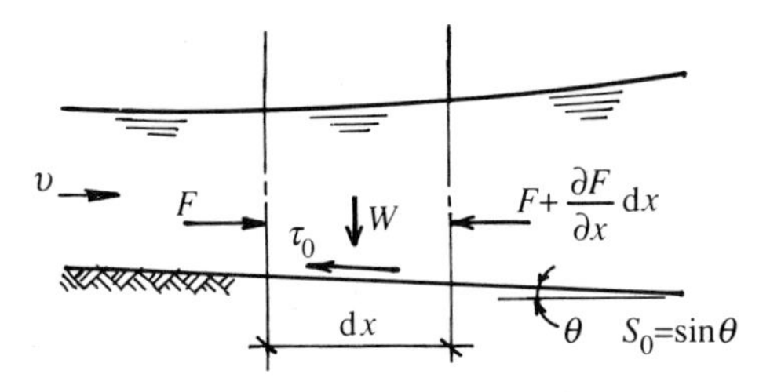

Fig. 10.2 Reference diagram for the momentum equation.

change:

$$F - \left(F + \frac{\partial F}{\partial x}\,\mathrm{d}x\right) + W \sin\theta - \tau_0 P_\mathrm{r}\,\mathrm{d}x = \rho A\,\mathrm{d}x\,\frac{\mathrm{d}v}{\mathrm{d}t} \tag{10.5}$$

where the pressure force $F = \rho g A\bar{y}$, the weight force $W = \rho g A\,\mathrm{d}x$, the wall shear stress $\tau_0 = \rho g R_\mathrm{h} S_\mathrm{f}$, and P_r is the perimeter length. Equation (10.5) may therefore be expressed in terms of more basic flow parameters:

$$-\rho g\,\frac{\partial}{\partial x}(A\bar{y})\,\mathrm{d}x + \rho g A S_0\,\mathrm{d}x - \rho g A S_\mathrm{f}\,\mathrm{d}x = \rho A\,\mathrm{d}x\,\frac{\mathrm{d}v}{\mathrm{d}t}.$$

Dividing across by $\rho g\,\mathrm{d}x$:

$$-\frac{\partial}{\partial x}(A\bar{y}) + AS_0 - AS_\mathrm{f} = \frac{A}{g}\left(v\,\frac{\partial v}{\partial x} + \frac{\partial v}{\partial t}\right). \tag{10.6}$$

For a rectangular section this simplifies to

$$-\frac{\partial y}{\partial x} + S_0 - S_\mathrm{f} = \frac{v}{g}\frac{\partial v}{\partial x} + \frac{1}{g}\frac{\partial v}{\partial t}$$

or

$$v\,\frac{\partial v}{\partial x} + \frac{\partial v}{\partial t} + g\,\frac{\partial y}{\partial x} + g(S_\mathrm{f} - S_0) = 0. \tag{10.7}$$

The terms of the momentum equation have the dimension of acceleration or force per unit mass. The first two terms on the left-hand side are the fluid acceleration terms, $g\,\mathrm{d}y/\mathrm{d}x$ represents the pressure force component, and gS_f and gS_0 represent the friction and gravity force components, respectively.

The forms of the continuity and momentum equations, represented in eqns (10.3) and (10.7), respectively, are known as the St Venant equations; they relate the dependent variables y and v to the independent space and time variables x and t, respectively.

10.3 Solution by the characteristics method

The same procedure as that used in Chapter 6 for the solution of the corresponding pair of equations for unsteady flow in pipes is applied here. Multiplying the continuity equation (10.3) by the factor λ and adding to the momentum equation (10.7):

$$\left[\frac{\partial v}{\partial x}(\lambda y + v) + \frac{\partial v}{\partial t}\right] + \lambda\left[\frac{\partial y}{\partial t} + \frac{\partial y}{\partial x}\left(v + \frac{g}{\lambda}\right)\right] + g(S_\mathrm{f} - S_0) = 0. \tag{10.8}$$

This partial differential equation can be converted to a total differential equation provided that

$$\frac{\mathrm{d}x}{\mathrm{d}t} = \lambda y + v = v + g/\lambda$$

or $\lambda y = g/\lambda$. Hence

$$\lambda = \pm\left(\frac{g}{y}\right)^{1/2}$$

and

$$\frac{\mathrm{d}x}{\mathrm{d}t} = v \pm (gy)^{1/2} = v \pm c \tag{10.9}$$

where c is the gravity wavespeed; hence $\lambda = \pm g/c$. Thus eqn (10.8) can be written in the equivalent total differential form:

$$\frac{\mathrm{d}v}{\mathrm{d}t} + \frac{g}{c}\frac{\mathrm{d}y}{\mathrm{d}t} + g(S_\mathrm{f} - S_0) = 0 \tag{10.10}$$

subject to:

$$\frac{\mathrm{d}x}{\mathrm{d}t} = v + c; \tag{10.11}$$

and

$$\frac{\mathrm{d}v}{\mathrm{d}t} - \frac{g}{c}\frac{\mathrm{d}y}{\mathrm{d}t} + g(S_\mathrm{f} - S_0) = 0 \tag{10.12}$$

subject to

$$\frac{\mathrm{d}x}{\mathrm{d}t} = v - c. \tag{10.13}$$

Thus, the two partial differential equations (10.3) and (10.7) have been converted to their characteristic form, that is, linked pairs of ordinary differential equations. On integration of the latter over the time interval Δt we get a pair of C^+ characteristic equations and a pair of C^- characteristic equations:

$$C^+ \quad \begin{cases} v_P - v_R + g\displaystyle\int_{y_R}^{y_P} \frac{1}{c}\,\mathrm{d}y + \int_{t_R}^{t_P} g(S_\mathrm{f} - S_0)\,\mathrm{d}t = 0 \\ x_P - x_R = \displaystyle\int_{t_R}^{t_P} (v + c)\,\mathrm{d}t \end{cases}$$

$$C^- \quad \begin{cases} v_P - v_S - g\displaystyle\int_{y_S}^{y_P} \frac{1}{c}\,\mathrm{d}y + \int_{t_S}^{t_P} g(S_\mathrm{f} - S_0)\,\mathrm{d}t = 0 \\ x_P - x_S = \displaystyle\int_{t_S}^{t_P} (v - c)\,\mathrm{d}t \end{cases}$$

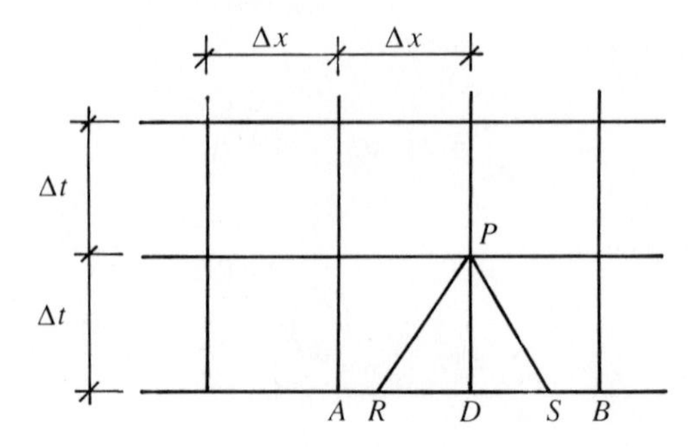

Fig. 10.3 The x–t plane.

where v_R and v_S are the interpolated values of v at x_R and x_S, respectively, as illustrated on the x–t plane in Fig. 10.3. The foregoing integrations may be approximated to a first-order accuracy by assigning their known values to v, c, and S_f, giving the characteristic equations the following format:

$$C^+ \begin{cases} v_P - v_R + \dfrac{g}{c_R}(y_P - y_R) + g(S_R - S_0)\Delta t = 0 & (10.14) \\ x_P - x_R = (v_R + c_R)\Delta t & (10.15) \end{cases}$$

$$C^- \begin{cases} v_P - v_S + \dfrac{g}{c_S}(y_P - y_S) + g(S_S - S_0)\Delta t = 0 & (10.16) \\ x_P - x_S = (v_S - c_S)\Delta t & (10.17) \end{cases}$$

where S_R and S_S are the values of S_f at R and S, respectively. The parameter values at R are found by linear interpolation in the interval AD and the parameter values at S are found by linear interpolation in the interval DB. Referring to the interval AD in Fig. 10.4:

$$x_D - x_R = \Delta t(v_R + c_R)$$

$$\frac{v_D - v_R}{v_D - v_A} = \frac{x_D - x_R}{x_D - x_A}$$

$$\frac{c_D - c_R}{c_D - c_A} = \frac{x_D - x_R}{x_D - x_A}.$$

Replacing x_D by x_P and $(x_D - x_A)$ by Δx, the following are the interpolated values at R:

$$v_R = \frac{v_D + \theta(-v_D c_A + c_D v_A)}{1 + \theta(v_D - v_A + c_D - c_A)} \qquad (10.18)$$

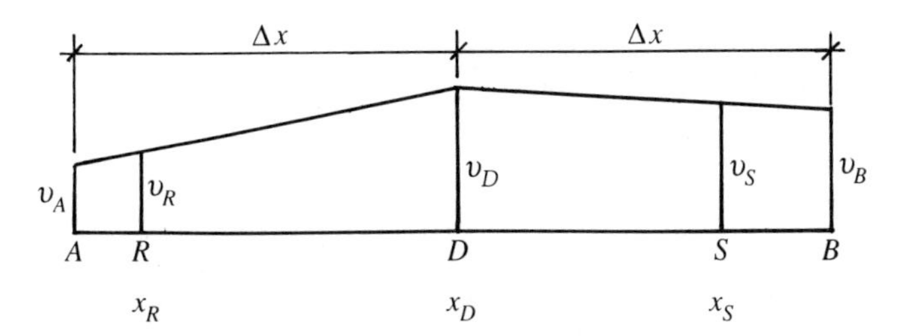

Fig. 10.4 Linear interpolation.

$$c_R = \frac{c_D - v_R\theta(c_D - c_A)}{1 + \theta(c_D - c_A)} \tag{10.19}$$

$$y_R = y_D - \theta(y_D - y_A)(v_R + c_R) \tag{10.20}$$

where $\theta = \Delta t/\Delta x$. Interpolated values are similarly established at S on the negative characteristic side of D:

$$x_D - x_S = \Delta t(v_S - c_S)$$

$$\frac{v_D - v_S}{v_D - v_B} = \frac{x_S - x_D}{x_B - x_D}$$

$$\frac{c_D - c_S}{c_D - c_B} = \frac{x_S - x_D}{x_B - x_D}.$$

Solution of these equations gives the following interpolated values at S:

$$v_S = \frac{v_D - \theta(v_D c_B - c_D v_B)}{1 - \theta(v_D - v_B - c_D + c_B)} \tag{10.21}$$

$$c_S = \frac{c_D + v_S\theta(c_D - c_B)}{1 + \theta(c_D - c_B)} \tag{10.22}$$

$$y_S = y_D + \theta(v_S - c_S)(y_D - y_B). \tag{10.23}$$

10.4 Numerical computation procedure: program OCUSF

The foregoing finite difference formulation of the characteristic form of the unsteady flow equations can be used where there are no abrupt changes in the water surface profile and where conditions are subcritical. The computational procedure adopted is similar to that outlined in Chapter 6 for the solution of the corresponding set of pipe flow equations. The channel length is divided into N reaches, each of length Δx. The corresponding value of the time step Δt is set by the so-called Courant condition:

$$\Delta t \leq \frac{\Delta x}{|v| + c}. \tag{10.24}$$

This ensures that the characteristic curves plotted on the x–t plane (Fig. 10.3) remain within a single x–t grid. At time zero the values of y and v are known at each channel node point. Their values at internal nodes, at one time interval Δt later, are found by solution of eqns (10.14) and (10.16) and are as follows:

$$y_P = \frac{1}{c_R + c_S}\left\{y_S c_R + y_R c_S + c_R c_S\left[\frac{v_R - v_S}{g} - \Delta t(S_R - S_S)\right]\right\} \tag{10.25}$$

$$v_P = v_R - \frac{g(y_P - y_R)}{c_R} - g\Delta t(S_R - S_0). \tag{10.26}$$

The updated values of y (y_p) and v (v_p) at the upstream end of the channel are governed by the negative characteristic equations (10.16) and (10.17) and the prevailing upstream boundary condition equation, which is typically in the form of a defined variation of either y or Q with time. Solution of eqn (10.17) and the boundary condition equation yields the required values for v_p and y_p.

The new values for v_p and y_p at the downstream end of the channel are found in the same manner as their corresponding values at the upstream end, the defining equations being the positive characteristic equation (10.18) and the prevailing downstream boundary condition equation.

Program OCUSF implements the foregoing numerical computational scheme for the computation of y and Q in open channel unsteady flow. The program caters for rectangular, trapezoidal, and circular channel sections. The computation of friction slope is based on the Manning equation. The upstream and downstream boundary condition options included in the program are linear time variations of y and Q. It should be noted that the foregoing analytical treatment and program OCUSF relate to tranquil flow only, that is, where the Froude number F_r is less than unity. As the flow depth approaches critical value ($F_r = 1$), the numerical computation becomes unstable. At critical depth, $v = c$ and hence the negative characteristic on the x–t plane becomes vertical, that is, points S and D are coincident.

It should be noted that the foregoing analysis assumes that the friction resistance under unsteady flow conditions is the same as under steady flow conditions. This may represent a significant underestimate of the actual frictional resistance.

Sample program run

As the program input data show, the problem being analysed is the unsteady flow resulting from closure of the outlet from a 4 m diameter culvert, thus linearly reducing the initial steady discharge rate of 4 $m^3 s^{-1}$ to zero over a period of 60 seconds. The culvert length is 500 m, bottom slope is 0.0005, and the Manning n-value is 0.015. The depth of water at the upstream end of the channel remains constant at its steady flow value.

The tabulated program output shows the variation with time of the discharge rate and flow depth at specified points along the channel length. The oscillatory nature of these variations can be more clearly seen in Fig. 10.5, which shows them in graphical form for the upstream and downstream ends of the channel.

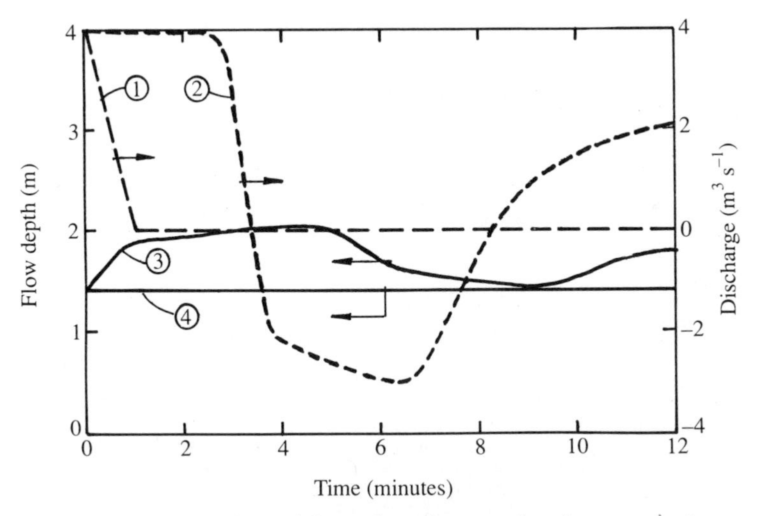

Fig. 10.5 Unsteady open channel flow: plotted output data from a sample program run with steady conditions at time zero. ① is the discharge at the downstream end; ② is the discharge at the upstream end; ③ is the flow depth at the downstream end; and ④ is the flow depth at the upstream end.

```
RUN
                         PROGRAM OCUSF

    This program computes the transient flow and water depth in
    open channels of rectangular, trapezoidal and circular cross-
    sections, using a numerical computation procedure based on the
    method of characteristics, as outlined in this chapter. The
    computation of frictional resistance is based on the Manning
    equation.

    Computation starts from a specified steady state at time zero.
    The program offers a choice of two initial steady states viz.
    steady uniform flow and zero flow.

    Boundary conditions:  The program caters for the following
    parameter variations at both ends of the channel:
       (a) linear variation of flow depth with time
       (b) linear variation of discharge rate with time

    A constant value for either boundary parameter is obtained
    by specifying a zero rate for the parameter variation.

Press the space bar to continue

ENTER CHANNEL DATA:

Is section:   1 CIRCULAR   2 RECTANGULAR   3 TRAPEZOIDAL ?

Enter 1, 2 or 3, as appropriate? 1

Diameter (m)? 3.0

Enter channel length (m)? 500.0
Enter Manning's n-value? 0.015
Enter channel bed slope (sin theta)? 0.0005

Enter initial steady flow rate (m**3/s)? 4.0

Select upstream boundary condition:
                    1. linear variation of discharge with time
                    2. linear variation of depth with time
Enter 1 or 2 as appropriate? 2
Enter rate of depth variation with time (m/s)? 0

Select downstream boundary condition:
                    1. linear variation of discharge with time
                    2. linear variation of depth with time
Enter 1 or 2 as appropriate? 1
Enter rate of discharge variation (m**3/s/s)? -0.0666
Enter final d/s discharge rate (m**3/s)? 0

Enter number of reaches into which the channel length is divided
for computational purposes (multiple of 10)? 10

Enter number of computation iterations? 100

DATA INPUT COMPLETED; COMPUTATION IN PROGRESS
```

```
**** Tabulation of computed values follows ****

TIME                 DISTANCE ALONG CHANNEL
(MIN)    0.0L     0.2L     0.4L     0.6L     0.8L     1.0L

0.00     4.000    4.000    4.000    4.000    4.000    4.000    Q (m**3/s)
         1.431    1.431    1.431    1.431    1.431    1.431    DEPTH (m)

0.37     4.004    4.004    4.004    4.004    4.004    2.521    Q (m**3/s)
         1.431    1.431    1.431    1.431    1.431    1.617    DEPTH (m)

0.74     4.007    4.007    4.007    4.007    3.474    1.043    Q (m**3/s)
         1.431    1.431    1.431    1.431    1.507    1.778    DEPTH (m)

1.11     4.010    4.010    4.010    3.858    1.947    0.000    Q (m**3/s)
         1.431    1.431    1.431    1.453    1.683    1.887    DEPTH (m)

1.48     4.012    4.012    3.975    3.024    0.406    0.000    Q (m**3/s)
         1.431    1.431    1.437    1.565    1.842    1.906    DEPTH (m)

1.85     4.014    4.002    3.676    1.068    0.246   -0.000    Q (m**3/s)
         1.431    1.433    1.482    1.772    1.874    1.925    DEPTH (m)

2.22     4.010    3.911    2.167    0.527    0.241    0.000    Q (m**3/s)
         1.431    1.447    1.660    1.840    1.893    1.943    DEPTH (m)

2.59     3.932    3.231    0.923    0.499    0.248    0.000    Q (m**3/s)
         1.431    1.540    1.796    1.862    1.912    1.962    DEPTH (m)

2.96     3.267    1.532    0.748    0.498    0.251    0.000    Q (m**3/s)
         1.431    1.718    1.829    1.881    1.932    1.982    DEPTH (m)

3.36     0.283    0.556    0.682    0.487    0.246    0.000    Q (m**3/s)
         1.431    1.754    1.845    1.901    1.952    2.002    DEPTH (m)

3.76    -1.881   -0.857    0.295    0.428    0.233    0.000    Q (m**3/s)
         1.431    1.649    1.833    1.917    1.972    2.023    DEPTH (m)

4.15    -2.240   -1.922   -0.875    0.066    0.182    0.000    Q (m**3/s)
         1.431    1.557    1.746    1.907    1.988    2.041    DEPTH (m)

4.55    -2.410   -2.351   -1.889   -0.956   -0.151    0.000    Q (m**3/s)
         1.431    1.528    1.663    1.834    1.978    2.053    DEPTH (m)

4.94    -2.565   -2.564   -2.420   -1.893   -1.020    0.000    Q (m**3/s)
         1.431    1.523    1.625    1.760    1.912    2.018    DEPTH (m)

5.33    -2.711   -2.706   -2.680   -2.422   -1.586    0.000    Q (m**3/s)
         1.431    1.523    1.613    1.715    1.821    1.891    DEPTH (m)

5.72    -2.839   -2.841   -2.804   -2.446   -1.405    0.000    Q (m**3/s)
         1.431    1.524    1.607    1.675    1.716    1.761    DEPTH (m)

6.11    -2.953   -2.915   -2.648   -1.931   -0.998    0.000    Q (m**3/s)
         1.431    1.522    1.586    1.612    1.629    1.664    DEPTH (m)

6.50    -3.007   -2.767   -2.125   -1.365   -0.647    0.000    Q (m**3/s)
         1.431    1.502    1.535    1.549    1.565    1.602    DEPTH (m)

6.90    -2.731   -2.285   -1.592   -0.938   -0.411    0.000    Q (m**3/s)
         1.431    1.455    1.478    1.496    1.522    1.563    DEPTH (m)

                etc. up to 100 iterations
```

10.5 Simplification of the St Venant equations

The St Venant equations can be made more amenable to solution by omitting selected terms from the momentum equation (10.2). The latter may be written in the form

$$S_0 = S_f + \frac{\partial y}{\partial x} + \frac{v}{g}\frac{\partial v}{\partial x} + \frac{1}{g}\frac{\partial v}{\partial t}. \tag{10.27}$$

Henderson (1966) has pointed out that the acceleration terms (the third and fourth on the right-hand side of (10.27)) are usually two orders of magnitude less than the gravity (S_0) and friction (S_f) terms and one or two orders of magnitude less than the remaining term $\partial y/\partial x$. This suggests that the solution of the simplified equation obtained by dropping the acceleration terms may provide a good approximation to the solution based on the full equations. The resulting simplified momentum equation becomes

$$\frac{\mathrm{d}y}{\mathrm{d}x} = S_0 - S_f. \tag{10.28}$$

In combination with the continuity equation (10.3) the resulting unsteady open channel flow equation for a rectangular channel has the form

$$\frac{y}{v}\frac{\partial v}{\partial x} + \frac{1}{v}\frac{\partial y}{\partial t} = S_f - S_0. \tag{10.29}$$

A further simplification of the momentum equation is obtained by omission of the $\mathrm{d}y/\mathrm{d}x$ term (this term represents the unbalanced pressure force component), reducing the momentum equation to its steady uniform flow form:

$$S_f = S_0. \tag{10.30}$$

Using the Manning expression of friction slope, eqn (10.30) becomes

$$S_0 = \left(\frac{nQ}{AR_h^{0.67}}\right)^2 \tag{10.31}$$

and hence we can write

$$Q = f(A) \qquad \text{and} \qquad \frac{\partial Q}{\partial x} = \frac{\partial Q}{\partial A}\frac{\partial A}{\partial x} = F(A)\frac{\partial A}{\partial x}.$$

Combining this form of simplified momentum equation with the continuity equation (10.2), where $q_1 = 0$, the resulting open channel unsteady flow equation becomes:

$$F(A)\frac{\partial A}{\partial x} + \frac{\partial A}{\partial t} = 0. \tag{10.32}$$

This equation is known as the kinematic wave equation because the dynamic terms of the momentum equation have been omitted in its development. The solution of eqn (10.32) is clearly of the form

$$A = \phi\left(t - \frac{x}{F(A)}\right) \tag{10.33}$$

where the form of the function ϕ is determined by the boundary condition for $x = 0$.

10.6 Rapidly varied unsteady flow

Rapidly varied unsteady flow gives rise to a surge or wave front, which moves as a step-change in water depth along the channel. A positive surge is defined as one which leaves an increased water depth in its wake as the wave front passes, while a negative surge is one which leaves a shallower depth in its wake as the wave front passes. In the following simplified analysis of surge front movement the effect of frictional resistance is neglected.

10.6.1 Upstream positive surge

An upstream positive surge may be created in channel flow, for example, by the rapid closure of a gate, resulting in a step reduction in flow rate. The effect of this on the upstream side of the gate is the development of a wave front which travels upstream, as illustrated in Fig. 10.6.

Referring to Fig. 10.6, the surge front is seen to leave in its wake an increased depth y_2, hence the description 'positive'. By superimposing a downstream velocity c on the flow system, the flow regime is converted to an equivalent steady state, in which the wave front is now stationary. Applying the continuity and momentum principles to the control volume between sections 1 and 2, under the transformed steady state conditions:

$$\text{Continuity:} \quad A_1(v_1 + c) = A_2(v_2 + c). \tag{10.34}$$

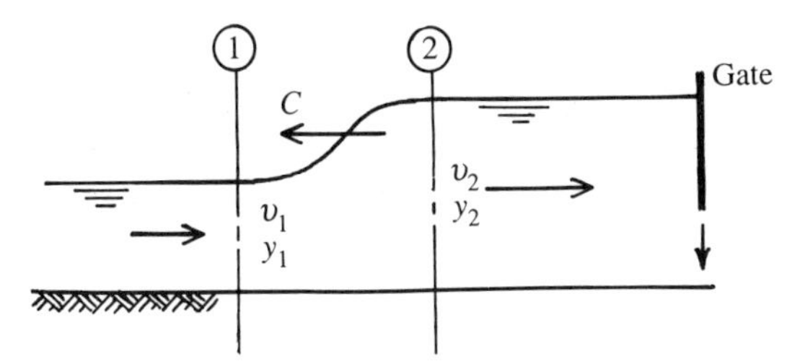

Fig. 10.6 Upstream positive surge.

Hence

$$v_2 = \frac{A_1 v_1 - c(A_2 - A_1)}{A_2}$$

and

$$c = \frac{Q_1 - Q_2}{A_2 - A_1}.$$

Neglecting the flow friction force:

$$\text{Momentum:} \qquad \rho g \bar{y}_1 A_1 - \rho g \bar{y}_2 A_2 = \rho A_1 (v_1 + c)(v_2 - v_1). \quad (10.35)$$

Solving equations (10.34) and (10.35) for c and v_2:

$$c = \left[g A_2 \frac{A_2 \bar{y}_2 - A_1 \bar{y}_1}{A_1 (A_2 - A_1)} \right]^{1/2} - v_1 \quad (10.36)$$

$$v_2 = v_1 - \left[\frac{g(A_2 - A_1)(A_2 \bar{y}_2 - A_1 \bar{y}_1)}{A_1 A_2} \right]^{1/2}. \quad (10.37)$$

For a rectangular channel:

$$c = \left[\frac{g y_2}{2} \frac{(y_2 + y_1)}{y_1} \right]^{1/2} - v_1. \quad (10.38)$$

If c is assumed equal to zero in eqn (10.38) the resulting relation between y_1 and y_2 is that for a hydraulic jump. Thus, the hydraulic jump can be considered to be a stationary surge. It should be noted that the continuity and momentum equations are not sufficient on their own to define the flow regime since there are three unknowns, c, y_2, and v_2 (or Q_2). One of these must therefore be known to enable computation of the remaining two parameters.

10.6.2 Downstream positive surge

A downstream positive surge is caused, for example, by the sudden opening of a gate, which results in an instantaneous increase in discharge and flow depth downstream of the gate, as illustrated in Fig. 10.7.

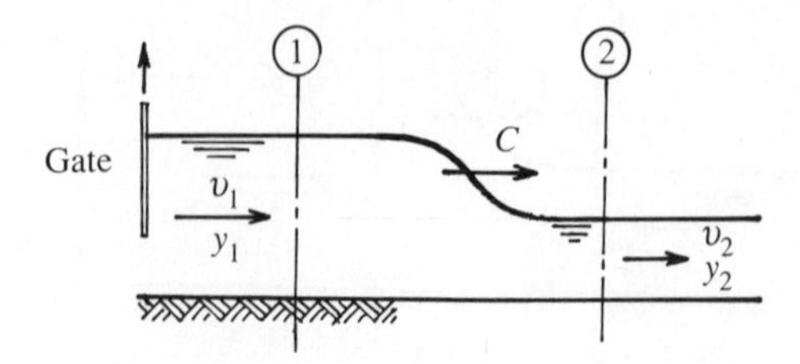

Fig. 10.7 Downstream positive surge.

Applying the same analytical approach as that used for the analysis of the upstream positive surge, the flow regime is transformed to an equivalent steady state by superimposing a backward velocity of magnitude c on the system. Thus, referring to Fig. 10.7, the continuity and momentum principles can be applied to the control volume defined by sections 1 and 2:

$$\text{Continuity:} \quad A_1(v_1 - c) = A_2(v_2 - c). \tag{10.39}$$

Hence

$$c = \frac{Q_1 - Q_2}{A_1 - A_2}. \tag{10.40}$$

Neglecting the boundary friction force:

$$\text{Momentum:} \quad \rho g(A_1\bar{y}_1 - A_2\bar{y}_2) = \rho A_2(v_2 - c)(v_2 - v_1). \tag{10.41}$$

Solving equations (10.39) and (10.41) for c and v_1:

$$c = \left[\frac{gA_1(A_1\bar{y}_1 - A_2\bar{y}_2)}{A_2(A_1 - A_2)}\right]^{1/2} + v_2 \tag{10.42}$$

$$v_1 = \left[\frac{g(A_1 - A_2)(A_1\bar{y}_1 - A_2\bar{y}_2)}{A_1A_2}\right]^{1/2} + v_2. \tag{10.43}$$

For a rectangular section:

$$c = \left[\frac{gy_1}{2y_2}(y_1 + y_2)\right]^{1/2} + v_2 \tag{10.44}$$

$$v_1 = \left[\frac{g(y_1 - y_2)^2(y_1 + y_2)}{2y_1y_2}\right]^{1/2} + v_2 \tag{10.45}$$

10.6.3 Upstream negative surge

A negative surge is seen by the observer as a wave front movement which leaves a lowered water surface level in its wake. An upstream negative surge may be caused, for example, upstream of a rapidly opened gate, as illustrated in Fig. 10.8. The wave front flattens as it travels along the channel, due to the top of the wave having a greater velocity than the bottom. It is necessary, therefore, to calculate two wavespeeds, one for the wave crest and the second for the wave trough.

Consider a small rapid disturbance giving rise to a small negative surge, moving upstream as illustrated in Fig. 10.9. Applying the continuity and

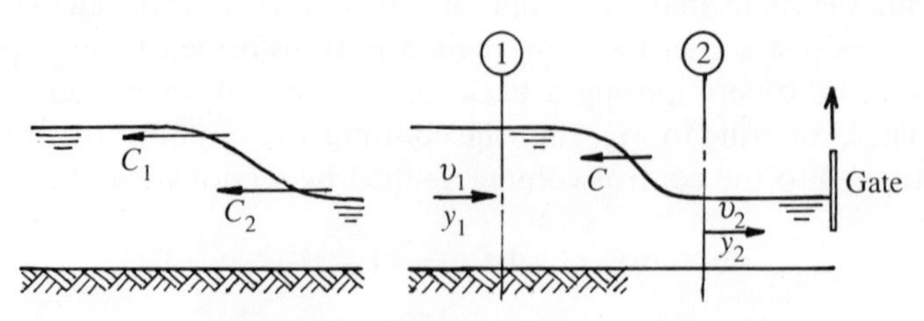

Fig. 10.8 Upstream negative surge.

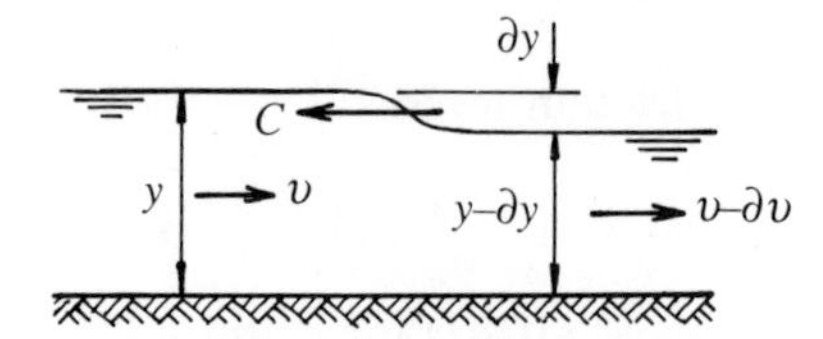

Fig. 10.9 Negative surge propagation.

momentum principles as before:

Continuity: $$(v + c)y = (y - \partial y)(v - \partial v + c) \quad (10.46)$$

Momentum: $$\rho g\left(\frac{y^2}{2} - \frac{(y - \partial y)^2}{2}\right) = \rho y(v + c)(-\partial v). \quad (10.47)$$

From eqn (10.46) $\partial y = -y\,\partial v/(v + c)$; from eqn (10.47) $\delta y = -\partial v(v + c)/g$. Hence

$$c = \sqrt{gy} - v \quad (10.48)$$

and also

$$\partial y = -\frac{\partial v}{g}\sqrt{gy}$$

which, as δy approaches zero, can be written

$$\frac{\mathrm{d}y}{\sqrt{y}} = -\frac{\mathrm{d}v}{\sqrt{g}}.$$

Integrating for a wave of finite height

$$v = -2\sqrt{gy} + \text{constant}. \quad (10.49)$$

For the upstream negative surge, illustrated in Fig. 10.8, we have the known boundary condition, $v = v_1$ when $y = y_1$; using these values in eqn (10.49), the integration constant is found to be $2\sqrt{gy_1} + v_1$. Hence, from eqn (10.49)

$$v_2 = 2\sqrt{gy_1} - 2\sqrt{gy_2} + v_1 \tag{10.50}$$

and from eqn (10.48)

$$c_2 = \sqrt{gy_2} - v_2.$$

Hence from (10.50)

$$c_2 = 3\sqrt{gy_2} - 2\sqrt{gy_1} + v_1. \tag{10.51}$$

10.6.4 Downstream negative surge

A downstream negative surge is propagated downstream of a rapidly closed gate, for example, as illustrated in Fig. 10.10. Using the same analytical

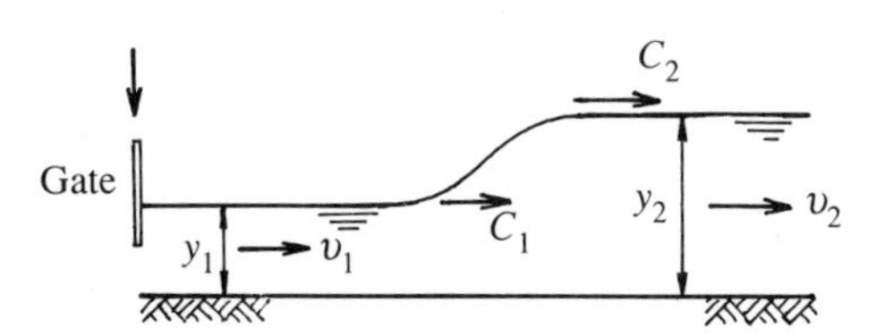

Fig. 10.10 Downstream negative surge.

procedure as that applied in the case of the upstream negative surge, the wave front velocity c can be shown, in the case of a downstream negative surge, to be

$$c = \sqrt{gy} + v \tag{10.52}$$

and the velocity is

$$v = 2\sqrt{gy} - 2\sqrt{gy_2} + v_2. \tag{10.53}$$

Hence the values of v_1 and c_1 are found to be:

$$v_1 = 2\sqrt{gy_1} - 2\sqrt{gy_2} + v_2 \tag{10.54}$$

$$c_1 = 3\sqrt{gy_1} - 2\sqrt{gy_2} + v_2. \tag{10.55}$$

Related reading

Abbott, M. B. (1966). *An Introduction to the Method of Characteristics*. Elsevier, New York.

Chaudhry, M. H. (1987). *Applied Hydraulic Transients*, (2nd edn). Van Nostrand Reinhold, New York.

Chow, Ven Te (1959). *Open Channel Hydraulics*. McGraw-Hill, New York.

Dooge, J. C. I. (1986). *Theory of Flood Routing, in River Flow Modelling and Forecasting*, (ed. D. A. Kraijenhoff and J. R. Moll). D. Reidel Publishing Co., Dordrecht.

Featherstone, R. E. and Nalluri, C. (1982). *Civil Engineering Hydraulics*. Collins, London.

Henderson, F. M. (1966). *Open Channel Flow*. Macmillan, New York.

Wylie, E. B. and Streeter, V. L. (1978). *Fluid Transients*. McGraw-Hill, New York.

11
Pumping installations

11.1 Introduction

Pumping installations are intrinsic elements of water supply and wastewater disposal systems. Essentially, they transfer energy to the through-flow of water by effecting a step-increase in head or pressure. In water supply applications, pumps are typically running continuously and, in most cases, the required step-increase in head is substantial. In wastewater applications, pumps are typically running in a start/stop mode and the required wastewater lift is usually not large, the pump installation being typically used to augment gravity flow systems.

11.2 Pump types

The following pump types, differentiated by their mode of pumping action, are in general use in the water industry.

(1) Positive displacement pumps, for example, helical rotor, diaphragm, and piston pumps. The use of positive displacement pumps is confined to specialist applications such as the dosing of chemical solutions where a high level of flow control is required, the pumping of viscous fluids such as sewage sludges, and high-pressure applications.

(2) Rotodynamic pumps, which may be classified as centrifugal, mixed flow, and axial flow. Rotodynamic pumps are by far the most widely used pump type for bulk-pumping of clean water and wastewaters.

(3) Air-lift pumps which may be of the submerged air injection type, in which compressed air is injected into a vertical riser pipe, thus air-lifting liquid in a two-phase flow, or the air ejector type, where compressed air is used to displace liquid from a closed vessel. The latter operates in a batch mode.

(3) The Archimedean screw pump is an inclined screw conveyer which is used for low-lift applications, particularly in the wastewater treatment field, where it is used for raw sewage and activated sludge pumping.

11.2.1 Positive displacement pumps

Positive displacement pumps can be divided into reciprocating and rotary types. Reciprocating pumps include piston, plunger (ram), and diaphragm types. They have the common characteristic of a discontinuous pulsed delivery, the magnitude of which is effectively independent of delivery head, hence the description positive displacement. The peak pressure generated during the delivery stroke can be reduced by the installation of an air vessel on the pump delivery. This will also have a smoothing effect on the pump discharge. The sequential filling and emptying of the pump body requires non-return valves on both the suction and delivery sides of the pump.

Rotary pumps may be of the gear, lobe, helical rotor, and sliding vane types. They combine the continuous discharge characteristic of rotodynamic pumps and the positive displacement characteristic of reciprocating pumps.

Positive displacement pumps are particularly suited to chemical dosing applications in water and wastewater treatment plants, where their accurate and controllable discharge characteristics can be used to advantage. Helical rotor or progressive cavity pumps and diaphragm pumps are widely used to pump viscous liquids, including sewage sludges.

11.2.2 Rotodynamic pumps

The active element of a rotodynamic pump is the rotating impeller or propeller, which imparts a momentum to the fluid, that, on deceleration, is converted to a pressure rise. If geometrically similar impellers of differing size are driven at appropriate speeds, such that the exiting fluid has the same flow direction in all cases, the impellers are described as a homologous set. Using the method of dimensional analysis, it has been shown in Chapter 9 that such a homologous set can be categorized by the compound non-dimensional parameter, known as the specific speed N_s, which is a function of the pump speed N, the pump head H, and the pump discharge Q:

$$N_s = \frac{NQ^{0.5}}{(gH)^{0.75}}. \tag{9.12}$$

In pump technology literature, the gravity constant is often omitted, resulting in the following form of the specific speed function:

$$N_s = \frac{NQ^{0.5}}{H^{0.75}}. \tag{9.13}$$

As N_s in this form is not dimensionless, its numerical value is dependent on the system of units used for N, H, and Q.

On the basis of their specific speed value, rotodynamic pumps can be broadly classified into the following types:

Pump type	Specific speed
Centrifugal	≤ 80
Mixed flow	80–150
Axial flow	150–300

based on parameter units: N (rpm), Q ($m^3\ s^{-1}$), and H (m).

The typical geometric form of the impeller component of each of these rotodynamic pump types is illustrated in Fig. 11.1.

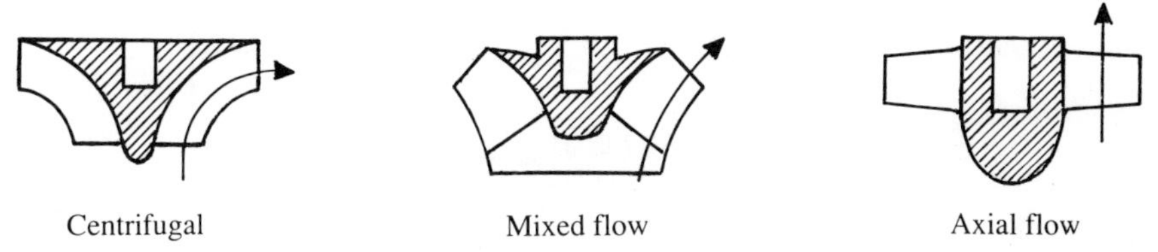

Fig. 11.1 Rotodynamic pump impeller types.

The manometric head H is defined as the step change in total head across the pump:

$$H = \left(\frac{p_{\mathrm{d}}}{\rho g} + \frac{v_{\mathrm{d}}^2}{2g}\right) - \left(\frac{p_{\mathrm{s}}}{\rho g} + \frac{v_{\mathrm{s}}^2}{2g}\right) \tag{11.1}$$

where the subscripts s and d relate to the pump suction and delivery sides, respectively. Pump efficiency η is defined as the ratio of the hydraulic power transferred to the fluid, to the shaft power, P:

$$\eta = \frac{\rho g H Q}{P}. \tag{11.2}$$

Efficiencies of up to 90 per cent can be achieved in large centrifugal pump units pumping clean water. The performance characteristics of individual pumps are generally presented graphically as plots of manometric head (H), power (P), and efficiency (η), as functions of discharge (Q), as illustrated in Fig. 11.2.

It will be noted that head and power increase sharply as the discharge approaches zero in axial flow pumps. For this reason, such pumps should

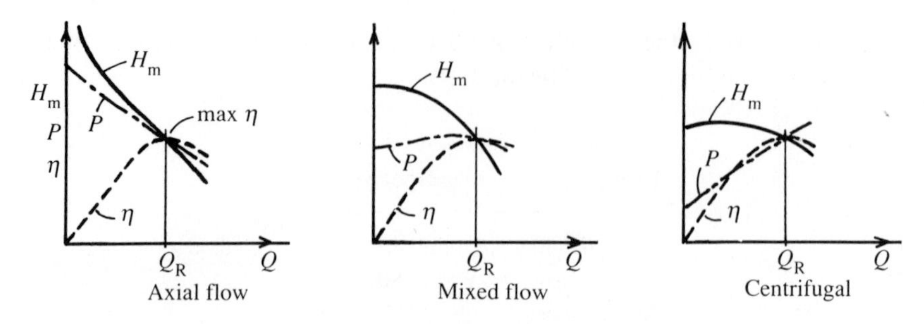

Fig. 11.2 Rotodynamic pump characteristics; H_m is the manometric head; P is the power; and η is the efficiency.

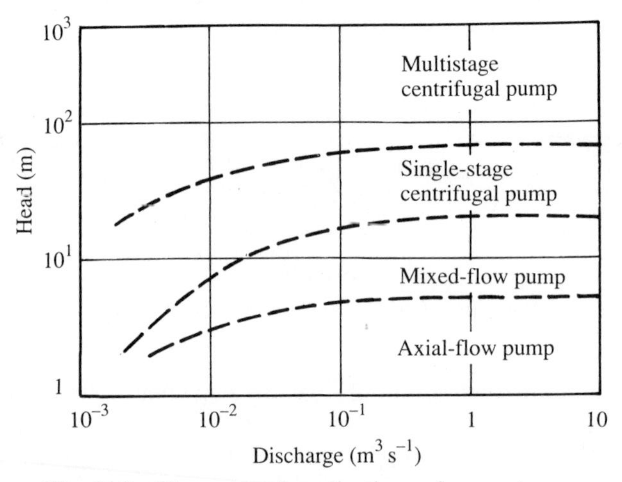

Fig. 11.3 The range of applications of pump types.

never be started up against a closed valve in the delivery line. A centrifugal pump, on the other hand, has its minimum power demand at zero discharge rate.

The respective ranges of practical application for the three types of rotodynamic pump, in terms of head/discharge capacity, are presented in Fig. 11.3.

As shown in Chapter 5, the H/Q relationship for rotodynamic pumps, driven at their rated speed N_R, may be expressed in quadratic equation form as follows:

$$H_R = A_0 + A_1 Q_R + A_2 Q_R^2 \tag{5.9}$$

where A_0 is the manometric head at zero flow, and A_1 and A_2 are constants.

As shown in Chapter 6, the H/Q equation for any other speed N, as derived from the rated speed, has the following form:

$$H_N = A_0\left(\frac{N}{N_R}\right)^2 + A_1\left(\frac{N}{N_R}\right)Q_N + A_2 Q_N^2. \tag{6.30}$$

This modification of eqn (5.9) is based on the fact that for a given impeller, $Q \propto N$ and $H \propto N^2$.

11.2.3 The air-lift pump

The mode of operation of the submerged air injection type of air-lift pump is illustrated in Fig. 11.4. Air is injected into the vertical pipe at a depth h_s below the free water surface. The resulting rising stream of air bubbles within the riser pipe produces a two-phase fluid with a lower composite density than the liquid on its own and thus enables the riser to discharge at a height h_L above the free water surface.

It is clear that the upward driving force at the bottom of the two-phase column is the static liquid pressure at that point and hence the average driving force per unit column length is the ratio $h_s/(h_s + h_L)$, which is defined as the submergence ratio S_r. The fluid efficiency η_a of an air-lift pump is expressed as the ratio of the work done in lifting the liquid to the work done by the expanding air stream:

$$\eta_a = \frac{\rho_L g h_L Q_L}{P_2 Q_{g2} \ln(P_0/P_2)} \tag{11.3}$$

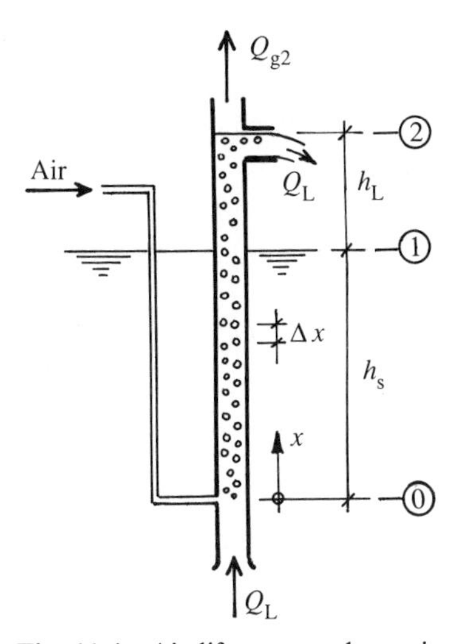

Fig. 11.4 Air-lift pump schematic.

where ρ_L is the liquid density, P_0 and P_2 are the absolute pressures at levels 0 and 2, respectively, and Q_L and Q_{g2} are the volumetric flow rates for liquid and gas, respectively. (Q_{g2} is the volumetric gas flow rate at the discharge pressure P_2, which would normally be atmospheric pressure, as in Fig. 11.4.)

Computation of air-lift pump discharge

The two-phase flow in an air-lift riser pipe can be analysed by application of the momentum principle (Nicklin 1963; Clark and Daybolt 1986). Consider a control volume of length Δx, as shown in Fig. 11.4. Since the flow is steady, the sum of the acting forces—pressure, gravity, and friction—must be zero. Neglecting the weight of the air phase, the momentum equation is written as

$$\Delta P\, A + \rho_L g A(1 - \varepsilon)\Delta x + F_w \Delta x = 0 \tag{11.4}$$

where ΔP is the pressure change over the length Δx, A is the riser cross-sectional area, ε is the void (air) fraction, and F_w is the frictional drag force per unit length.

The void fraction is influenced by the relative motion or 'slip' of the air past the liquid in the riser. This is minimized if the injected air remains as dispersed bubbles. In air-lift pipes the air bubbles typically coalesce to form air 'slugs' or elongated round-nosed bubbles. Nicklin has shown that the average rise velocity of air slugs within a riser can be represented as

$$\frac{Q_g}{\varepsilon A} = \frac{1.2(Q_g + Q_L)}{A} + 0.35\sqrt{gD} \tag{11.5}$$

where D is the pipe diameter. ($0.35\sqrt{gD}$ is the theoretical rise velocity under still water conditions.) The variation in gas volume with reduction in pressure over the column height is given by the isothermal relation

$$Q_g = Q_{g2}\frac{P_2}{P}. \tag{11.6}$$

Clarke and Daybolt have shown that the friction loss in the two-phase air-lift flow (F_w) can be approximately related to the friction loss $F_{w(L)}$ of the liquid, flowing on its own, as follows:

$$F_w = F_{w(L)}(1 + 1.5\varepsilon). \tag{11.7}$$

The liquid flow frictional resistance per unit length can be expressed in accordance with the Darcy–Weisbach equation, as

$$F_{w(L)} = \frac{\rho_L f Q_L^2}{2AD}. \tag{11.8}$$

Hence

$$F_w = \frac{\rho_L f Q_L^2}{2AD}(1 + 1.5\varepsilon). \tag{11.9}$$

Inserting the foregoing expression for F_w into the momentum equation (11.4) and combining with eqns (11.5) and (11.6) leads to the following expression in P and x:

$$-\Delta P = \left[\rho_L g\left\{1 - \frac{Q_{g2}P_2}{1.2(Q_{g2}P_2 + Q_L P) + Q_D P)}\right\} + K_f\left\{1 + \frac{1.5Q_{g2}P_2}{1.2(Q_{g2}P_2 + Q_L P) + Q_D P)}\right\}\right]\Delta x \tag{11.10}$$

where

$$K_f = \frac{\rho_L f Q_L^2}{2A^2 D} \quad \text{and} \quad Q_D = 0.35(gD)^{0.5}A.$$

Integrating eqn (11.10) between the limits $P = P_2$ and $P = P_0$ and $x = (h_s + h_L)$ and $x = 0$ yields the following:

$$\frac{P_2 - P_0}{\rho_L g + K_f} - \left(\frac{(1.5K_f - \rho_L g)Q_{g2}P_2}{(\rho_L g + K_f)^2 S}\right)\ln\left(\frac{R + (\rho_L g + K_f)SP_2}{R + (\rho_L g + K_f)SP_0}\right) + (h_s + h_L) = 0 \tag{11.11}$$

where $R = Q_{g2}P_2(0.2\rho_L g + 2.7K_f)$ and $S = 1.2Q_L + Q_D$.

In typical water engineering applications the pump outlet discharges to the atmosphere and hence the outlet pressure P_{g2} is atmospheric pressure. In practice, the injection pressure P_0 will be less than the local external hydrostatic value by an amount equal to the total head loss in the pipe between the pipe inlet and the point of air injection into the pipe. In general, the major energy loss in air-lift pumps is due to slippage between the air and water flows and hence the friction loss component tends not to be significant, except in circumstances where the lift is small and the suction pipe is long.

The efficiency of air-lift pumps, as earlier defined, varies with the submergence ratio S_r, the liquid and gas flow rates, and the riser pipe diameter. Nicklin has shown that the best achievable efficiency, based on the foregoing theory, is in the approximate range of 50–60 per cent, depending on pipe diameter, the magnitude increasing marginally with increase in diameter. The following may be used as an approximate guide to the most efficient operating regions for liquid and gas flow rates:

$$\text{Liquid:} \qquad \frac{Q_L}{A\sqrt{gD}} = 0.5\text{–}1.5$$

$$\text{Gas:} \qquad \frac{Q_{g2}}{A\sqrt{gD}} = 0.1\text{–}1.0.$$

The foregoing theory is based on the assumption that the two-phase flow remains within the slug flow mode. For practical design purposes, this can be assumed to be the case provided that the superficial air velocity does not exceed twice the superficial liquid velocity.

Program AIRLIFT solves eqn (11.11), using the interval-halving method (Appendix B), to determine the required air flow rate, or the required submergence, given the remaining parameters as input data. It also computes the pump efficiency η_a. The program assumes free discharge to atmosphere, that is, $P_{g2} = P_a$, where P_a is atmospheric pressure. In computing the air injection point pressure P_0, account is taken of the entry head loss, the velocity head, and the friction loss in the suction pipe upstream of the injection point.

A program listing is given in Appendix A. The use of the program is illustrated in the following program run.

Sample program run: program AIRLIFT

```
RUN
This program computes the following airlift pump design
parameters, given the values for the remaining parameters
as input data:

   1.  REQUIRED AIR INPUT
   2.  REQUIRED SUBMERGENCE

ENTER 1 OR 2, AS APPROPRIATE? 1

ENTER PIPE DIAMETER (mm)? 200
ENTER SUBMERGENCE OF AIR INJECTION PT. (m)? 2.0
ENTER STATIC LIFT (m)? 0.5
ENTER LENGTH OF SUCTION PIPE U/S OF INJECTION PT. (m)? 0.1
ENTER PIPE WALL ROUGHNESS (mm)? 0.1
ENTER REQUIRED LIQUID PUMPING RATE (m**3/s)? 0.04
ENTER LIQUID DENSITY (kg/m**3)? 1000
ENTER LIQUID VISCOSITY (Ns/m**2)? 0.001

Computed output values:

REQUIRED AIR INPUT RATE (m**3/s) =  .021875
SUPERFICIAL AIR VEL. (m/s) =  .696656
SUPERFICIAL WATER VELOCITY (m/s) =  1.273885
SUBMERGENCE RATIO =  .8
EFFICIENCY =  .5100837
Ok
```

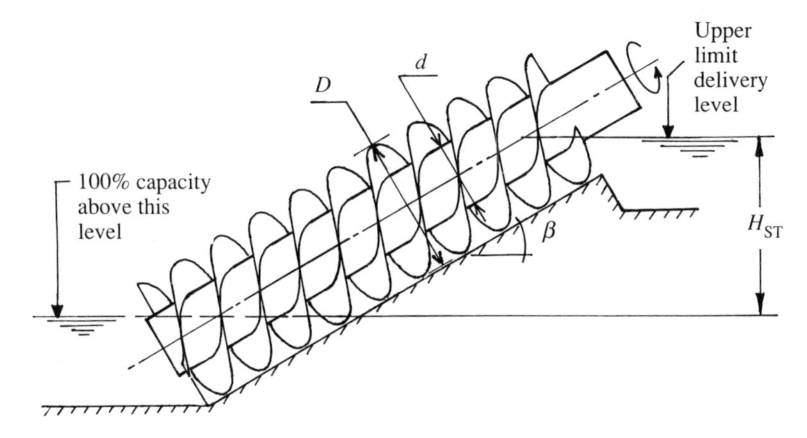

Fig. 11.5 Schematic outline of the Archimedean screw pump (courtesy of Spaans Babcock, Hoofddorp, The Netherlands). β is in the range 22–38 degrees.

The **air-ejector** type of air-lift pump is used in association with a closed vessel sump which fills by gravity inflow. Its contents are displaced into a rising main by the admission of compressed air. This mode of operation requires non-return valves on the inflow and discharge lines. Air ejectors are suitable for low-volume wastewater pumping duties.

11.2.4 The Archimedean screw pump

The Archimedean screw pump is illustrated in Fig. 11.5. It consists of an inclined screw conveyer shaft rotating within a close-fitting open-top cylindrical conduit. In rotation, the helical screw blades force the liquid up along the inclined conduit, achieving a liquid lift equal to the vertical projection of the inclined rotor shaft. The discharge capacity is a function of the rotor diameter and its rotational speed. The operational blade tip speed is usually in the range of 2.0–3.5 m s^{-1}, with a corresponding diameter (D) range of 0.4–3.0 m and a pumping capacity range of 0.02–3.2 $m^3\ s^{-1}$. Archimedean screw pumps retain their pumping efficiency over a wide flow range. A best overall efficiency of about 75 per cent is claimed for pumping at rated capacity, reducing to about 65 per cent at 30 per cent of rated capacity.

11.3 Hydraulics of rotodynamic pump/rising main systems

Figure 11.6 illustrates a typical pump/rising main system. As already defined by eqn (11.1), the pump manometric head is the total differential head across

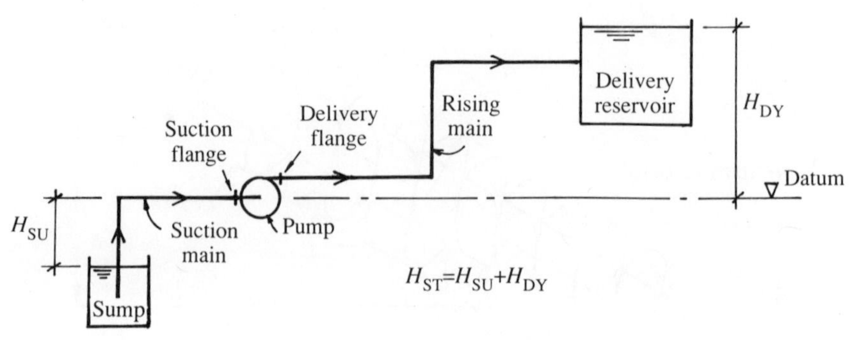

Fig. 11.6 Pump and pipe system.

the pump. In steady state pump operation the manometric head H is equal to the pipe system head:

$$H = H_{ST} + h_L \tag{11.12}$$

where H_{ST} is the static head and is the sum of the suction lift H_{SU} and the delivery lift H_{DY}; h_L is the total head loss between suction and delivery reservoirs.

The maximum lift for a rotodynamic pump is limited by the necessity to avoid cavitation. This phenomenon is caused by a drop in fluid pressure to vapour level and the consequent formation of vapour cavities; hence the term cavitation. Cavitation is manifested by noise and vibration as the vapour pockets implode on moving to regions of higher pressure. It reduces pump efficiency and may cause pitting and erosion of the pump impeller and housing. It is avoided by setting a lower limit to the allowable pressure at the suction flange. This limit is conventionally expressed as the 'net positive suction head' or NPSH for a pump. NPSH is defined as

$$\text{NPSH} = \left(\frac{P_s}{\rho g} - \frac{p_v}{\rho g}\right) \tag{11.13}$$

where P_s is the absolute pressure at the suction flange and p_v is the prevailing vapour pressure. The NPSH value can be related to the specific speed N_s through the Thoma cavitation number σ, which is defined as the ratio of NPSH to manometric head H:

$$\sigma = \frac{\text{NPSH}}{H}. \tag{11.14}$$

Based on empirical evidence (Wijdieks 1971), σ can be approximately correlated with the specific speed N_s, as follows:

(1) for single suction pumps: $\sigma = 0.001 N_s^{1.36}$;

(2) for double suction pumps: $\sigma = 0.0006 N_s^{1.36}$.

While such correlations may be used as a general guide for preliminary design, the individual pump NPSH specification should be used in final design computations.

Pumps may be operated in parallel or in series. The resulting H/Q characteristics are shown in Fig. 11.7(a) and (b), respectively.

The duty point at which a pumped system operates is determined by solution of the pump equation (6.33) and the system equation (11.12) for H and Q. This solution is illustrated graphically in Fig. 11.8. The duty point may be computed using program PUMP. This program (a) determines the pump equation coefficients, given three sets of H/Q values from the rated characteristic pump curve, and (b) solves the system and pump equations to find the duty point value for the operating speed.

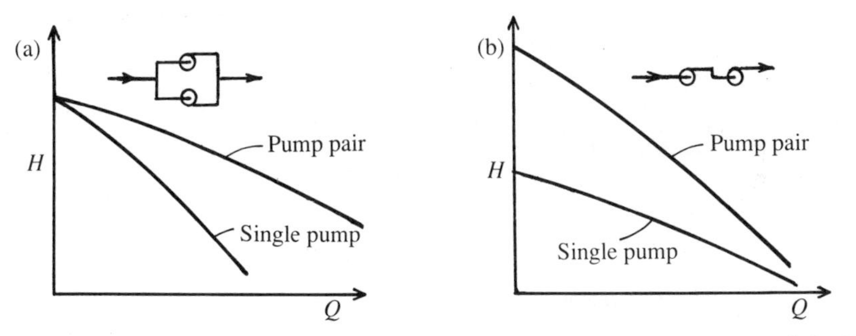

Fig. 11.7 Characteristic H/Q curves for pump combinations: (a) in parallel; (b) in series.

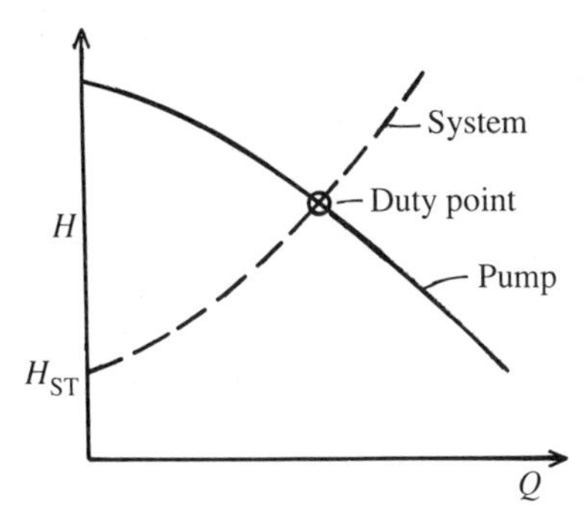

Fig. 11.8 Determination of pump duty point.

The program listing is given in Appendix A. The following is a sample program run.

Sample program run: program PUMP

```
RUN
This program computes the coefficients Al and A2 in the rotodynamic
pump equation: H = A0+A1*Q+A2*Q*Q, given 3 points on the H/Q standard
speed curve, including the shut-off head A0.

ENTER THE VALUE OF A0 (m)? 34.0
ENTER VALUES FOR H (m) AND Q(m**/s), SEPARATED BY A COMMA
ENTER FIRST PAIR OF VALUES? 23.5,0.075
ENTER SECOND PAIR OF VALUES? 0,0.125
ENTER VALUE OF STANDARD PUMP SPEED (rpm)? 1450
ENTER VALUE OF DUTY SPEED (rpm)? 1450

Coefficient values at standard speed 1450  rpm are:

      A0 =  34
      Al =  58.00003
      A2 = -2640

ENTER PUMP SUMP WATER LEVEL (mOD)? 5.0
ENTER RISING MAIN DISCHARGE LEVEL (mOD)? 25.0
ENTER SUCTION MAIN LENGTH (m)? 5.0
ENTER SUCTION MAIN DIAMETER (mm)? 250
ENTER SUCTION MAIN WALL ROUGHNESS (mm)? 0.1
ENTER DELIVERY MAIN LENGTH (m)? 2500.0
ENTER DELIVERY MAIN DIAMETER (mm)? 250
ENTER DELIVERY MAIN WALL ROUGHNESS (mm)? 0.1
ENTER ESTIMATE OF DUTY POINT DISCHARGE (m**3/S)? 0.05

Computed duty point values are:

      DUTY POINT DISCHARGE (m**3/s) =  5.137079E-02
      DUTY POINT HEAD (m) =  30.01266
      VALUES RELATE TO PUMP SPEED (rpm) =  1450
Ok
```

11.4 Economics of pump/rising main systems

The cost of water pumping includes the capital cost of the water transportation system, including structures, pumps, power supply, pipelines, reservoirs, and also the operational cost. In selecting the type and size of rising mains, particularly where the friction head represents a significant fraction of the total head, the influence of pipe size on the unit cost of pumping merits examination. A larger pipe incurs less friction head loss and hence requires smaller pumps and lower-power input; thus the problem is one of trading off increased capital costs against reduced running costs or vice versa.

Assuming the following unit costs

Pipeline:	£ C_{pi}/m length/m diameter
Pumps:	£ C_{pu}/installed kW
Energy:	£ C_e/kWy (kWy = kW × 1 year)
Capital:	£ P% annual charge (interest + capital repayment)

and using the following system variables

Maximum pumping rate:	Q (m^3 s^{-1})
Pipe length:	L (m)
Pipe diameter:	D (m)
Pump + motor efficiency:	η

then, allowing for some standby capacity, the installed pump power P_i can be expressed in the form

$$P_i = \frac{S\rho gHQ}{\eta} \text{ (W)} = \frac{SgHQ}{\eta} \text{ (kW)} \tag{11.15}$$

where S is a standby factor. For example, where three pumps of equal size are installed, one of which is a standby unit, the value of S is 1.5.

Using the foregoing data, an annual cost function C_a can be written as follows:

$$C_a = \frac{P}{100}\left[C_{pu}\left(\frac{SgHQ}{\eta}\right) + C_{pi}LD\right] + C_e\frac{gHQ}{\eta}. \tag{11.16}$$

For minimum cost:

$$\frac{dC_a}{dD} = 0 = \frac{P}{100}\left[C_{pu}\left(\frac{SgQ}{\eta}\frac{dH}{dD}\right) + C_{pi}L\right] + C_e\frac{gQ}{\eta}\frac{dH}{dD}. \tag{11.17}$$

The duty point head H can be written in the form

$$H = H_{ST} + h_f$$

where H_{ST} is the static head and $h_f = KQ^2/D^5$ is the system friction head, where K is a system constant. Then

$$\frac{dH}{dD} = \frac{dh_f}{dD} = -\frac{5KQ^2}{D^6}.$$

Insertion of this value for dH/dD in eqn (11.17) gives the following expression for the minimum cost diameter D:

$$D = \left\{\frac{5KQ^2[(C_e gQ) + (C_{pu}SgQP/100)]}{P/100(C_{pi}L\eta)}\right\}^{1/6}. \tag{11.18}$$

11.5 Pumping station design

The stages in pumping station design include:

(1) selection of type, size, and number of pumps;

(2) design of general layout;

(3) design of the wet well or sump;

(4) selection of a pumping control system.

11.5.1 Pump selection

Invariably, rotodynamic pumps are used for pumping clean water and wastewaters. In general, pump type selection is largely determined by the required duty, from which the specific speed can be calculated, thus indicating whether the pump should be of the centrifugal, mixed flow, or axial flow type. The number and size of pumps is normally selected to match the pattern of flow variation. Where the pumping demand is more or less constant, there is an obvious maintenance advantage in using a single size and type of pump. Except in very high head installations, pumps are used in parallel configuration. The pumping efficiency will vary, depending on whether one, two, or more pumps are working. Energy efficiency is obviously an important consideration in pump set selection, the normal design objective being the achievement of the highest possible operational efficiency, taking all modes of operation into account. While pump units of best efficiency in a particular category may be chosen for clean water pumping, the designer has generally to settle for a lower efficiency in pumping wastewater, where the necessity to pass suspended solids calls for large flow passages and clog-free impeller geometry. The provision of standby capacity is essential. The minimum standby capacity is that which will allow the station to operate at design load with any one of its pumps down for maintenance.

11.5.2 General layout

The underground structure of a pumping station typically consists of a wet well or sump, the volume and shape of which are determined by the factors outlined in the following section, and a dry well in which the pump set is housed, as illustrated schematically in Fig. 11.9(a) and (b). It will be noted that the use of a vertical shaft pump allows a more compact dry well and has the environmental advantage of drive motor placement at ground level. The use of submersible pumps eliminates the necessity for a separate dry well, as shown in Fig. 11.9(c).

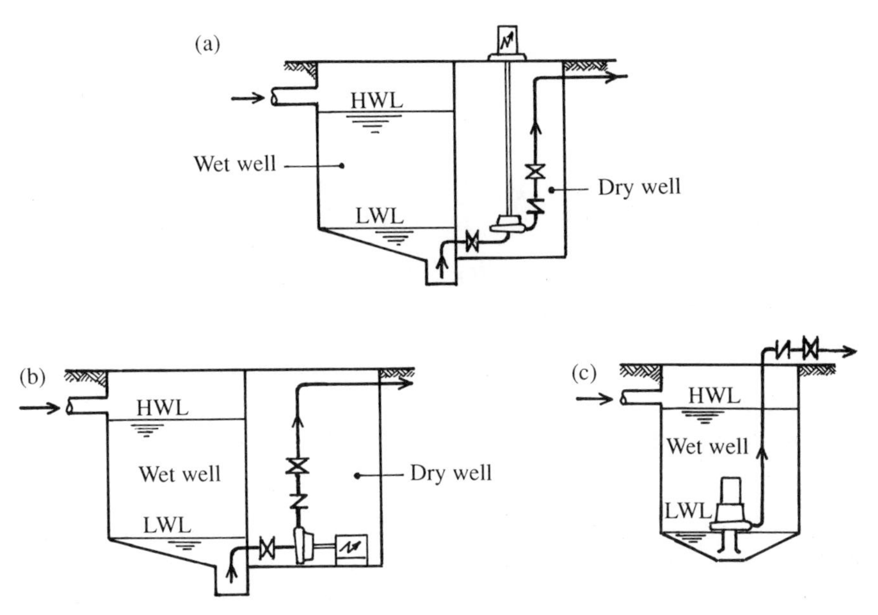

Fig. 11.9 Schematic of underground pump structure.

11.5.3 Pump sump design

Pump sumps are preferably designed to provide a flooded pump suction (positive gauge pressure at the suction flange) at start-up and thus obviate priming problems. Also, as discussed later, the geometry of the sump and submergence of the inlet should be such as to avoid vortex formation and air intake.

Where the inflow rate to the sump is variable, as in sewage and stormwater pumping, it is necessary to provide a storage volume in the sump to avoid too frequent pump starting, which would lead to motor-starter burn-out. At the same time, it is also desirable to minimize solids deposition in pump sumps and the attendant septicity problems; hence there is a need to avoid unduly large sumps. The sump volume required to satisfy starting frequency criteria may be calculated on the following basis, where P is the pumping rate, Q is the inflow rate to the sump, and V is the effective sump volume:

$$\text{Time to empty sump} = \frac{V}{P - Q}$$

$$\text{Time to fill sump} = \frac{V}{Q}$$

Interval between starts: I = time to empty + time to fill. Hence

$$I = \frac{PV}{Q(P-Q)} \tag{11.19}$$

$$N = \frac{1}{I} = \frac{Q(P-Q)}{PV} \tag{11.20}$$

where N is the frequency of starting. To obtain a maximum value for N:

$$\frac{dN}{dQ} = \frac{P-2Q}{PV} = 0$$

giving $N = N_{max}$ when $Q = P/2$. Inserting this value for Q in eqn (11.20), the corresponding value for V, which is the minimum sump volume, is found to be

$$V_{min} = \frac{P}{4N_{max}}. \tag{11.21}$$

For a typical value for N_{max} of $15\,h^{-1}$, $V_{min} = P/60$, or a volume corresponding to a pumping duration of one minute.

The foregoing minimum sump volume refers to the volume between pump cut-in level and pump cut-out level, as illustrated in Fig. 11.10. It is normal practice to separate the cut-in and cut-out levels for individual pumps, as shown in Fig. 11.10. Where this is the case, the appropriate value for P in eqn (11.21) is the output capacity of one pump. It should also be noted that the separation of cut-in and cut-out levels for pumps is desirable on the grounds of minimization of waterhammer effects associated with starting and stopping.

The selected pump cut-out level is significant in the separate contexts of cavitation and air entrainment, both of which reduce pumping efficiency. Cavitation problems are avoided by meeting the NPSH pump specification.

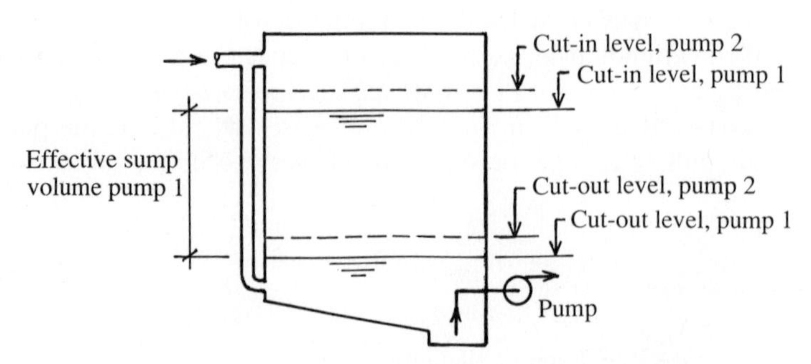

Fig. 11.10 Minimum sump volume for start/stop pumping.

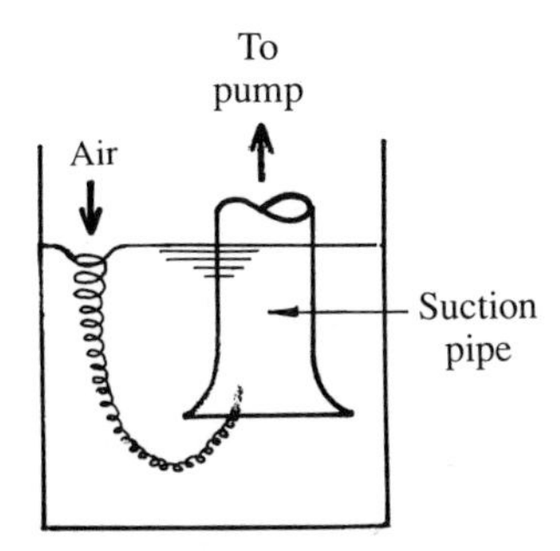

Fig. 11.11 Air-entrainment vortex.

Air entrainment adversely affects pump performance. Denny (1956) reported that 1 per cent free air could reduce the efficiency of a centrifugal pump by 5–15 per cent. Air entrainment may result from the formation of air-entraining vortices in the vicinity of the pump suction, as illustrated in Fig. 11.11, or may be caused by a cascading inflow to the pump sump resulting in the dispersion of air bubbles in the fluid bulk.

The tendency towards vortex formation can be effectively controlled by appropriate geometric design of pump sumps. Boundary discontinuities, which cause boundary layer separation or flow obstructions which give rise to vortex shedding, should be avoided. Where feasible, flows should be symmetrically directed to individual pump intakes. For detailed discussion related to the geometric aspects of pump sump design the reader is referred to the publications of Hattersley (1965), Prosser (1977), and Sweeney *et al.* (1982).

Design guidelines (Prosser 1977) for a single pump intake are presented in Fig. 11.12. For this type of intake layout a length of straight approach channel of at least $10D_b$ is recommended. Iversen (1953) has shown that pump efficiency is adversely affected if the bellmouth floor clearance is less than about $0.5D_b$.

For large multiple intake sumps and sumps of unusual geometry the design process can be greatly aided by physical model studies.

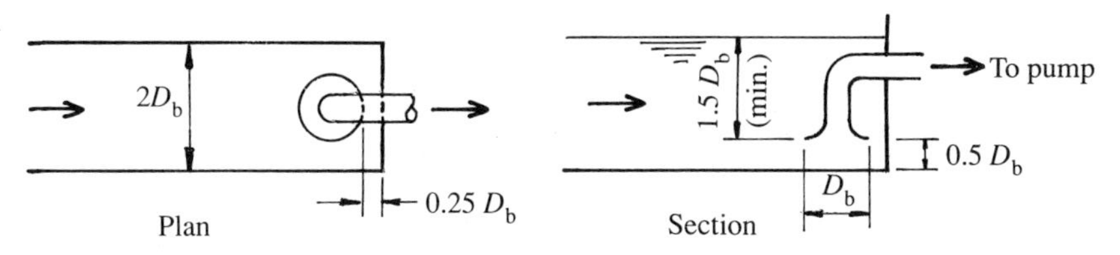

Fig. 11.12 Single pump intake layout.

11.6 Control of pumping

Pumping systems are normally automatically controlled, responding to signals generated by the particular duty regime. In wastewater pumping installations, the control signal is normally generated by the water surface level in the pump sump. A variety of level-sensing devices is available, including electrode rods, floats, ultrasonic devices, pressure sensors, and so on. These sensors produce a signal which switches on a pump when the water level in the sump reaches the upper set point for that pump or switches off a pump when the water level drops to the cut-out level for that pump. Typically, each pump has its own cut-in and cut-out level, as illustrated in Fig. 11.10. This means that only one pump is involved in any switching operation, thus minimizing the resulting step-change in flow in the rising main. The minimum height difference between two adjacent water surface control levels should be such as to avoid simultaneous switching of these pumps due to water surface wave action.

Pump control may also be exercised in response to signals generated on the delivery side. This type of control is typical of water supply installations, where the requirement may, for example, be the maintenance of a fixed pressure at a specified location in the distribution pipe system. The use of variable-speed electric motor pump drives offers considerable control flexibility in this category of application, allowing a close matching of supply and demand and thus reducing energy costs.

Pump installations are typically fitted with a valve set, as illustrated in Fig. 11.9, that is, with a gate or sluice valve on the suction side and with a gate valve plus a non-return valve on the delivery side. The gate valves allow the isolation of the pump and non-return valve for repair and maintenance purposes, while the non-return valve prevents the emptying of the rising main when the pump is not operating.

All pumping installations should be checked at the design stage for waterhammer effects and provided with appropriate control devices, if required; see Chapter 6.

References

Clark, N. N. and Dabolt, R. J. (1986). A general design equation for air-lift pumps operating in slug flow. *AIChE J.*, **32**, No. 1, 56–64.

Denny, D. F. (1956). An experimental study of air-entraining vortices in pump sumps. *Proc. I. Mech. E.*, **170**, No. 2, 106–16.

Hattersley, R. T. (1965). Hydraulic design of pump intakes. *Proc. ASCE*, **91**, No. HY2, 223–48.

Iversen, H. W. (1953). Studies of the submergence requirements of high specific speed pumps. *Trans. ASME*, **75**, No. 4, 635–41.
Nicklin, D. J. (1963). The air-lift pump: theory and optimisation. *Trans. IChemE*, **41**, 29–39.
Prosser, M. J. (1977). *The hydraulic design of pump sumps and intakes.* British Hydromechanics Research Association/Ciria, Cranfield, UK.
Sweeney, C. E., Elder, R. A., and Hay, D. (1982). Pump sump design experience: summary. *Proc. ASCE*, **108**, No. HY3, 361–76.
Wijdieks, J. (1971). Hydraulic aspects of the design of pump installations. *Land and Water International*, **12**, 1–7.

Related reading

Bartlett, R. E. (1974). *Pumping stations for water and sewage.* Applied Science Publishers.
Kristal, F. A. (1953). *Pumps.* McGraw-Hill, New York.

Appendix A
Computer program listings

Title	Function
FLUPROPS	Fluid properties data base
FRICTF	Pipe friction factor computation
PIPFLO	Steady pipe flow computation
SSFLO	Computation of sewage sludge flow in pipes
MANIFLO	Analysis of flow distribution in pipe manifolds
PNA	Computation of steady flow in pipe networks
WATHAM	Waterhammer computation
OCSF	Open channel steady uniform flow computation
HJUMP	Hydraulic jump computation
GVF	Gradually varied steady flow in open channels
BROAD	Design of broad-crested weirs
FLUME	Design of long-throated critical depth flumes
SHARP1	Design of rectangular sharp-crested weirs
SHARP2	Design of V-notch weirs
SUTRO	Design of proportional flow weir
OCUSF	Computation of unsteady flow in open channels
AIRLIFT	Air-lift pump flow computation
PUMP	Rotodynamic pump H/Q relation and duty point computation.

Supply of programs on disk

The author will be pleased to supply readers with copies of the program listings on disk. Please apply to the address below stating preferred disk format ($3\frac{1}{2}''$ or $5\frac{1}{4}''$, single or double density, and so on). Please also enclose £5 (Irish or sterling) to cover the costs of disks, postage, etc.

Professor T. J. Casey
University College Dublin
Civil Engineering Department
Earlsfort Terrace
Dublin 2
Eire

Listing of program **FLUPROPS.BAS**

```
10      REM   PROGRAM FLUPROPS
20      REM   THIS PROGRAM PROVIDES A DATA BASE OF FLUID PROPERTIES
30      DIM DENS(30),VISC(30),STEN(30),SVP(30)
40      FOR I=1 TO 21:READ DENS(I),VISC(I),STEN(I),SVP(I): NEXT I
50      READ AICP,AICV,AIA,AIB,AIC,AID,AIS
60      READ OXCP,OXCV,OXA,OXB,OXC,OXD,OXS
70      READ NICP,NICV,NIA,NIB,NIC,NID,NIS
80      READ MECP,MECV,MEA,MEB,MEC,MED,MES
90      READ CDCP,CDCV,CDA,CDB,CDC,CDD,CDS
100     CLS:PRINT TAB(5) "PROGRAM FLUPROPS":PRINT
110     PRINT TAB(5) "TO ACCESS WATER DATA, ENTER 1"
120     PRINT TAB(5) "TO ACCESS GASES DATA, ENTER 2"
130     PRINT: PRINT TAB(5)"ENTER 1 OR 2, AS APPROPRIATE"
140     INPUT SEL: IF SEL<1 OR SEL>2 THEN GOTO 130
150     ON SEL GOTO 170,340
160     REM PHYSICAL PROPERTIES OF WATER
170     PRINT: INPUT"ENTER THE WATER TEMPERATURE (deg C) ";T
180     I=INT(T/5)+1
190     DENS=DENS(I)+(T-(I-1)*5)*(DENS(I+1)-DENS(I))/5
200     VISC=VISC(I)+(T-(I-1)*5)*(VISC(I+1)-VISC(I))/5
210     STEN=STEN(I)+(T-(I-1)*5)*(STEN(I+1)-STEN(I))/5
220     SVP = SVP(I)+(T-(I-1)*5)*(SVP(I+1)-SVP(I))/5
230     PRINT:PRINT  "PHYSICAL PROPERTIES OF WATER AT ";T;
240     PRINT " deg C ARE AS FOLLOWS:": PRINT
250     PRINT "DENSITY (kg/m**3) = ";DENS
260     PRINT "DYNAMIC VISCOSITY (Ns/m**2) = ";VISC
270     PRINT "SURFACE TENSION (N/m) = ";STEN
280     PRINT "SATURATION VAPOUR PRESSURE (N/m**2) = ";SVP
290     PRINT: PRINT"PRESS THE SPACE BAR TO CONTINUE":Y$=INPUT$(1)
300     PRINT: INPUT"DO YOU WISH TO OBTAIN FURTHER DATA (Y/N)";ANS$
310     IF ANS$="Y" THEN GOTO 100
320     IF ANS$<>"N" THEN GOTO 300 ELSE GOTO 1040
330     REM   PHYSICAL PROPERTIES OF GASES
340     PRINT TAB(20) "1.  AIR"
350     PRINT TAB(20) "2.  OXYGEN"
360     PRINT TAB(20) "3.  NITROGEN"
370     PRINT TAB(20) "4.  METHANE"
380     PRINT TAB(20) "5.  CARBON DIOXIDE"
390     PRINT: INPUT"SELECT GAS BY TYPING ITS NUMBER";Z
400     PRINT: IF Z<1 OR Z>5 THEN GOTO 390
410     INPUT "INPUT GAS TEMPERATURE (deg C)";T
420     INPUT "INPUT GAS ABSOLUTE PRESSURE (N/m**2)";P
430     ON Z GOTO 440,560,680,800,920
440     REM   AIR PROPERTIES
450     T=T+273.2: AIR=AICP-AICV: RHO=P/(AIR*T)
460     VISC=AIS*2.71828^(AIA*LOG(T)+(AIB/T)+AIC/(T*T)+AID)
470     PRINT:PRINT "DENSITY OF AIR (kg/m**3) = ";RHO
480     PRINT "DYNAMIC VISCOSITY OF AIR (Ns/m**2) = ";VISC
490     PRINT "SPECIFIC HEAT AT CONSTANT PRESSURE (J/kg.K) = ";AICP
500     PRINT "SPECIFIC HEAT AT CONSTANT VOLUME (J/kg.K) = ";AICV
510     PRINT "SPECIFIC GAS CONSTANT (J/kg.K) = ";AIR
520     PRINT:PRINT"PRESS THE SPACE BAR TO CONTINUE": Y$=INPUT$(1)
530     PRINT:INPUT"DO YOU WISH TO OBTAIN FURTHER DATA (Y/N)";ANS$
540     IF ANS$="Y" THEN GOTO 100
550     IF ANS$<>"N" THEN GOTO 530 ELSE GOTO 1040
560     REM   OXYGEN PROPERTIES
570     T=T+273.2: OXR=OXCP-OXCV: RHO=P/(OXR*T)
580     VISC=OXS*2.71828^(OXA*LOG(T)+(OXB/T)+OXC/(T*T)+OXD)
590     PRINT:PRINT"DENSITY OF OXYGEN (kg/m**3) = ";RHO
```

```
600    PRINT"DYNAMIC VISCOSITY OF OXYGEN (Ns/m**2) = ";VISC
610    PRINT"SPECIFIC HEAT AT CONSTANT PRESSURE (J/kg.K) = ";OXCP
620    PRINT"SPECIFIC HEAT AT CONSTANT VOLUME (J/kg.K) = ";OXCV
630    PRINT"SPECIFIC GAS CONSTANT (J/kg.K) = ";OXR
640    PRINT: PRINT"PRESS THE SPACE BAR TO CONTINUE": Y$=INPUT$(1)
650    PRINT:INPUT"DO YOU WISH TO OBTAIN FURTHER DATA (Y/N)";ANS$
660    IF ANS$="Y" THEN GOTO 100
670    IF ANS$<>"N" THEN GOTO 650 ELSE GOTO 1040
680    REM  NITROGEN PROPERTIES
690    T=T+273.2: NIR=NICP-NICV: RHO=P/(NIR*T)
700    VISC=NIS*2.71828^(NIA*LOG(T)+(NIB/T)+NIC/(T*T)+NID)
710    PRINT:PRINT"DENSITY OF NITROGEN (kg/m**3) = ";RHO
720    PRINT"DYNAMIC VISCOSITY OF NITROGEN (Ns/m**2) = ";VISC
730    PRINT"SPECIFIC HEAT AT CONSTANT PRESSURE (J/kg.K) = ";NICP
740    PRINT"SPECIFIC HEAT AT CONSTANT VOLUME (J/kg.K) = ";NICV
750    PRINT"SPECIFIC GAS CONSTANT (J/kg.K) = ";NIR
760    PRINT: PRINT"PRESS THE SPACE BAR TO CONTINUE": Y$=INPUT$(1)
770    PRINT:INPUT"DO YOU WISH TO OBTAIN FURTHER DATA (Y/N)";ANS$
780    IF ANS$="Y" THEN GOTO 100
790    IF ANS$<>"N" THEN GOTO 770 ELSE GOTO 1040
800    REM  METHANE PROPERTIES
810    T=T+273.2: MER=MECP-MECV: RHO=P/(MER*T)
820    VISC=MES*2.71828^(MEA*LOG(T)+(MEB/T)+MEC/(T*T)+MED)
830    PRINT:PRINT"DENSITY OF METHANE (kg/m**3) = ";RHO
840    PRINT"DYNAMIC VISCOSITY OF METHANE (Ns/m**2) = ";VISC
850    PRINT"SPECIFIC HEAT AT CONSTANT PRESSURE (J/kg.K) = ";MECP
860    PRINT"SPECIFIC HEAT AT CONSTANT VOLUME (J/kg.K) = ";MECV
870    PRINT"SPECIFIC GAS CONSTANT (J/kg.K) = ";MER
880    PRINT"PRESS THE SPACE BAR TO CONTINUE": Y$=INPUT$(1)
890    PRINT:INPUT"DO YOU WISH TO OBTAIN FURTHER DATA (Y/N)";ANS$
900    IF ANS$="Y" THEN GOTO 100
910    IF ANS$<>"N" THEN GOTO 890 ELSE GOTO 1040
920    REM   CARBON DIOXIDE PROPERTIES
930    T=T+273.2: CDR=CDCP-CDCV: RHO=P/(CDR*T)
940    VISC=CDS*2.71828^(CDA*LOG(T)+(CDB/T)+CDC/(T*T)+CDD)
950    PRINT:PRINT"DENSITY OF CARBON DIOXIDE (kg/m**3) = ";RHO
960    PRINT"DYNAMIC VISCOSITY OF CARBON DIOXIDE (Ns/m**2) = ";VISC
970    PRINT"SPECIFIC HEAT AT CONSTANT PRESSURE (J/kg.K) = ";CDCP
980    PRINT"SPECIFIC HEAT AT CONSTANT VOLUME (J/kg.K) = ";CDCV
990    PRINT"SPECIFIC GAS CONSTANT (J/kg.K) = ";CDR
1000    PRINT:PRINT"PRESS THE SPACE BAR TO CONTINUE": Y$=INPUT$(1)
1010    PRINT:INPUT"DO YOU WISH TO OBTAIN FURTHER DATA (Y/N)";ANS$
1020    IF ANS$="Y" THEN GOTO 100
1030    IF ANS$<>"N" THEN GOTO 1010 ELSE GOTO 1040
1040    END
1050    DATA 999.968,1.787E-3,75.6,0.6107E3
1060    DATA 999.992,1.519E-3,74.9,0.8721E3
1070    DATA 999.728,1.307E-3,74.22,1.2277E3
1080    DATA 999.129,1.139E-3,73.49,1.7049E3
1090    DATA 998.234,1.002E-3,72.75,2.3378E3
1100    DATA 997.075,0.8904E-3,71.97,3.1676E3
1110    DATA 995.678,0.7975E-3,71.18,4.2433E3
1120    DATA 994.064,0.7194E-3,70.37,5.6237E3
1130    DATA 992.247,0.6529E-3,69.56,7.3774E3
1140    DATA 990.24,0.5960E-3,68.74,9.5848E3
1150    DATA 998.07,0.5468E-3,67.91,12.338E3
1160    DATA 985.73,0.5040E-3,67.05,15.745E3
1170    DATA 983.24,0.4665E-3,66.18,19.924E3
1180    DATA 980.59,0.4335E-3,65.29,25.013E3
1190    DATA 977.81,0.4042E-3,64.40,31.166E3
1200    DATA 974.89,0.3781E-3,63.50,38.553E3
1210    DATA 971.83,0.3547E-3,62.60,47.364E3
```

```
1220    DATA 968.65,0.3337E-3,61.68,57.808E3
1230    DATA 965.34,0.3147E-3,60.76,70.112E3
1240    DATA 961.92,0.2975E-3,59.84,84.528E3
1250    DATA 958.38,0.2818E-3,58.90,101.325E3
1260    DATA 1005.0,717.9,0.63404,-45.638,380.87,-3.4505,182.0E-7
1270    DATA 920.0,657.1,0.52662,-97.589,2650.7,-2.6892,203.2E-7
1280    DATA 1040.0,742.9,0.60097,-57.005,1029.1,-3.2322,175.7E-7
1290    DATA 2260.0,1725.2,0.54188,-127.57,4700.8,-2.6952,109.3E-7
1300    DATA 876.0,673.8,0.44037,-288.4,19312.0,-1.7418,146.7E-7
```

Listing of program **FRICTF.BAS**

```
10      PRINT"Program FRICTF.BAS":PRINT
20      PRINT"THIS PROGRAM COMPUTES THE FRICTION FACTOR F"
30      PRINT"FOR FLUID FLOW IN PIPES":PRINT
40      PRINT:INPUT"ENTER FLUID DENSITY (kg/m**3)";RHO
50      PRINT:INPUT"ENTER FLUID DYNAMIC VISCOSITY (Ns/m**2)";MU
60      PRINT:INPUT"ENTER PIPE VELOCITY (m/s)";V
70      PRINT:INPUT"ENTER PIPE INTERNAL DIAMETER (mm)";D
80      PRINT:INPUT"ENTER PIPE WALL ROUGHNESS (mm)";K
90      KVISC=MU/RHO
100     RE=ABS(V)*.001*D/KVISC
110     IF RE>2000 THEN GOTO 120 ELSE F=64/RE: GOTO 250
120     UPV=.5
130     LOV=0
140     F=(UPV+LOV)/2
150     Y=.5/SQR(F)
160     X=((K/(3.7*D))+2.51/(RE*SQR(F)))
170     W=Y+LOG(X)/LOG(10)
180     IF W<0 THEN UPV=F
190     IF W>0 THEN LOV=F
200     Z=(UPV+LOV)/2
210     E=ABS(Z-F)
220     IF E<.0001 THEN GOTO 240
230     F=Z: GOTO 150
240     F=Z
250     PRINT:PRINT TAB(5)"Friction factor f = ";F
260     PRINT TAB(7)"Reynolds number = ";RE
270     PRINT:INPUT"DO YOU WISH TO COMPUTE ANOTHER VALUE (Y/N)";ANS$
280     IF ANS$="Y" THEN GOTO 40
290     END
```

Listing of program **PIPFLO.BAS**

```
10      REM PROGRAM PIPFLO
20      REM PIPFLO COMPUTES VELOCITY AND HEAD LOSS IN PIPES
30      PRINT "Program PIPFLO.BAS": PRINT
40      PRINT "THIS PROGRAM RELATES TO FLOW OF CLEAN WATER"
50      PRINT "(TEMP 10 deg C) IN PIPES.  IT COMPUTES:":PRINT
60      PRINT TAB(20) "1. HEAD LOSS FOR GIVEN FLOW"
70      PRINT TAB(20) "2. FLOW AT GIVEN HEAD LOSS":PRINT
80      INPUT "ENTER 1 OR 2, AS APPROPRIATE";SEL:PRINT
90      IF SEL>2 THEN GOTO 80
100     IF SEL=1 THEN INPUT "ENTER PIPE DISCHARGE RATE (m**3/h)";Q:PRINT
110     IF SEL=2 THEN INPUT "ENTER HYDRAULIC GRADIENT H/L";HG:PRINT
120     PRINT "WHICH OF THE FOLLOWING FORMULAE DO YOU WISH TO USE ?":PRINT
130     PRINT TAB(20) "1. COLEBROOK-WHITE"
140     PRINT TAB(20) "2. HAZEN-WILLIAMS"
150     PRINT TAB(20) "3. MANNING":PRINT
160     INPUT "ENTER YOUR SELECTED FORMULA NUMBER";FMUL:PRINT
170     IF FMUL>3 THEN GOTO 160
180     IF FMUL=1 THEN INPUT "ENTER VALUE OF PIPE ROUGHNESS, K (m)";K
190     IF FMUL=2 THEN INPUT "ENTER HAZEN-WILLIAMS C-VALUE";C
200     IF FMUL=3 THEN INPUT "ENTER MANNING n-VALUE";N
210     INPUT "ENTER PIPE INTERNAL DIAMETER (m)";D
220     IF SEL=1 THEN IF FMUL=1 THEN GOTO 280
230     IF SEL=1 THEN IF FMUL=2 THEN GOTO 340
240     IF SEL=1 THEN IF FMUL=3 THEN GOTO 370
250     IF SEL=2 THEN IF FMUL=1 THEN GOTO 410
260     IF SEL=2 THEN IF FMUL=2 THEN GOTO 440
270     IF SEL=2 THEN IF FMUL=3 THEN GOTO 460
280     V=Q/(3600*.785*D*D): GOSUB 510
290     HG=F*V*V/(19.62*D)
300     PRINT: PRINT TAB(8) "Friction factor = ";F
310     PRINT TAB(5)"Hydraulic gradient = ";HG
320     PRINT TAB(8)"Reynolds number = ";RE
330     GOTO 480
340     V=Q/(3600*.785*D*D)
350     HG=6.81*V^1.852/(C^1.852*D^1.167)
360     GOTO 390
370     V=Q/(3600*.785*D*D)
380     HG=6.35*N*N*V*V/D^1.33
390     PRINT "HYDRAULIC GRADIENT = ";HG
400     GOTO 480
410     FC=(19.62*D*HG)^.5
420     V=-.88*FC*LOG(K/(3.7*D)+3.2675E-06/(D*FC))
430     GOTO 470
440     V=.355*C*D^.63*HG^.54
450     GOTO 470
460     V=.397*D^.67*HG^.5/N
470     PRINT "VELOCITY (m/s) = ";V
480     PRINT: INPUT "DO YOU WISH TO MAKE ANOTHER COMPUTATION Y/N";ANS$
490     IF ANS$="Y" THEN GOTO 30
500     END
510     REM ********* SUBROUTINE FRICTION FACTOR *********
520     KVISC=1.307E-06
530     RE=V*D/KVISC
540     UPV=.5
550     LOV=0
560     F=(UPV+LOV)/2
570     Y=.5/SQR(F)
580     X=((K/(3.7*D))+2.51/(RE*SQR(F)))
```

```
590     W=Y+LOG(X)/LOG(10)
600     IF W<0 THEN UPV=F
610     IF W>0 THEN LOV=F
620     Z=(UPV+LOV)/2
630     E=ABS((Z-F)/F)
640     IF E<.005 THEN GOTO 660
650     F=Z: GOTO 570
660     F=Z
670     RETURN
```

Listing of program **SSFLO.BAS**

```
10      REM PROGRAM SSFLO
20      REM SSFLO COMPUTES HEAD LOSS FOR SLUDGE FLOW IN PIPES
30      PRINT  "Program SSFLO.BAS" :PRINT
40      PRINT "THIS PROGRAM COMPUTES THE HEAD LOSS DUE TO FLOW"
50      PRINT "OF PRIMARY, ACTIVATED, DIGESTED AND HUMUS SLUDGES"
60      PRINT "IN PIPES.":PRINT
70      PRINT TAB(10) "1. PRIMARY"
80      PRINT TAB(10) "2. ACTIVATED"
90      PRINT TAB(10) "3. ANAEROBICALLY DIGESTED"
100     PRINT TAB(10) "4. HUMUS": PRINT
110     INPUT  "ENTER YOUR SELECTED NUMBER";SEL:PRINT
120     IF SEL>4 THEN GOTO 110
130     INPUT "INPUT THE SUSPENDED SOLIDS CONC. (kg/m**3)";C
140     INPUT "INPUT THE PIPE INTERNAL DIAMETER (m)";D
150     INPUT "INPUT THE PIPE ROUGHNESS (m)"; KP
160     INPUT "INPUT THE SLUDGE DISCHARGE RATE (m**3/h)";Q
170     V=Q/(3600*.785*D*D)
180     ON SEL GOTO 190,210,230,250
190     K=.000056*C^2.8: N=.79/C^.172: TY=.00013*C^2.72: HLR=1.5
200     GOTO 260
210     K=.000086*C^3: N=1.7/C^.45: TY=.00013*C^3: HLR=.88+.024*C
220     GOTO 260
230     K=.0000057*C^3.48: N=.9/C^.24: TY=.000014*C^3.37: HLR=.8+.016*C
240     GOTO 260
250     K=.000017*C^3: N=1.9/C^.45: TY=.000016*C^3: HLR=.8+.02*C
260     RE=1000*V*D/(K*((3*N+1)/(4*N))^N*(8*V/D)^(N-1))
270     PRINT: PRINT TAB(10)"Velocity (m/s) = ";V
280     PRINT TAB(10)"Reynolds number = ";RE: PRINT
290     IF RE<2300 THEN PRINT "HENCE, FLOW IS LAMINAR": GOSUB 360
300     IF RE>4100 THEN PRINT "HENCE, FLOW IS TURBULENT": GOSUB 490
310     IF RE>2300 THEN IF RE<4100 THEN PRINT "HENCE, FLOW IS IN THE
TRANSITION REGION BETWEEN LAMINAR AND TURBULENT. HEAD LOSS IS COMPUTED AS FOR
TURBULENT FLOW": GOSUB 490
320     PRINT: PRINT TAB(10) "Hydraulic gradient = ";HG
330     PRINT: INPUT "DO YOU WISH TO MAKE ANOTHER COMPUTATION (Y/N)";ANS$
340     IF ANS$="Y" THEN GOTO 40
350     END
360     REM SUBROUTINE FOR LAMINAR FLOW
370     Q=Q/3600
380     UPV=5000: LOV=0
390     TW=(UPV+LOV)/2
400     XX=TY/(TW*(2*N+1))*(1+(2*N*TY)/((N+1)*TW)*(1+N*TY/TW))
410     RES=(3.14*D^3/8)*(N/(3*N+1))*((TW-TY)/K)^(1/N)*(1-XX)-Q
```

```
420    IF RES>0 THEN UPV=TW
430    IF RES<0 THEN LOV=TW
440    IF ABS(UPV-LOV)<10 THEN GOTO 470
450    TW=(UPV+LOV)/2
460    GOTO 400
470    HG=4*TW/(9810*D)
480    RETURN
490    REM ********* SUBROUTINE FRICTION FACTOR *********
500    KVISC=.000001
510    RE=V*D/KVISC
520    UPV=.5
530    LOV=0
540    F=(UPV+LOV)/2
550    Y=.5/SQR(F)
560    X=((KP/(3.7*D))+2.51/(RE*SQR(F)))
570    W=Y+LOG(X)/LOG(10)
580    IF W<0 THEN UPV=F
590    IF W>0 THEN LOV=F
600    Z=(UPV+LOV)/2
610    E=ABS((Z-F)/F)
620    IF E<.005 THEN GOTO 640
630    F=Z: GOTO 550
640    F=Z
650    IF HLR<1 THEN HLR=1
660    HG=HLR*F*V*V/(19.62*D)
670    RETURN
```

Listing of program **MANIFLO.BAS**

```
10      REM PROGRAM MANIFLO
20      DIM QO(100), QL(100)
30      PRINT"Program MANIFLO.BAS":PRINT
40      PRINT "THIS PROGRAM COMPUTES THE DISTRIBUTION OF FLOW IN A"
50      PRINT"PIPE MANIFOLD SYSTEM WITH SQUARE-EDGED LATERAL PIPES"
60      PRINT"HAVING UNIFORMLY SPACED SHARP-EDGED ORIFICES":PRINT
70      PRINT "INPUT THE FOLLOWING DATA IN THE SPECIFIED UNITS:"
80      PRINT "(pipes sloping upwards in flow dir. have pos. slope)":PRINT
90      INPUT "Enter orifice diameter (mm)";OD
100     INPUT "Enter orifice spacing (m)";OS
110     INPUT "Enter number of orifices per lateral";NOPL
120     INPUT "Enter lateral diameter (mm)";LD
130     INPUT "Enter lateral slope (sin theta)";SLO
140     INPUT "Enter lateral wall roughness (mm)";ROL
150     INPUT "Enter lateral spacing (m)";LS
160     PRINT:INPUT "Enter manifold diameter (mm)";MD
170     INPUT "Enter manifold slope (sin theta)";SMO
180     INPUT "Enter manifold wall roughness (mm)";ROM
190     INPUT "Enter number of laterals (total both sides)";NL
200     INPUT "Are laterals on both sides of manifold (Y/N)";ANS$
210     PRINT:INPUT "Enter flowrate (m**3/h)";Q
220     INPUT "Enter density of manifold fluid (kg/m**3)";RHOM
230     INPUT "Enter viscosity of manifold fluid (Ns/m**2)";MU
240     KVISC=MU/RHOM
250     INPUT "Enter density of external fluid (kg/m**3)";RHO
260     PRINT: PRINT"     COMPUTATION IN PROGRESS"
270     PRINT"                 ------- please wait ------- "
280     IF ANS$="Y" THEN FACT=2 ELSE FACT=1
290     G=9.809999:QO(1)=Q/(3600*NL*NOPL):SUMQ=0
300     NLL=NL/FACT
310     LD=LD/1000:OD=OD/1000:MD=MD/1000:ROL=ROL/1000:ROM=ROM/1000
320     OA=.785*OD^2:LA=.785*LD^2:MA=.785*MD^2
330     OV(1)=QO(1)/(OA*.66):E=OV(1)^2/(2*G)
340     FOR I = 1 TO 3
350     LV=QO(1)/LA: CD=.66-.75*(LV^2)/(2*G*E)
360     QO(1)=CD*OA*(2*G*E)^.5: NEXT I
370     V=LV:K=ROL:D=LD
380     GOSUB 980
390     SF=(F*LV^2)/(2*G*LD):E=E+OS*(SF+SLO*(1-RHO/RHOM))
400     SUMQ=QO(1)
410     FOR J=2 TO NOPL
420     LLV=LV
430     FOR I=1 TO 3
440     ER=(LV^2/(2*G))/E:CD=.66-.75*ER:QO(J)=CD*OA*(2*G*E)^.5
450     LV=LLV+QO(J)/LA: NEXT I
460     V=LV: K=ROL: D=LD
470     GOSUB 980
480     SF=(F*LV^2)/(2*G*LD)
490     E=E+OS*(SF+SLO*(1-RHO/RHOM))
500     SUMQ=SUMQ+QO(J)
510     NEXT J
520     EL=E:CDL=SUMQ/(LA*(2*G*EL)^.5)
530     LV=SUMQ/LA: MV=FACT*SUMQ/MA: EHL=(LV^2/(2*G))*(.9*(MV/LV)^2+.4)
540     E=E+EHL
550     QL(1)=SUMQ
560     FOR I = 2 TO NLL
```

```
570     MMV=MV
580     V=MV :K=ROM: D=MD
590     GOSUB 980
600     SF=(F*MV^2)/(2*G*MD):E=E+LS*(SF+SMO*(1-RHO/RHOM))
610     FOR J= 1 TO 3: EL=E-EHL: QL(I)=CDL*LA*(2*G*EL)^.5
620     MV=MMV+FACT*QL(I)/MA: LV=QL(I)/LA
630     EHL=(LV^2/(2*G))*(.9*(MV/LV)^2+.4)
640     NEXT J
650     NEXT I
670     QLE=QL(1)
680     FOR I=1 TO NLL
690     QL(I)=QL(I)/QLE
700     NEXT I
710     QOE=QO(1)
720     FOR I =1 TO NOPL
730     QO(I)=QO(I)/QOE
740     NEXT I
750     QR=QL(NLL)*QO(NOPL):LD=LD*1000:MD=MD*1000:OD=OD*1000
770     REM             Print results to screen
790     PRINT TAB(25) "Manifold Hydraulics"
800     PRINT:PRINT "Orifice" TAB(10) "Orifice" TAB(20)"No per" TAB(30)
"Lateral";
810     PRINT TAB(40) "Lateral" TAB(50) "No of" TAB(60) "Manifold";
820     PRINT TAB(70) "Discharge"
830     PRINT "Diameter" TAB(10) "Spacing" TAB(20) "lateral" TAB(30)
"diameter";
840     PRINT TAB(40) "spacing" TAB(50) "laterals" TAB(60) "diameter"
850     PRINT "(mm)" TAB(10) "(m)"  TAB(30) "(mm)" TAB(40) "(m)";
860     PRINT TAB(60) "(mm)" TAD(70) "(m3/h)"
870     PRINT OD TAB(10) OS TAB(20) NOPL TAB(30) LD TAB(40) LS;
880     PRINT TAB(50) NL TAB(60) MD TAB(70) Q
890     PRINT:PRINT "DISCHARGE RELATIVE TO DEAD END DISCHARGE:":PRINT
900     PRINT "IN LATERALS: ";
910     FOR I=1 TO NOPL:PRINT USING"#.## ";QO(I);:NEXT I:PRINT
920     PRINT:PRINT "IN MANIFOLD: ";
930     FOR I=1 TO NLL:PRINT USING"#.## ";QL(I);:NEXT I:PRINT
940     PRINT:PRINT "Min orifice discharge/max orifice discharge = "QR
950     PRINT:PRINT "System headloss (m) =  "E
960     PRINT:PRINT "Manifold inlet end velocity (m/s) = "MV
970     END
980     REM ********* SUBROUTINE FRICTION FACTOR *********
990     RE=V*D/KVISC
1000    IF RE>2000 THEN GOTO 1010 ELSE F=64/RE: RETURN
1010    UPV=.5
1020    LOV=0
1030    F=(UPV+LOV)/2
1040    Y=.5/SQR(F)
1050    X=((K/(3.7*D))+2.51/(RE*SQR(F)))
1060    W=Y+LOG(X)/LOG(10)
1070    IF W<0 THEN UPV=F
1080    IF W>0 THEN LOV=F
1090    Z=(UPV+LOV)/2
1100    EE=ABS(Z-F)
1110    IF EE<.0001 THEN GOTO 1130
1120    F=Z: GOTO 1040
1130    F=Z
1140    RETURN
```

Listing of program **PNA.BAS**

```
10   REM PROGRAM PNA ** PIPE NETWORK ANALYSIS **
20   CLEAR ,,8000:CLS
30   PRINT "          ******************************************          "
40   PRINT "          *       PIPE NETWORK ANALYSIS PROGRAM      *          "
50   PRINT "          *                                        *          "
60   PRINT "          *                 PNA.BAS                *          "
70   PRINT "          *                                        *          "
80   PRINT "          ******************************************          "
90   PRINT "          * DEPT. OF CIVIL ENGINEERING, UCD, 1991  *          "
100  PRINT "          ******************************************          "
110  PRINT "                                                              "
120  FOR I =1 TO 4000 STEP 1
130  NEXT I
140  CLS
150  PRINT"THIS PROGRAM ANALYSES FLOW IN WATER PIPE NETWORKS": PRINT
160  PRINT"DO YOU NEED INSTRUCTIONS ON HOW TO USE THIS PROGRAM? (Y/N)"
170  INPUT Q$
180  IF Q$="Y" THEN GOTO 7480: REM CALL INSTRUCTION SECTION OF PROGRAM
190  IF Q$="N" THEN GOTO 210
200  GOTO 150
210  DIM N1(125),N2(125),LENGTH(125),DIA(125),CVALUE(125),C(125),L(125),P(125),
T(125),HEIGHT(125),Q(125),LOOP(500),VALVE(50),DEVICE(150),VEL(125),X(125),
Y(125),Z(125),DEMAND(125),HEAD(125),SR(125)
220  PRINT"DO YOU WISH TO ANALYZE A NEW NETWORK (NN) OR A NETWORK FOR WHICH
DATA":PRINT"IS ON FILE (FN) ?; TYPE FN OR NN AS APPROPRIATE"
230  INPUT REP$
240  IF REP$="NN" THEN GOTO 280
250  IF REP$="FN" THEN GOTO 7730
260  GOTO 220
270  REM     #### MENU DISPLAY####
280  PRINT"ENTER TITLE OF NETWORK AND DATE FOR RECORDING ON OUTPUT"
290  INPUT"TITLE (up to 80 characters) ";TITLE$:PRINT
300  PRINT TITLE$;"  ";DATE$
310  PRINT:PRINT "PROGRAM MENU": PRINT
320  PRINT" 1.    NETWORK                  ENTER FOR EACH NEW NETWORK"
330  PRINT" 2.    PIPE                     ENTERS/DISPLAYS PIPE DATA"
340  PRINT" 3.    PUMP                     ENTERS/DISPLAYS PUMP DATA"
350  PRINT" 4.    DEMAND/SUPPLY            ENTERS/DISPLAYS SUPPLY/DEMAND DATA"
360  PRINT" 5.    VALVE N-R                ENTERS/DISPLAYS NR VALVE DATA"
370  PRINT" 6.    VALVE P-R                ENTERS/DISPLAYS PR VALVE DATA"
380  PRINT" 7.    HEAD                     ENTERS/DISPLAYS FIXED HEAD VALUES"
390  PRINT" 8.    ELEVATION                ENTERS/DISPLAYS NODE ELEVATION DATA"
400  PRINT" 9.    FILE                     WRITES NETWORK DATA TO A FILE"
410  PRINT"10.   *ANALYSIS*                RUNS ANALYSES ON DATA"
420  PRINT"11.    END PROGRAM"                                  :PRINT " "
430  INPUT "ENTER THE NUMBER OF YOUR CHOICE"; CHOICE:PRINT
440  ON CHOICE GOSUB 470,790,1510,1840,2060,2370,2680,2880,3100,3210,7920
450  CLS
460  LOCATE 4,15: GOTO 310
470  REM     ##### THIS PART INITIALISES NETWORK ####
480  PRINT "NETWORK IS BEING INITIALISED":PRINT
490  IF REP$="FN" THEN RETURN
500  PIPENO=0:HEADNO=0:CONTRL=0:PSR=0
510  FOR I= 1 TO 125 STEP 1
520  N1(I)=0:N2(I)=0:LENGTH(I)=0:DIA(I)=0:CVALUE(I)=0:C(I)=0:SR(I)=0
530  HEIGHT(I)=1000
540  NEXT I
550  REM     #### INPUT UNITS OF COMPUTATION ####
```

```
560  DIML$="m":DIMD$="mm":DIMV$="m/s":METRIC=1000:CONST=10.67372
570  PRINT "INPUT UNITS"
580  PRINT "SPECIFY FLOW COMPUTATION UNITS:":PRINT
590  PRINT "                1. m**3/s - CUBIC METRES PER SECOND"
600  PRINT "                2. LPS    - LITRES PER SECOND"
610  PRINT " "
620  INPUT "ENTER 1 OR 2, AS APPROPRIATE";NUMB
630  IF NUMB=1 THEN FACTOR=1 : DIMQ$="m**3/s":S=1
640  IF NUMB=2 THEN FACTOR=.001 :DIMQ$="  LPS " : S=1
650  IF S<>1 THEN PRINT "ILLEGAL UNITS RESPECIFY":GOTO 570
660  PRINT:PRINT "ENTER ACCURACY OF FLOW COMPUTATION"
670  INPUT "i.e. INTEGRATED LOOP HEAD LOSS TOLERANCE (m)";ACRACY:PRINT
680  PRINT"SPECIFY PIPE FLOW FORMULA TO BE USED:":PRINT
690  PRINT"                 1. HAZEN-WILLIAMS (C-VALUE) 0R"
700  PRINT"                 2. DARCY-WEISBACH (SURFACE ROUGHNESS)"
710  PRINT:INPUT"ENTER 1 OR 2, AS APPROPRIATE";NUM
720  IF NUM=1 THEN POW=1.85
730  IF NUM=2 THEN PSR=100:POW=2
740  NET$="NET"+LEFT$(TITLE$,4)
750  OPEN "O",1,NET$
760  WRITE# 1,FACTOR,ACRACY,CONST,CONTRL,POW,METRIC,DIMD$,DIMV$,DIMQ$,DIML$,S,
PSR
770  CLOSE 1
780  RETURN
790  REM        #### THIS PART INPUTS PIPE DATA ####
800  CLS
810  IF REP$="FN" THEN GOTO 940
820  PRINT"INPUT OF PIPE DATA": PRINT
830  INPUT"NUMBER OF PIPES IN NETWORK";PIPENO:PRINT
840  SFULL=0:PRINT"ENTER NUMERICAL VALUES, SEPARATED BY COMMAS":PRINT
850  FOR I=1 TO PIPENO STEP 1
860  IF PSR=100 THEN GOTO 880
870  IF SFULL=0 THEN PRINT"NODE1, NODE2, LENGTH(";DIML$;"), DIAMETER(";DIMD$;"),
C-VALUE":SFULL=1:GOTO 900
880  IF SFULL=0 THEN PRINT"NODE1, NODE2, LENGTH(";DIML$;"), DIAMETER(";DIMD$;"),
ROUGHNESS(";DIMD$;")":SFULL=1:GOTO 910
890  IF PSR>0 THEN GOTO 910
900  INPUT N1(I),N2(I),LENGTH(I),DIA(I),CVALUE(I):GOTO 920
910  INPUT N1(I),N2(I),LENGTH(I),DIA(I),SR(I)
920  IF I\15=I/15 THEN SFULL=0
930  NEXT I
940  CLS: PRINT"YOU HAVE ENTERED THE FOLLOWING PIPE DATA; MAKE A NOTE OF ANY
ERRORS": SFULL=0: PRINT
950  FOR I=1 TO PIPENO
960  IF PSR>0 THEN GOTO 980
970  IF SFULL=0 THEN PRINT"PIPE NO  NODE 1  NODE 2  LENGTH(";DIML$;")  DIAMETER
(";DIMD$;")  C-VALUE": SFULL=1: PRINT:GOTO 990
980  IF SFULL=0 THEN PRINT"PIPE NO  NODE 1  NODE 2  LENGTH(";DIML$;")  DIAMETER
(";DIMD$;") ROUGHNESS(";DIMD$;")": SFULL=1: PRINT
990  LET C$="  ###        ###      ###      #####.#      ####.#      ###.###"
1000 IF PSR>0 THEN GOTO 1020
1010 PRINT USING C$;I;N1(I);N2(I);LENGTH(I);DIA(I);CVALUE(I):GOTO 1030
1020 PRINT USING C$;I;N1(I);N2(I);LENGTH(I);DIA(I);SR(I)
1030 IF I\15 = I/15 THEN SFULL=0 ELSE GOTO 1060
1040 PRINT: PRINT "PRESS THE SPACE BAR TO CONTINUE"
1050 ZZ$=INPUT$(1): CLS
1060 NEXT I
1070 PRINT: INPUT"DO YOU WISH TO REVISE ABOVE DATA (Y/N)";ANS$
1080 IF ANS$="Y" THEN GOTO 1110
1090 IF ANS$="N" THEN GOTO 1250
1100 GOTO 1070
```

```
1110 PRINT: PRINT: PRINT
1120 INPUT"ENTER NO OF PIPE FOR WHICH YOU WISH TO CHANGE DATA";I
1130 PRINT: PRINT"CURRENT VALUE OF PIPE LENGTH IS "; LENGTH(I)
1140 INPUT"ENTER NEW VALUE OF PIPE LENGTH "; LENGTH(I)
1150 PRINT: PRINT"CURRENT VALUE OF PIPE DIAMETER IS ";DIA(I)
1160 INPUT"ENTER NEW VALUE OF PIPE DIAMETER"; DIA(I)
1170 IF PSR=100 THEN GOTO 1200
1180 PRINT: PRINT"CURRENT C-VALUE FOR PIPE IS ";CVALUE(I)
1190 INPUT"ENTER NEW C-VALUE";CVALUE(I):GOTO 1220
1200 PRINT:PRINT"CURRENT ROUGHNESS VALUE IS (mm) ";SR(I)
1210 INPUT"ENTER NEW ROUGHNESS VALUE (mm) ";SR(I)
1220 PRINT: PRINT"CURRENT NODE NUMBERS OF THIS PIPE ARE ";N1(I);" ";N2(I)
1230 INPUT"ENTER NEW NODE NUMBERS (N1,N2)";N1(I),N2(I)
1240 GOTO 1070
1250 PRINT:INPUT"DO YOU WISH TO ADD NEW PIPES (Y/N)";ANS$
1260 IF ANS$="N" THEN GOTO 1350
1270 PIPENO=PIPENO+1 :PN=PIPENO
1280 IF PSR=100 THEN GOTO 1310
1290 PRINT"ENTER NODE 1, NODE 2, LENGTH, DIAM., C-VALUE"
1300 INPUT"N1(PN),N2(PN),LENGTH(PN),DIA(PN),CVALUE(PN):GOTO 1320
1310 PRINT"ENTER NODE 1, NODE 2, LENGTH, DIAM., ROUGHNESS"
1320 INPUT"N1(PN),N2(PN),LENGTH(PN),DIA(PN),SR(PN)
1330 INPUT" DO YOU WISH TO ADD ANOTHER PIPE (Y/N)";ANS$
1340 IF ANS$="Y" THEN GOTO 1270
1350 PRINT: INPUT" DO YOU WISH TO DELETE PIPES (Y/N)";ANS$
1360 IF ANS$="N" THEN GOTO 1450
1370 INPUT" DELETE PIPE NO.:";I
1380  IF PSR=100 THEN GOTO 1400
1390 LENGTH(I)=LENGTH(PIPENO):DIA(I)=DIA(PIPENO):CVALUE(I)=CVALUE(PIPENO)
1400 LENGTH(I)=LENGTH(PIPENO):DIA(I)=DIA(PIPENO):CVALUE(I)=SR(PIPENO)
1410 PIPENO=PIPENO-1
1420 INPUT" DO YOU WISH TO DELETE ANOTHER PIPE (Y/N)";ANS$
1430 IF ANS$="Y" THEN GOTO 1370
1440 REM  *** CALCULATE NODENO, THE NUMBER OF NODES ***
1450 NODENO=0
1460 FOR I=1 TO PIPENO
1470 IF N1(I)>NODENO THEN NODENO=N1(I)
1480 IF N2(I)>NODENO THEN NODENO=N2(I)
1490 NEXT I
1500 RETURN
1510 REM      #### THIS PART ENTERS PUMP DATA ####
1520 CLS
1530 PRINT"                      ENTRY OF PUMP DATA": PRINT
1540 IF REP$="FN" THEN GOTO 1690
1550 INPUT" DO YOU WISH TO ENTER PUMP DATA (Y/N)";ANS$
1560 IF ANS$="N" THEN GOTO 1830 ELSE IF ANS$<>"Y" THEN GOTO 1550
1570 PRINT "ENTER PUMP DATA:  NODE 1, NODE 2, A, B, C (values sep. by commas)"
1580 PRINT: INPUT "PUMP DATA:";A,B,F,G,R
1590 FOR I=1 TO PIPENO
1600 IF A=N2(I)  THEN IF  B=N1(I)  THEN D=N1(I) : N1(I)=N2(I):N2(I)=D:GOTO 1630
1610 IF A=N1(I) THEN IF B=N2(I) THEN GOTO 1630
1620 GOTO 1650
1630 DEVICE(I)=20
1640 X(I)=F:Y(I)=G: Z(I)=R: GOTO 1660
1650 NEXT I
1660 INPUT"DO YOU WISH TO ENTER DATA FOR ANOTHER PUMP (Y/N)";ANS$
1670 IF ANS$="Y" THEN GOTO 1570: IF ANS$<>"N" THEN GOTO 1660
1680 REM  *******SCREEN DISPLAY OF PUMP DATA*********
1690 PRINT"                     PUMP DATA": PRINT
1700 PRINT "PIPE NO.   NODE 1  NODE 2    A          B      C (coeffs)"
1710 FOR I =1 TO PIPENO
```

```
1720 IF DEVICE(I)<>20 THEN GOTO 1750
1730 FORM$= " ###        ###      ###  ####.###  ####.###   ###.#"
1740 PRINT USING FORM$; I,N1(I),N2(I),X(I),Y(I),Z(I)
1750 NEXT I
1760 PRINT:INPUT" DO YOU WISH TO DELETE A PUMP (Y/N)"; ANS$
1770 IF ANS$="N" THEN GOTO 1810
1780 INPUT" ENTER THE NUMBER OF THE PIPE CONTAINING THE PUMP";I:DEVICE(I)=0
1790 PRINT:INPUT" DO YOU WISH TO DELETE ANOTHER PUMP (Y/N)";ANS$
1800 IF ANS$="Y" THEN GOTO 1780
1810 PRINT: INPUT" DO YOU WISH TO ENTER NEW PUMP DATA (Y/N)";ANS$
1820 IF ANS$="Y" THEN GOTO 1570
1830 RETURN
1840 REM    #### THIS PART ENTERS DEMANDS/SUPPLIES ####
1850 CLS
1860 PRINT"                 ENTRY OF SUPPLY/DEMAND DATA" :PRINT
1870 IF REP$="FN" THEN GOTO 1960
1880 INPUT" DO YOU WISH TO ENTER SUPPLY/DEMAND DATA (Y/N)";ANS$
1890 IF ANS$="N" THEN GOTO 2050 ELSE PRINT
1900 PRINT"DEMAND IS ENTERED AS A POSITIVE NUMBER, SUPPLY AS A NEGATIVE NUMBER"
1910 INPUT"HOW MANY SUPPLY/DEMAND VALUES DO YOU WISH TO ENTER?";NSD
1920 FOR J=1 TO NSD
1930 INPUT "node ,amount ";F,G
1940 DEMAND(F)=G
1950 NEXT J
1960 PRINT:PRINT"                 SUPPLY/DEMAND DATA":PRINT
1970 PRINT"             NODE NO.       SUPPLY/DEMAND":PRINT
1980 FORM$="              ####           ####.####"
1990 FOR I=1 TO NODENO
2000 IF DEMAND(I)=0 THEN GOTO 2020
2010 PRINT USING FORM$;I,DEMAND(I)
2020 NEXT I
2030 PRINT:INPUT" DO YOU WISH TO REVISE SUPPLY/DEMAND DATA (Y/N)";ANS$
2040 IF ANS$="Y" THEN GOTO 1910
2050 RETURN
2060 '                       #### NON RETURN VALVE ENTERED HERE ####
2070 CLS
2080 PRINT"              ENTRY OF NRV DATA":PRINT
2090 IF REP$="FN" THEN GOTO 2210
2100 INPUT"DO YOU WISH TO ENTER NRV DATA (Y/N)";ANS$
2110 IF ANS$="N" THEN GOTO 2360 ELSE IF ANS$<>"Y" THEN GOTO 2100
2120 PRINT"ENTER NRV LOCATION: NODE 1,  NODE 2"
2130 INPUT" PIPE NODES";A,B
2140 FOR I=1 TO PIPENO
2150 IF(A=N2(I)) AND (B=N1(I)) THEN D=N1(I):N1(I)=N2(I):N2(I)=D:GOTO 2170
2160 IF (A<>N1(I)) AND (B<>N2(I)) THEN GOTO 2180
2170 DEVICE(I)=10: GOTO 2190
2180 NEXT I
2190 INPUT"DO YOU WISH TO ENTER DATA FOR ANOTHER NRV (Y/N)";ANS$
2200 IF ANS$="Y" THEN GOTO 2120: IF ANS$<>"N" THEN GOTO 2190
2210 REM    ***** SCREEN DISPLAY OF NRV LOCATIONS *****
2220 PRINT"NR VALVES ARE LOCATED IN THE FOLLOWING PIPES:"
2230 PRINT" PIPE NO.   NODE 1    NODE 2"
2240 FORM$="######     ######    ######"
2250 FOR I=1 TO PIPENO
2260 IF DEVICE(I)<>10 THEN GOTO 2280
2270 PRINT USING FORM$;I,N1(I),N2(I)
2280 NEXT I
2290 PRINT:INPUT"DO YOU WISH TO DELETE AN NRV (Y/N)";ANS$
2300 IF ANS$="N" THEN GOTO 2360: IF ANS$<>"Y" THEN GOTO 2290
2310 INPUT"ENTER THE NUMBER OF THE PIPE CONTAINING THE NRV";I:DEVICE(I)=0
2320 PRINT:INPUT"DO YOU WISH TO DELETE ANOTHER NRV (Y/N)";ANS$
```

```
2330 IF ANS$="Y" THEN GOTO 2310
2340 PRINT:INPUT"DO YOU WISH TO ADD NEW NRVs (Y/N)";ANS$
2350 IF ANS$="Y" THEN GOTO 2120
2360 RETURN
2370 REM    #### PRESURE REDUCING VALVE ####
2380 CLS
2390 PRINT"                    ENTRY OF PRV DATA":PRINT
2400 IF REP$="FN" THEN GOTO 2530
2410 INPUT"DO YOU WISH TO ENTER PRV DATA (Y/N)";ANS$
2420 IF ANS$="N" THEN GOTO 2670 ELSE IF ANS$<>"Y" THEN GOTO 2410
2430 PRINT"ENTER NRV LOCATION:  NODE 1,  NODE 2,  CONSTANT
2440 INPUT " DATA:";A,B,C
2450 FOR I=1 TO PIPENO
2460 IF(A=N2(I)) AND (B=N1(I)) THEN D=N1(I):N1(I)=N2(I):N2(I)=D:GOTO 2480
2470 IF(A<>N1(I)) AND (B<>N2(I)) THEN GOTO 2490
2480 DEVICE(I)=1:X(I)=C:GOTO 2500
2490 NEXT I
2500 INPUT"DO YOU WISH TO ENTER DATA FOR ANOTHER PRV (Y/N)";ANS$
2510 IF ANS$="Y" THEN GOTO 2430: IF ANS$<>"N" THEN GOTO 2500
2520 REM        *****   SCREEN DISPLAY OF PRV DATA  ****
2530 PRINT"THE FOLLOWING PRV DATA HAS BEEN ENTERED:":PRINT
2540 PRINT"  PIPE NO.    NODE 1     NODE 2    CONSTANT"
2550 FORM$="   ####        ####        ####      ###.#####"
2560 FOR I=1 TO PIPENO
2570 IF DEVICE(I)<>1 THEN GOTO 2590
2580 PRINT USING FORM$; I,N1(I),N2(I),X(I)
2590 NEXT I
2600 PRINT:INPUT" DO YOU WISH TO DELETE A PRV (Y/N)";ANS$
2610 IF ANS$="N" THEN GOTO 2670
2620 PRINT:INPUT"ENTER THE NUMBER OF THE PIPE CONTAINING THE PRV";I:DEVICE(I)=0
2630 INPUT" DO YOU WISH TO DELETE ANOTHER PRV (Y/N)";ANS$
2640 IF ANS$="Y" THEN GOTO 2620
2650 PRINT:INPUT" DO YOU WISH TO ADD NEW PRVs (Y/N)";ANS$
2660 IF ANS$="Y" THEN GOTO 2430
2670 RETURN
2680 REM  **** ENTRY OF OF FIXED HEAD DATA ****
2690 CLS
2700 IF REP$="FN" THEN GOTO 2760
2710 PRINT:INPUT"HOW MANY FIXED HEADS DO YOU WISH TO ENTER";HEADNO:PRINT
2720 FOR J=1 TO HEADNO
2730 INPUT" NODE NO., HEAD:";F,G
2740 HEAD(F)=G
2750 NEXT J
2760 PRINT: PRINT"                FIXED HEAD DATA":PRINT
2770 PRINT"                NODE NO.        HEAD (m)":PRINT
2780 FORM$="                ####           ####.#"
2790 FOR I=1 TO NODENO
2800 IF HEAD(I)=0 THEN GOTO 2820
2810 PRINT USING FORM$;I,HEAD(I)
2820 NEXT I
2830 PRINT:INPUT" DO YOU WISH TO RE-ENTER HEAD DATA (Y/N)";ANS$
2840 IF ANS$="N" THEN GOTO 2870
2850 FOR I=1 TO NODENO: HEAD(I)=0:NEXT I
2860 GOTO 2710
2870 RETURN
2880 REM  **** ENTRY OF NODE ELEVATIONS ****
2890 CLS: IF REP$="FN" THEN GOTO 2990
2900 PRINT"                   ENTRY OF NODE ELEVATION DATA": PRINT
2910 FOR I=1 TO 125:HEIGHT(I)=1000:NEXT I
2920 PRINT"A DEFAULT LEVEL OF 1000m IS ASSUMED FOR ALL NODES"
2930 PRINT"IN THE ABSENCE OF INPUT DATA FOR NODE ELEVATIONS"
2940 PRINT: INPUT"DO YOU WISH TO ENTER NODE ELEVATIONS? (Y/N)";ANS$
```

```
2950 IF ANS$="N" THEN GOTO 3090: IF ANS$<>"Y" THEN GOTO 2910
2960 FOR I= 1 TO NODENO STEP 1
2970 PRINT "NODE  ";I;" HAS ELEVATION";:INPUT HEIGHT (I)
2980 NEXT I
2990 PRINT:PRINT"          NODE ELEVATION DATA":PRINT
3000 PRINT"               NODE NO.       ELEVATION (m)":PRINT
3010 FORM$="                ####          ####.#"
3020 FOR I=1 TO NODENO
3030 PRINT USING FORM$; I,HEIGHT(I): NEXT I
3040 PRINT:INPUT" DO YOU WISH TO REVISE NODE ELEVATION DATA (Y/N)";ANS$
3050 IF ANS$="N" THEN GOTO 3090
3060 PRINT:INPUT" NODE NO.,ELEVATION";I,HEIGHT(I)
3070 INPUT"DO YOU WISH TO REVISE ANOTHER VALUE (Y/N)";ANS$
3080 IF ANS$="Y" THEN GOTO 3060
3090 RETURN
3100 REM **** TRANSFER OF NETWORK DATA TO DISK FILE ****
3110 PIP$="PIP"+LEFT$(TITLE$,4)
3120 OPEN "O",1,PIP$
3130 WRITE# 1,PIPENO,NODENO,HEADNO
3140 FOR I=1 TO PIPENO
3150 IF PSR>0 THEN GOTO 3170
3160 WRITE# 1,N1(I),N2(I),LENGTH(I),DIA(I),CVALUE(I),DEVICE(I),X(I),Y(I),Z(I),
DEMAND(I),HEAD(I),HEIGHT(I):GOTO 3180
3170 WRITE# 1,N1(I),N2(I),LENGTH(I),DIA(I),SR(I),DEVICE(I),X(I),Y(I),Z(I),
DEMAND(I),HEAD(I),HEIGHT(I)
3180 NEXT I
3190 CLOSE 1
3200 RETURN
3210 REM   *** COMPUTATION OF PIPE RESISTANCE COEFFICIENTS C(I) ***
3220 FOR I=1 TO PIPENO STEP 1
3230 IF PSR>0 THEN GOTO 3270
3240 M=(DIA(I)/METRIC)^4.87
3250 M=M*CVALUE(I)^1.85
3260 C(I)=LENGTH(I)*CONST/M:GOTO 3290
3270 V=1:D=DIA(I):L=LENGTH(I):K=SR(I):GOSUB 7960
3280 C(I)=CPP
3290 NEXT I
3300 REM    **** FACTORING PUMP COEFFS TO m^3/s ****
3310 FOR I=1 TO PIPENO STEP 1
3320 F1=1/FACTOR:X(I)=X(I)*F1*F1:Y(I)=Y(I)*F1
3330 IF DEVICE(I)>0 THEN CONTRL=CONTRL+1
3340 NEXT I
3350 REM****MOVING PIPES CONTAINING CONTROL DEVICES TO END OF PIPE ARRAY****
3360 K1=PIPENO-CONTRL
3370 K2=K1+1
3380 IF CONTRL=0 GOTO 3520
3390 FOR I=1 TO K1 STEP 1
3400 IF DEVICE(I)=0 GOTO 3510
3410 ' GO TO CATCHER
3420 IF DEVICE(K2)<>0 THEN K2=K2+1: GOTO 3410
3430 K3=N1(I)
3440 N1(I)=N1(K2):N1(K2)=K3
3450 K3=N2(I): N2(I)=N2(K2): N2(K2)=K3
3460 TEMP =LENGTH(I):LENGTH(I)=LENGTH(K2):LENGTH(K2)=TEMP
3470 TEMP=DIA(I):DIA(I)=DIA(K2):DIA(K2)=TEMP
3480 TEMP=CVALUE(I):CVALUE(I)=CVALUE(K2):CVALUE(K2)=TEMP
3490 TEMP=C(I):C(I)=C(K2):C(K2)=TEMP
3500 DEVICE(K2)=DEVICE(I):DEVICE(I)=0:X(K2)=X(I):X(I)=0:Y(K2)=Y(I):Y(I)=0:
Z(K2)=Z(I):Z(I)=0
3510 NEXT I
3520 REM****SORTING REMAINDER OF PIPES IN ORDER OF RESISTANCE****
```

```
3530 PRINT "   ......SORTING THE PIPES NOW......
                                        ........please wait ........      "
3540 K1=K1-1
3550 FOR I=1 TO K1 STEP 1
3560 K2=PIPENO-CONTRL-I
3570 FOR J=1 TO K2 STEP 1
3580 IF C(J)<=C(J+1) GOTO 3650
3590 TEMP =C(J):C(J)=C(J+1):C(J+1)=TEMP
3600 K3=N1(J):N1(J)=N1(J+1):N1(J+1)=K3
3610 K3=N2(J):N2(J)=N2(J+1):N2(J+1)=K3
3620 TEMP=LENGTH(J):LENGTH(J)=LENGTH(J+1):LENGTH(J+1)=TEMP
3630 TEMP=DIA(J):DIA(J)=DIA(J+1):DIA(J+1)=TEMP
3640 TEMP=CVALUE(J):CVALUE(J)=CVALUE(J+1):CVALUE(J+1)=TEMP
3650 NEXT J
3660 NEXT I
3670 REM  *** THIS PART OF THE PROGRAM ANALYSES THE INPUT DATA ***
3605 REM  *** USING THE TRAVERS ALGORITHM  AND LOOP EQUATIONS  ***
3680 REM  *** INPUT DATA IS NOW PRINTED AS AN ECHO CHECK       ***
3690 PRINT CHR$(12)
3700 PRINT "           ************************************   "
3710 PRINT "           ***      INPUT    DATA          ***   "
3720 PRINT "           ************************************   "
3730 SFULL=0
3740 FORM$="###### ###### ###### #######     #####  ###.##   ########"
3750 FOR I=1 TO PIPENO STEP 1
3760 IF PSR>0 THEN GOTO 3780
3770 IF SFULL=0 THEN PRINT"PIPE NO  NODE1  NODE2  LENGTH  DIAMETER  CVALUE
RESISTANCE":SFULL=1:PRINT"                              ";DIML$;"
";DIMD$:PRINT:GOTO 3800
3780 IF SFULL=0 THEN PRINT"PIPE NO  NODE1  NODE2  LENGTH  DIAMETER ROUGHNESS
RESISTANCE":SFULL=1:PRINT"                               ";DIML$;"       ";DIMD$;"
   ";DIMD$:PRINT:GOTO 3810
3790 IF PSR>0 THEN GOTO 3810
3800 PRINT USING FORM$;I,N1(I),N2(I),LENGTH(I),DIA(I),CVALUE(I),C(I):GOTO 3820
3810 PRINT USING FORM$;I,N1(I),N2(I),LENGTH(I),DIA(I),SR(I),C(I)
3820 IF I\15=I/15 THEN SFULL=0 ELSE GOTO 3850
3830 PRINT: PRINT"PRESS THE SPACE BAR TO CONTINUE"
3840 ZZ$=INPUT$(1)
3850 NEXT I
3860 PRINT: PRINT "PRESS THE SPACE BAR TO CONTINUE"
3870 ZZ$=INPUT$(1)
3880 PRINT:PRINT:PRINT
3890 REM   ****SCREEN PRINT OF SUPPLY/DEMAND/HEAD/ELEVATION DATA****
3900 SFULL=0
3910 FOR I=1 TO NODENO STEP 1
3920 IF SFULL=0 THEN PRINT"NODE NO   DEMAND   HEAD   ELEVATION":SFULL=1:PRINT"
      ";DIMQ$;"     ";"m";"       ";"m":PRINT
3930 FORM$="#####   ####.##    ####    ####"
3940 PRINT USING FORM$;I,DEMAND(I),HEAD(I),HEIGHT(I)
3950 IF I\15=I/15 THEN SFULL=0 ELSE GOTO 3980
3960 PRINT:PRINT"PRESS THE SPACE BAR TO CONTINUE"
3970 ZZ$=INPUT$(1)
3980 NEXT I
3990 PRINT:PRINT: PRINT"PRESS THE SPACE BAR TO CONTINUE"
4000 ZZ$=INPUT$(1)
4010 REM            **** INPUT CHECK FINISHED ****
4020 PRINT:PRINT
4030 PRINT".....LOOP SELECTION NOW IN PROGRESS....."
4040 PRINT"                                  .....please wait....."
4050 REM
4060 REM *** SELECTION OF LOOPS; THE PIPES IN EACH LOOP ARE LISTED ***
```

```
4070 REM *** IN LOOP ORDER IN THE ARRAY LOOP(I), A ZERO ARRAY VALUE ***
4080 REM *** SEPARATING LOOP SETS.                                  ***
4090 REM *** SET NODE T-VALUES TO ZERO ***
4100 FOR I = 1 TO NODENO
4110 T(I) = 0
4120 NEXT I
4130 TREE = 0
4140 NEXTLP = 0
4150 ITER = 0
4160 LIMIT = PIPENO
4170 REM  ##### POINTS OF FIXED HEAD #####
4180 IF HEADNO < 2 GOTO 4400
4190 EXPECT = HEADNO - 1
4200 LIMIT = LIMIT + EXPECT
4210 FOR I = 1 TO NODENO
4220 IF HEAD(I) = 0  GOTO  4280
4230 K1 = PIPENO + 1
4240 EXTRA = 0
4250 N1(K1) = I
4260 K2 = I + 1
4270 GOTO 4300
4280 NEXT I
4290 REM  ##### ASCEND THE NODES MARKING POINTS OF FIXED HEAD #####
4300 FOR I = K2 TO NODENO
4310 IF HEAD(I) = 0  GOTO  4380
4320 N2(K1) = I
4330 EXTRA = EXTRA + 1
4340 DEVICE(K1) - -1
4350 IF EXTRA = EXPECT  GOTO 4400
4360 K1 = K1 + 1
4370 N1(K1) = I
4380 NEXT I
4390 REM  ##### CONSTRUCT THE TREE #####
4400 FOR I = 1 TO LIMIT
4410 IF T(N1(I)) = 0  GOTO  4450
4420 IF T(N2(I)) = 0  GOTO  4470
4430 IF T(N1(I)) = T(N2(I))  GOTO 4520  ELSE GOTO 4880
4440 REM  ##### INSERT NODE1 AND NODE2 INTO TREE #####
4450 IF T(N2(I)) = 0  THEN  TREE=TREE+1:T(N1(I))=TREE:T(N2(I))=TREE:L(N1(I))=0:
L(N2(I))=1:P(N1(I))=0:P(N2(I))=I:GOTO 5110 : ELSE T(N1(I))=T(N2(I)):
L(N1(I))=L(N2(I))+1:P(N1(I))=I: GOTO 5110
4460 REM    ##### INSERT NODE2 INTO TREE #####
4470 T(N2(I)) = T(N1(I))
4480 L(N2(I)) = L(N1(I)) + 1
4490 P(N2(I)) = I
4500 GOTO  5110
4510 REM   ##### BOTH NODES IN TREE, CLOSING PIPE OR NEW LOOP #####
4520 NEXTLP = NEXTLP + 1
4530 LOOP(NEXTLP) = I
4540 K1 = N1(I)
4550 K2 = N2(I)
4560 IF L(K2) = L(K1)  GOTO  4680
4570 IF L(K2) < L(K1)  THEN LOOP(NEXTLP)=-I:K1=K2:K2=N1(I)
4580 REM
4590 NXPIPE = P(K2)
4600 NEXTLP = NEXTLP + 1
4610 LOOP(NEXTLP) = NXPIPE
4620 K3 = N1(NXPIPE)
4630 K4 = N2(NXPIPE)
4640 IF K2 <> K3  THEN LOOP(NEXTLP)=-NXPIPE:K4=K3
4650 K2 = K4
```

```
4660 IF L(K1) <> L(K2)  GOTO  4590
4670 REM
4680 IF K1 = K2  GOTO 4830
4690 NXPIPE = P(K1)
4700 NEXTLP = NEXTLP + 1
4710 LOOP(NEXTLP) = NXPIPE
4720 K3 = N1(NXPIPE)
4730 K4 = N2(NXPIPE)
4740 IF K1 = K4  THEN K1=K3 ELSE LOOP(NEXTLP)=-NXPIPE:K1=K4
4750 NXPIPE = P(K2)
4760 NEXTLP = NEXTLP + 1
4770 LOOP(NEXTLP) = NXPIPE
4780 K3 = N1(NXPIPE)
4790 K4 = N2(NXPIPE)
4800 IF K2 = K3  THEN  K2=K4 ELSE LOOP(NEXTLP)=-NXPIPE:K2=K3
4810 GOTO 4680
4820 REM   ##### LOOP IS CREATED , BEGIN A NEW LOOP #####
4830 NEXTLP = NEXTLP + 1
4840 LOOP(NEXTLP) = 0
4850 ITER = ITER + 1
4860 GOTO  5110
4870 REM   ##### NODES OF PIPE ON DIFFERENT TREES #####
4880 K1 = N1(I)
4890 K2 = N2(I)
4900 TNEW = T(K1)
4910 TOLD = T(K2)
4920 IF TOLD <= TNEW  THEN K1=N2(I):K2=N1(I):TNEW=T(K1):TOLD=T(K2)
4930 T(K2) = TNEW
4940 L(K2) = L(K1) + 1
4950 P(K2) = I
4960 NO = I - 1
4970 REM     ##### JOIN TWO SEPARATE TREES #####
4980 LEVEL = 0
4990 FOR J = 1 TO NO
5000 K3 = N1(J)
5010 K4 = N2(J)
5020 IF T(K3) = TNEW  THEN GOSUB 5490  ELSE GOSUB 5510
5030 ON S GOTO 5050,5090,5090,5050,5090
5040 REM    ##### ASCEND ONE PIPE AND INCREASE LEVEL #####
5050 T(K4) = TNEW
5060 L(K4) = L(K3) + 1
5070 P(K4) = J
5080 LEVEL = 1
5090 NEXT J
5100 IF LEVEL = 1  GOTO 4980
5110 NEXT I
5120 REM    ##### ERROR CHECK FOR UNCONNECTED NODE #####
5130 FOR I = 1 TO NODENO
5140 IF T(I) = 1 THEN GOTO 5180
5150 PRINT "******UNCONNECTED NODE NO. = "  I  "*******"
5160 PRINT:PRINT"PRESS THE SPACE BAR TO RETURN TO MAIN MENU
5170 Y$=INPUT$(1):GOTO 310
5180 NEXT I
5190 REM    ##### ERROR CHECK FOR CORRECT NO OF LOOPS #####
5200 NLOOP = LIMIT - NODENO + 1
5210 IF NLOOP <> ITER  THEN PRINT "******** INCORRECT NO. OF LOOPS   EXPECTED
NO " NLOOP " FOUND  NO " ITER "********" :GOTO 5230
5220 GOTO 5250
5230 PRINT:PRINT"PRESS THE SPACE BAR TO RETURN TO MAIN MENU"
5240 Y$=INPUT$(1):GOTO 310
5250 PRINT: PRINT"   LOOP CONSTRUCTION NOW COMPLETE,"
```

```
5260 PRINT"   STARTING ITERATIVE COMPUTATION....."
5270 PRINT"                                          .....please wait....."
5280 FOR I = 1 TO NODENO
5290 DEMAND(I) = DEMAND(I) * FACTOR
5300 IF LEVEL < L(I)  THEN  LEVEL = L(I)
5310 NEXT
5320 FOR I = 1 TO PIPENO
5330 Q(I) = 0
5340 NEXT
5350 REM      ##### DESCEND THE TREE ACCUMULATING DEMANDS IN EACH PIPE #####
5360 FOR I = 1 TO NODENO
5370 IF T(I) = 0  GOTO  5410
5380 IF L(I) = LEVEL  THEN NO=P(I):K1=N1(NO):K2=N2(NO):K3=1:Q(NO)=DEMAND(I)
5390 IF (K2<>I) AND (L(I)=LEVEL) THEN K1=N2(NO):K2=N1(NO):K3=-1:Q(NO)=-Q(NO)
5400 IF L(I)=LEVEL THEN DEMAND(K1)=DEMAND(K1)+Q(NO)*K3
5410 NEXT I
5420 LEVEL = LEVEL - 1
5430 IF LEVEL > 0    GOTO  5360
5440 REM     ##### ERROR CHECK FOR FLOW BALANCE AT EACH NODE #####
5450 IF  ABS (DEMAND(K1))> 8.999999E-03  THEN  TEMP=DEMAND(K1)/FACTOR:PRINT
"DIFFERFNCE BETWEEN SUPPLY AND DEMAND OF "  TEMP,DIMQ$
5460 PRINT
5470 GOTO 5560
5480 REM  *** SUBROUTINE TO JOIN TWO TREES ***
5490 IF T(K4)=TOLD  THEN S = 1 ELSE S = 2
5500 RETURN
5510 IF T(K3)=TOLD  THEN IF T(K4) <> TNEW THEN S = 3 ELSE K3=N2(J):K4=N1(J):S =
4 ELSE S = 5
5520 RETURN
5530 REM  ***************************************
5540 REM            LOOP  FLOW  CORRECTION
5550 REM  ***************************************
5560 ITER = 0
5570 REM      #####  START NEW ITERATION  #####
5580 K1 = 1:ERRMAX=0
5590 ITER = ITER + 1
5600 REM     ##### START NEW LOOP #####
5610 H = 0
5620 HDOT = 0
5630 FOR I = K1 TO NEXTLP
5640 SIGN = 1
5650 J = LOOP(I)
5660 IF J = 0 THEN GOTO 5790
5670 IF J < 0  THEN  SIGN=-SIGN:J=-J
5680 IF ABS(Q(J)) < 1E-15  THEN Q(J)=0
5690 IF PSR<100 THEN GOTO 5740
5695 IF J>PIPENO THEN GOTO 5740
5700 IF Q(J)<>0 THEN V=ABS(Q(J))/(.785*(DIA(J)/METRIC)^2)
5710 D=DIA(J):L=LENGTH(J):K=SR(J):GOSUB 7960
5720 C(J)=CPP
5730 REM      #####  TWO POINTS OF FIXED HEAD IN A LOOP #####
5740 IF DEVICE(J) = -1  THEN  K2=N2(J):K3=N1(J):HDLSS=(HEAD(K2)-HEAD(K3))*SIGN:
H=H+HDLSS:GOTO 5770
5750 IF Q(J)<>0 THEN TEMP=ABS(Q(J)):HDLSS=SIGN*C(J)*Q(J)*(TEMP^(POW-1)):
HDLSSD=POW*C(J)*(TEMP^(POW-1)):H=H-HDLSS:HDOT=HDOT+HDLSSD
5760 IF DEVICE(J) > 0  THEN GOSUB 7210
5770 NEXT I
5780 REM      ##### CORRECT FLOW IN LOOP TO BALANCE HEAD #####
5790 DELTAQ = -H/HDOT
5800 IF ERRMAX < ABS(H)  THEN ERRMAX = ABS(H)
5810 FOR I = K1 TO NEXTLP
```

```
5820 J = LOOP(I)
5830 IF J = 0  THEN  GOTO 5880
5840 IF J < 0  THEN J=-J:Q(J)=Q(J)+2*DELTAQ
5850 Q(J) = Q(J) - DELTAQ
5860 NEXT I
5870 REM      ##### START NEW LOOP #####
5880 K1 = I + 1
5890 IF K1 < NEXTLP GOTO 5610
5900 IF ITER = 500  GOTO 5950
5910 PRINT "ITER = ";ITER;:PRINT"   ERRMAX (m) = ";ERRMAX
5920 IF ERRMAX > ACRACY  GOTO 5580
5930 IF ITER < 20 THEN GOTO 5580
5940 REM      ##### PRINT MAX ERROR AND NO OF ITERATIONS #####
5950 PRINT "**** NO OF ITERATIONS = " ITER
5960 PRINT " MAXIMUM ERROR = " ERRMAX "******"
5970 REM        #####  PRINT TITLE BLOCKS #####
5980 PRINT CHR$(12)
5990 PRINT "           ***********************************    "
6000 PRINT "           ***     OUTPUT  OF RESULTS       ***   "
6010 PRINT "           ***********************************    "
6020 PRINT "     "
6030 SFULL=0
6040 FOR I = 1 TO PIPENO
6050 IF PSR>0 THEN GOTO 6070
6060 IF SFULL=0 THEN PRINT"PIPE NO  NODE1  NODE2   FLOW   LENGTH  DIAMETER
C-VALUE  CONTROL":SFULL=1:PRINT"                              ";DIMQ$;"    ";DIML$;"
    ";DIMD$:PRINT:GOTO 6080
6070 IF SFULL=0 THEN PRINT"PIPE NO  NODE1  NODE2   FLOW   LENGTH  DIAMETER
ROUGHNESS CONTROL":SFULL=1:PRINT"                             ";DIMQ$;"     ";DIML$;"
     ";DIMD$;"          ";DIMD$:PRINT
6080 FORM$="#####    ####   #### ####.###  #####   ####.#     ###.## "
6090 TEMP = Q(I) / FACTOR
6100 VEL = Q(I) * 1.27324 / ((DIA(I)/METRIC)^2)
6110 IF VEL < 0  THEN  VEL = ABS(VEL)
6120 IF DEVICE(I) = 0  THEN CNLOP$="      ":GOTO 6160
6130 IF DEVICE(I) = 20 THEN CNLOP$= "    PUMP":GOTO 6160
6140 IF DEVICE(I) = 10 THEN CNLOP$= "    NR VALVE" :GOTO 6160
6150 IF DEVICE(I) = 1 THEN CNLOP$ = "    PR VALVE"
6160 IF PSR>0 THEN GOTO 6180
6170 PRINT USING FORM$;I,N1(I),N2(I),TEMP,LENGTH(I),DIA(I),CVALUE(I);:PRINT
CNLOP$:GOTO 6190
6180 PRINT USING FORM$;I,N1(I),N2(I),TEMP,LENGTH(I),DIA(I),SR(I);:PRINT CNLOP$
6190 IF I\15=I/15 THEN SFULL=0 ELSE GOTO 6220
6200 PRINT:PRINT"PRESS THE SPACE BAR TO CONTINUE"
6210 ZZ$=INPUT$(1):CLS
6220 NEXT I
6230 PRINT:PRINT:PRINT
6240 INPUT" DO YOU WISH TO PRINT THE ABOVE OUTPUT ON PAPER (Y/N)";ANS$
6250 IF ANS$="Y" THEN GOSUB 6890
6260 REM  ***************************************************
6270 REM  * COMPUTE SUPPLIES/DEMANDS FROM COMPUTED PIPE FLOWS *
6280 REM  ***************************************************
6290 FOR I = 1 TO NODENO
6300 DEMAND(I) = 0
6310 NEXT I
6320 FOR I = 1 TO PIPENO
6330 K1 = N1(I)
6340 K2 = N2(I)
6350 TEMP = Q(I) / FACTOR
6360 DEMAND(K1) = DEMAND(K1) - TEMP
6370 DEMAND(K2) = DEMAND(K2) + TEMP
```

```
6380 NEXT I
6390 PRINT
6400 REM     *****************************
6410 REM     * COMPUTATION OF NODE HEADS *
6420 REM     *****************************
6430 REM     ##### NO FIXED HEAD GIVEN,ASSUME FIXED HEAD OF 1000 #####
6440 IF HEADNO>0 THEN GOTO 6480
6450 FOR I=1 TO NODENO
6460 IF T(I)=1 THEN HEAD(I)=1000:GOTO 6480
6470 NEXT I
6480 LEVEL=0
6490 SFULL=0
6500 FOR I =1 TO PIPENO STEP 1
6510 K1=N1(I)
6520 K2=N2(I)
6530 HDOT=Q(I)
6540 IF HEAD(K1)<>0 THEN GOTO 6560
6550 IF HEAD(K2)=0 THEN LEVEL=1:GOTO 6660 ELSE K3=K1:K1=K2:K2=K3:HDOT=-HDOT:GOTO
6580
6560 IF HEAD(K2)<>0 THEN GOTO 6660
6570 REM       ##### APPLY HEADLOSS/HEADRISE FOR PUMPS AND VALVES #####
6580 SIGN=1
6590 IF HDOT<0  THEN SIGN=-SIGN:HDOT=-HDOT
6600 TEMP=ABS(Q(I))
6610 H=-SIGN*C(I)*HDOT*(TEMP^(POW-1))
6620 IF DEVICE(I)=0 THEN GOTO 6650
6630 K3=DEVICE(I)
6640 IF DEVICE(I)=20 THEN HDRISE=X(I)*Q(I)*Q(I)+Y(I)*Q(I)+Z(I):H=H+HDRISE*SIGN
6650 HEAD(K2)=HEAD(K1)+H
6660 NEXT I
6690 IF LEVEL=1 THEN GOTO 6480
6700 REM        ##### CALCULATE GUAGE HEIGHT #####
6710 FOR I=1 TO NODENO STEP 1
6720 TEMP=HEAD(I)-HEIGHT(I)
6730 IF SFULL=1 THEN GOTO 6770
6740 PRINT"NODE NO   DEMAND   GAUGE   ELEVATION   TOTAL HEAD":SFULL=1
6750 PRINT"           ";DIMQ$;"     m          m            m":PRINT
6760 FORM$="#####   ####.###  ###.##  #####.##    #####.##"
6770 PRINT USING FORM$;I;DEMAND(I);TEMP;HEIGHT(I);HEAD(I)
6780 IF I\15=I/15 THEN SFULL=0 ELSE GOTO 6810
6790 PRINT"PRESS THE SPACE BAR TO CONTINUE"
6800 ZZ$=INPUT$(1)
6810 NEXT I
6820 PRINT:INPUT" DO YOU WISH TO PRINT THE ABOVE DATA ON PAPER (Y/N)";ANS$
6830 IF ANS$="Y" THEN GOSUB 7100
6840 PRINT:PRINT"PRESS THE SPACE BAR TO CONTINUE"
6850 ZZ$=INPUT$(1)
6860 REM       ### RETURNS TO MAIN MENU ####
6870 CLS
6880 GOTO 310
6890 REM      ****** THIS SECTION PRINTS FLOW OUTPUT ON PRINTER ******
6900 IF PSR>0 THEN GOTO 6930
6910 LPRINT"          PIPE NO  NODE1  NODE2   FLOW   LENGTH  DIAMETER  C-VALUE
CONTROL"
6920 LPRINT"                                       ";DIMQ$;"      ";DIML$;"
";DIMD$:GOTO 6950
6930 LPRINT"          PIPE NO  NODE1  NODE2   FLOW   LENGTH  DIAMETER  ROUGHNESS
CONTROL"
6940 LPRINT"                                       ";DIMQ$;"      ";DIML$;"
";DIMD$;"      ";DIMD$
6950 LPRINT
```

```
6960 FORM$="          #####     ####    ####  ####.##  #####   ####.#     ### "
6970 FOR I = 1 TO PIPENO
6980 TEMP = Q(I) / FACTOR
6990 VEL = Q(I) * 1.27324 / ((DIA(I)/METRIC)^2)
7000 IF VEL < 0  THEN  VEL = ABS(VEL)
7010 IF DEVICE(I) = 0  THEN CNLOP$="      ":GOTO 7060
7020 IF DEVICE(I) = 20  THEN CNLOP$= "PUMP":GOTO 7060
7030 IF DEVICE(I) = 10 THEN CNLOP$= "NR VALVE" :GOTO 7060
7040 IF DEVICE(I) = 1 THEN CNLOP$ = "PR VALVE"
7050 IF PSR=100 THEN GOTO 7070
7060 LPRINT USING FORM$;I,N1(I),N2(I),TEMP,LENGTH(I),DIA(I),CVALUE(I);:LPRINT
CNLOP$:GOTO 7080
7070 LPRINT USING FORM$;I,N1(I),N2(I),TEMP,LENGTH(I),DIA(I),SR(I);:LPRINT CNLOP$
7080 NEXT I
7090 RETURN
7100 REM     ***** THIS SECTION PRINTS NODE OUTPUT ON PAPER *****
7110 LPRINT"          NODE NO   DEMAND   GAUGE   ELEVATION   TOTAL HEAD"
7120 LPRINT"                     ";DIMQ$;"      m          m            m"
7130 LPRINT
7140 FORM$="          #####    #####.#  ####.#   #####.#     #####.#"
7150 FOR I=1 TO NODENO STEP 1
7160 TEMP=HEAD(I)-HEIGHT(I)
7170 LPRINT USING FORM$;I;DEMAND(I);TEMP;HEIGHT(I);HEAD(I)
7180 NEXT I
7190 RETURN
7200 REM   *** SUBROUTINE FOR CONTROL DEVICES: PUMPS, NRVs, PRVs ***
7210 REM   *** PUMPS AND PRVs ARE TREATED AS NON-RETURN DEVICES ***
7220 IF ITER < 10 THEN RETURN
7230 IF Q(J) < 0 THEN DELTAQ=Q(J)*SIGN*.6: GOTO 5800
7240 IF DEVICE(J) = 20 THEN GOTO 7270
7250 IF DEVICE(J)=1 THEN GOTO 7350
7260 RETURN
7270 GOSUB 7390
7280 IF Q(J)>QH0 THEN DELTAQ=Q(J)-QH0:GOTO 5800
7290 HDRISE=X(J)*Q(J)^2+Y(J)*Q(J)+Z(J)
7300 HDRISE = HDRISE * SIGN
7310 HRDOT = 2*X(J)*Q(J) + Y(J)
7320 H = H + HDRISE
7330 HDOT = HDOT - HRDOT
7340 RETURN
7350 HDLSS=SIGN*X(J)*Q(J)*ABS(Q(J))
7360 HDLSSD=2*X(J)*ABS(Q(J)): H=H-HDLSS: HDOT=HDOT+HDLSSD
7370 RETURN
7380 REM*****CALCULATION OF Q(J) FOR ZERO HDRISE*******
7390 QH0=0
7400 FOR II=1 TO 10
7410 FP=X(J)*QH0^2+Y(J)*QH0+Z(J)
7420 DFP=2*X(J)*QH0+Y(J)
7430 DEL=-FP/DFP
7440 QH0=QH0+DEL
7450 NEXT II
7460 HDRISE=X(J)*QH0^2+Y(J)*QH0+Z(J)
7470 RETURN
7480 CLS
7490 PRINT:PRINT"                          NOTES ON PROGRAM USE":PRINT
7500 PRINT" This program calculates the flow distribution in pipe networks using
the"
7510 PRINT" Hardy Cross loop flow correction procedure. The program selects "
7520 PRINT" a set of loops for the network and also an initial estimate of flow"
7530 PRINT" distribution. Thus, only that information which is required to
define the"
```

```
7540 PRINT" problem, needs to be input by the user."
7550 PRINT" First, make a sketch of your pipe network and assign a number to
each node"
7560 PRINT" (pipe junction), numbering the nodes consecutively 1,2,3.....NN.
Then draw"
7570 PRINT" up a pipe data table, which has the following data for each pipe:
head node"
7580 PRINT" number, tail node number, length, diameter and C-value or roughness.
When"
7590 PRINT" this table is completed, number the pipes consecutively 1,2,3...NP."
7600 PRINT" Booster pump data is input by assigning values to the constants in
the"
7610 PRINT" characteristic eqn.: H=A*Q**2+B*Q+C. Pressure-reducing valve data is
input"
7620 PRINT" by assigning a value to the constant C in the relation H=C*Q**2."
7630 PRINT" PROGRAM OPERATION: The program is operated through a main menu which
offers"
7640 PRINT" the user a choice of operations. Each start-up of the program for a
non-"
7650 PRINT" filed network is initiated by selecting operation 1 (NETWORK) from
the"
7660 PRINT" main menu. The input data is filed on disk for future use by
selecting"
7670 PRINT" operation 9 (FILE) from the main menu."
7680 PRINT
7690 PRINT"                    PRESS THE SPACE BAR TO CONTINUE"
7700 ZI$=INPUT$(1)
7710 CLS
7720 GOTO 210
7730 REM recovery of data from files
7740 CLS
7750 PRINT:PRINT"THE DATA STORED ON FILE INCLUDES PIPE, HEAD, ELEVATION, "
7760 PRINT"SUPPLY/DEMAND AND OTHER NETWORK DATA REQUIRED FOR ANALYSIS. IT"
7770 PRINT"ALSO INCLUDES DATA ON PUMPS, NR AND PR VALVES."
7780 PRINT"TO IDENTIFY THIS DATA FOR RETRIEVAL YOU WILL BE ASKED TO ENTER"
7790 PRINT"THE TITLE OF THE NETWORK.": PRINT:PRINT
7800 INPUT "WHAT IS THE NETWORK TITLE";TITLE$
7810 NET$="NET"+LEFT$(TITLE$,4):PIP$="PIP"+LEFT$(TITLE$,4)
7820 OPEN "I",1,NET$:INPUT# 1,FACTOR,ACRACY,CONST,CONTROL,POW,METRIC,DIMD$,
DIMV$,DIMQ$,DIML$,S,PSR: CLOSE 1
7830 OPEN "I",1,PIP$
7840 INPUT# 1,PIPENO,NODENO,HEADNO
7850 FOR I=1 TO PIPENO
7860 IF PSR>0 THEN GOTO 7880
7870 INPUT# 1,N1(I),N2(I),LENGTH(I),DIA(I),CVALUE(I),DEVICE(I),X(I),Y(I),Z(I),
DEMAND(I),HEAD(I),HEIGHT(I):GOTO 7890
7880 INPUT# 1,N1(I),N2(I),LENGTH(I),DIA(I),SR(I),DEVICE(I),X(I),Y(I),Z(I),
DEMAND(I),HEAD(I),HEIGHT(I)
7890 NEXT I
7900 CLOSE 1
7910 GOTO 310
7920 REM            #### THIS ENDS PNA ####
7930 PRINT:PRINT
7940 PRINT"END PNA  THANK YOU"
7950 END
7960 REM*** SUBROUTINE TO CALCULATE C(I), BASED ON THE DARCY-WEISBACH EQN.***
7970 RHO=1000: MU=.001307
7980 KVISC=MU/RHO
7990 RE=ABS(V)*.001*D/KVISC
8000 IF RE>2000 THEN GOTO 8010 ELSE F=64/RE: GOTO 8140
8010 UPV=.5
```

```
8020 LOV=0
8030 F=(UPV+LOV)/2
8040 Y=.5/SQR(F)
8050 X=((K/(3.7*D))+2.51/(RE*SQR(F)))
8060 W=Y+LOG(X)/LOG(10)
8070 IF W<0 THEN UPV=F
8080 IF W>0 THEN LOV=F
8090 Z=(UPV+LOV)/2
8100 E=ABS(Z-F)
8110 IF E<.0001 THEN GOTO 8130
8120 F=Z: GOTO 8040
8130 F=Z
8140 CPP=8*F*L/(96.82*(D/METRIC)^5)
8150 RETURN
```

Listing of program **WATHAM .BAS**

```
10     REM PROGRAM WATHAM.BAS
20     CLS
30     PRINT"             ********************************              "
40     PRINT"             * WATERHAMMER ANALYSIS PROGRAM *              "
50     PRINT"             *                              *              "
60     PRINT"             *          WATHAM.BAS          *              "
70     PRINT"             ********************************              "
80     PRINT
90     FOR I=1 TO 3000:NEXT I
100    PRINT TAB(5)"This program computes the transient flow and pressure"
110    PRINT TAB(5)"conditions in a suction/rising main system due to pump"
120    PRINT TAB(5)"cut-out. It relates to a system bounded by reservoirs or"
130    PRINT TAB(5)"points of fixed head at its upstream and downstream ends"
140    PRINT TAB(5)"and which includes a non-return valve on the pump delivery."
150    PRINT
160    PRINT TAB(5)"It also offers the option of analysis of the above system"
170    PRINT TAB(5)"protected by an air vessel connected to the rising main on"
180    PRINT TAB(5)"the downstream side of the non-return valve.":PRINT
190    PRINT TAB(5)"The program computes the initial steady flow in the system"
200    PRINT TAB(5)"from the given pump and system data.":PRINT
210    PRINT TAB(5)"Pump performance is defined by quadratic expressions"
220    PRINT TAB(5)"for head H and torque T, as functions of discharge Q:"
230    PRINT
240    PRINT TAB(5)"      HEAD         H = A0 + A1*Q + A2*Q*Q"
250    PRINT TAB(5)"      TORQUE       T = B0 + B1*Q + B2*Q*Q"
260    PRINT:PRINT TAB(13)"PRESS THE SPACE BAR TO CONTINUE":PRINT
270    Z$=INPUT$(1)
280    PRINT TAB(5)"It should be noted that program computations are based"
290    PRINT TAB(5)"on piezometric head values and, hence, locations at"
300    PRINT TAB(5)"which the gauge pressure may have dropped to vapour"
310    PRINT TAB(5)"level are not sign-posted by the program.":PRINT
320    PRINT TAB(5)"The piezometric head datum is that to which the specified"
330    PRINT TAB(5)"upstream and downstream reservoir levels relate."
340    PRINT:PRINT TAB(10)"PRESS THE SPACE BAR TO CONTINUE"
350    Z$=INPUT$(1)
360    CLS
370    DIM QS(110),HS(110),QD(110),HD(110),QSP(110),HSP(110),QDP(110),HDP(110)
```

```
380   DIM HSMIN(110),HSMAX(110),HDMIN(110),HDMAX(110),WH(110),WB(110),HP(2000),
T(2000),PQ(2000)
390   PRINT:PRINT "WHICH SYSTEM DO YOU WISH TO ANALYSE ?":PRINT
400   PRINT TAB(10) "1.  PUMP/RISING MAIN SYSTEM"
410   PRINT TAB(10) "2.  PUMP/AIR VESSEL/RISING MAIN SYSTEM":PRINT
420   INPUT "ENTER 1 OR 2, AS APPROPRIATE";SEL
430   IF SEL>2 THEN GOTO 420
440   PRINT:PRINT "INPUT OF PUMP CHARACTERISTICS"
450   PRINT "THE PUMP HEAD/DISCHARGE CORRELATION IS AS FOLLOWS:":PRINT
460   PRINT TAB(5) "H = A0 + A1*Q + A2*Q*Q (Units: H(m), Q(m**3/s))":PRINT
470   INPUT"ENTER VALUES OF A0, A1 AND A2, SEPARATED BY COMMAS";A0R,A1R,A2
480   IF SEL=2 THEN GOTO 520
490   PRINT:PRINT"THE PUMP TORQUE/DISCHARGE CORRELATION IS AS FOLLOWS:":PRINT
500   PRINT TAB(5)"T = B0 + B1*Q + B2*Q*Q  (Units: T (Nm), Q (m**3/s))":PRINT
510   INPUT"ENTER VALUES OF B0,B1 AND B2, SEPARATED BY COMMAS";B0R,B1R,B2:PRINT
520   INPUT"ENTER DUTY POINT HEAD (m)";PHR
530   INPUT"ENTER DUTY POINT DISCHARGE (m**3/s)";PQR
540   IF SEL=2 GOTO 580
550   INPUT"ENTER DUTY POINT TORQUE (Nm)";PTR
560   INPUT"ENTER PUMP SPEED (rpm)";PS0
570   INPUT"ENTER MOMENT OF INERTIA OF PUMP SET (kg*m*m)";SECM
580   PRINT: PRINT"PIPE DATA INPUT"
590   INPUT"ENTER SUCTION PIPE LENGTH (m),DIAM(m),WALL THICKNESS (m),WALL
ROUGHNESS (m),      YOUNG'S MOD (N/m**2):";LS,DS,TS,ZS,ES
600   INPUT"ENTER DELIVERY PIPE LENGTH (m), DIAM (m), WALL THICKNESS (m), WALL
ROUGHNESS(m),  YOUNG'S MOD (N/m**2):";LD,DD,TD,ZD,ED
610   PRINT:INPUT"ENTER  SUCTION RESERVOIR LEVEL (mOD)";SRL
620   INPUT"DELIVERY RESERVOIR LEVEL (mOD)";DRL
630   IF SEL=2 THEN GOSUB 3260
640   PRINT:PRINT"ENTER NO OF SEGMENTS INTO WHICH DELIVERY MAIN IS DIVIDED"
650   INPUT"THIS SHOULD BE 10 OR A MULTIPLE THEREOF";ND:PRINT
660   INPUT"ENTER NO OF COMPUTATION ITERATIONS:";NITER
670   PRINT:PRINT"DATA ENTRY NOW COMPLETE"
680   PRINT:PRINT TAB(10)"....computation in progress, please wait....":PRINT
690   REM COMPUTATION OF PUMP DUTY POINT
700   AS=.785*DS*DS: AD=.785*DD*DD: PS=PS0
710   Y=1.31E-06: DE=1000: FK=2.2E+09: G=9.810001: BLF=1.35: TCUM=0:PI=3.142
720   KK=0: PQS=PQR
730   A0=A0R:A1=A1R:B0=B0R:B1=B1R
740   REM COMPUTATION OF PUMP DUTY POINT
750   FOR KI=1 TO 10
760   VS=PQS/AS: VD=PQS/AD
770   V=VS: D=DS: Z=ZS: GOSUB 2800
780   KP=X*LS/(19.62*DS*AS*AS)
790   V=VD:D=DD:Z=ZD:GOSUB 2800
800   KP=KP+X*LD/(19.62*DD*AD*AD)
810   F1=KP*PQS*PQS+DRL-SRL-A0-A1*PQS-A2*PQS*PQS
820   DF=2*KP*PQS-A1-2*A2*PQS
830   PQS=PQS-F1/DF
840   NEXT KI
850   PHS=A0+A1*PQS+A2*PQS*PQS
860   IF SEL=2 THEN GOTO 880
870   PTS=B0+B1*PQS+B2*PQS*PQS: PT=PTS
880   PRINT"PUMP STEADY DISCHARGE (m**3/s) IS "PQS
890   PRINT"PUMP STEADY HEAD (m) IS "PHS:PRINT
900   REM SEGMENTAL DIVISION FOR NUMERICAL COMPUTATION
910   ND1=ND+1
920   IF LD/ND>(5*LS) THEN SMAIN$="SHORT"
930   WVS=SQR(1/(DE*(1/FK+DS/(TS*ES))))
940   WVD=SQR(1/(DE*(1/FK+DD/(TD*ED))))
950   PRINT"SUCTION MAIN WAVE SPEED (m/s)= "WVS
```

```
960     PRINT"DELIVERY MAIN WAVE SPEED (m/s)= "WVD:PRINT
970     DELT=LD/(WVD*ND)
980     IF SMAIN$="SHORT" THEN NS1=1: GOTO 1010
990     SON=LS/(WVS*DELT)
1000    NS=INT(SON)+1: NS1=NS+1
1010    FOR I=1 TO NS1
1020    HSMIN(I)=5000: NEXT I
1030    FOR I=1 TO ND1
1040    HDMIN(I)=5000:NEXT I
1050    REM STEADY STATE CONDITIONS
1060    IF SMAIN$="SHORT" GOTO 1120
1070    V=PQS/AS: D=DS: Z=ZS: GOSUB 2800
1080    FOR I=1 TO NS1
1090    QS(I)=PQS
1100    HS(I)=SRL-(I-1)*X*(LS/NS)*PQS*PQS/(19.62*DS*AS*AS)
1110    NEXT I
1120    HS(NS1)=SRL:V=PQS/AD: D=DD: Z=ZD: GOSUB 2800
1130    HD(1)=DRL+X*LD*PQS*PQS/(19.62*DD*AD*AD): HDMAX(1)=HD(1):PHD=HD(1)
1140    QD(1)=PQS
1150    FOR I=2 TO ND1
1160    QD(I)=PQS
1170    HD(I)=HD(I-1)-X*(LD/ND)*PQS*PQS/(19.62*DD*AD*AD)
1180    NEXT I
1190    HP(1)=HD(1):PQ(1)=PQS:T(1)=0!
1200    IF SEL=2 GOTO 1250
1210    PQT=PQS
1220    TFACT=(30*DELT)/(PI*SECM)
1230    PRINT:GOTO 2140
1240    KK=KK+1
1250    REM TRANSIENT FLOW CALCS
1260    IF SMAIN$="SHORT" GOTO 1280
1270    BS=WVS/(G*AS): FS=(LS/NS)/(19.62*DS*AS*AS)
1280    BD=WVD/(G*AD): FD=(LD/ND)/(19.62*DD*AD*AD)
1290    REM PUMP BOUNDARY CONDITION
1300    TOL=.0002
1310    FOR COUNT =1 TO NITER
1320    IF SEL=1 THEN GOTO 1330 ELSE GOTO 2990
1330    IF PQT<0 GOTO 1620
1340    IF SMAIN$="SHORT" GOTO 1370
1350    V=QS(NS1)/AS: D=DS: Z=ZS: GOSUB 2800
1360    RS=X*FS
1370    V=QD(2)/AD: D=DD: Z=ZD: GOSUB 2800
1380    RD=X*FD
1390    IF SMAIN$="SHORT" GOTO 1410
1400    HCP=HS(NS)+BS*QS(NS)-RS*QS(NS)*ABS(QS(NS))
1410    HCM=HD(2)-BD*QD(2)+RD*QD(2)*ABS(QD(2))
1420    PS=PS-TFACT*PT: SF=PS/PS0
1430    A0=A0R*SF*SF:A1=A1R*SF
1440    B0=B0R*SF*SF:B1=B1R*SF
1450    IF SMAIN$="SHORT" GOTO 1520
1460    FOR KI=1 TO 10
1470    F1=(HCM-HCP-A0)+(BS+BD-A1)*PQT-A2*PQT*PQT
1480    DF=(BS+BD-A1)-2*A2*PQT
1490    PQT=PQT-F1/DF
1500    NEXT KI
1510    GOTO 1570
1520    FOR KI=1 TO 10
1530    F1=(HCM-SRL-A0)+(BD-A1)*PQT-A2*PQT*PQT
1540    DF=(BD+A1)-2*A2*PQT
1550    PQT=PQT-F1/DF
1560    NEXT KI
```

```
1570  IF SMAIN$="SHORT" THEN HSP(NS1)=SRL:QSP(NS1)=PQT: GOTO 1590
1580  QSP(NS1)=PQT:HSP(NS1)=HCP-BS*QSP(NS1)
1590  QDP(1)=PQT:HDP(1)=HCM+BD*QDP(1)
1600  PT=B0+B1*PQT+B2*PQT*PQT
1610  GOTO 1670
1620  QSP(NS1)=0: QDP(1)=0
1630  V=QD(2)/AD: D=DD: Z=ZD: GOSUB 2800
1640  RD=X*FD
1650  HCM=HD(2)-BD*QD(2)+RD*QD(2)*ABS(QD(2))
1660  HDP(1)=HCM:  HSP(NS1)=HS(NS1)
1670  REM SUCTION RESERVOIR END
1680  IF SMAIN$="SHORT" GOTO 1740
1690  HSP(1)=SRL
1700  V=QS(1)/AS: D=DS: Z=ZS: GOSUB 2800
1710  RS=X*FS
1720  HCM=HS(2)-BS*QS(2)+RS*QS(2)*ABS(QS(2))
1730  QSP(1)=(HSP(1)-HCM)/BS
1740  REM DELIVERY RESERVOIR END
1750  HDP(ND1)=DRL
1760  V=QD(ND1)/AD: D=DD: Z=ZD: GOSUB 2800
1770  RD=X*FD
1780  HCP=HD(ND1-1)+BD*QD(ND1-1)-RD*QD(ND1-1)*ABS(QD(ND1-1))
1790  QDP(ND1)=(HCP-HDP(ND1))/BD
1800  IF SEL =2 THEN GOTO 1910
1810  REM INTERNAL NODES SUCTION MAIN
1820  IF SMAIN$="SHORT" GOTO 1910
1830  FOR I=2 TO (NS1-1)
1840  V=QS(I)/AS: D=DS: Z=ZS: GOSUB 2800
1850  RS=X*FS
1860  CP=HS(I-1)+BS*QS(I-1)-RS*QS(I-1)*ABS(QS(I-1))
1870  CM=HS(I+1)-BS*QS(I+1)+RS*QS(I+1)*ABS(QS(I+1))
1880  HSP(I)=(CP+CM)/2
1890  QSP(I)=(HSP(I)-CM)/BS
1900  NEXT I
1910  REM INTERNAL NODES DELIVERY MAIN
1920  FOR I=2 TO (ND1-1)
1930  V=QD(I)/AD: D=DD: Z=ZD: GOSUB 2800
1940  RD=X*FD
1950  CP=HD(I-1)+BD*QD(I-1)-RD*QD(I-1)*ABS(QD(I-1))
1960  CM=HD(I+1)-BD*QD(I+1)+RD*QD(I+1)*ABS(QD(I+1))
1970  HDP(I)=(CP+CM)/2
1980  QDP(I)=(HDP(I)-CM)/BD
1990  NEXT I
2000  IF SEL =2 THEN GOTO 2060
2010  REM RESET H AND Q VALUES
2020  FOR I=1 TO NS1
2030  HS(I)=HSP(I)
2040  QS(I)=QSP(I)
2050  NEXT I
2060  TCUM=TCUM+DELT
2070  FOR I=1 TO ND1
2080  HD(I)=HDP(I)
2090  QD(I)=QDP(I)
2100  NEXT I
2110  IF SEL=2 THEN GOTO 2220
2120  IF COUNT\20=COUNT/20 THEN GOTO 2140 ELSE GOTO 2160
2130  IF COUNT>1 GOTO 2160
2140  PRINT"PUMP DISCHARGE   PUMP SPEED   HEAD U/S PUMP  HEAD D/S PUMP   TIME"
2150  PRINT"   (m**3/s)          (rpm)          (m)            (m)        (s)"
2160  PRINT USING"    +#.####  ";QD(1);
2170  PRINT USING"      +####.#";PS;
```

```
2180  PRINT USING"         +###.# ";HS(NS1),HD(1);
2190  PRINT USING"   ###.## ";TCUM
2200  IF KK=0 THEN GOTO 1240
2210  IF SEL=1 THEN GOTO 2350
2220  REM PRINTING AIR VESSEL SYSTEM OUTPUT ON SCREEN
2230  IF COUNT\20=COUNT/20 THEN GOTO 2250
2240  IF COUNT>1 GOTO 2280
2250  PRINT"TIME   WAT VOL   HD1   HD2   HD3   HD4   HD5   HD6   HD7   HD8
HD9   HD10"
2260  PRINT" (s)   IN AV     (m)   (m)   (m)   (m)   (m)   (m)   (m)   (m)
(m)   (m)"
2270  PRINT"        (m^3)"
2280  PRINT USING"##.##  ";TCUM;
2290  PRINT USING"###.##  ";(VAV-TAV);
2300  FOR I=1 TO 9
2310  PRINT USING" ###.#";HD(I);
2320  NEXT I
2330  PRINT USING" ###.#";HD(10)
2340  IF SEL=2 THEN GOTO 2390
2350  FOR I=1 TO NS1
2360  IF HSMIN(I)>HS(I) THEN HSMIN(I)=HS(I)
2370  IF HSMAX(I)<HS(I) THEN HSMAX(I)=HS(I)
2380  NEXT I
2390  FOR I=1 TO ND1
2400  IF HDMIN(I)>HD(I) THEN HDMIN(I)=HD(I)
2410  IF HDMAX(I)<HD(I) THEN HDMAX(I)=HD(I)
2420  NEXT I
2430  HP(COUNT)=HD(1):T(COUNT)=TCUM:PQ(COUNT)=QD(1)
2440  NEXT COUNT
2450  INC=ND/10: K=1
2460  IF SEL=2 THEN GOTO 2610
2470  PRINT:PRINT"NODE            RISING MAIN           SUCTION MAIN"
2480  PRINT"      MAX PRESSURE  MIN PRESSURE  MAX PRESSURE  MIN PRESSURE"
2490  PRINT"            (m)            (m)            (m)          (m)"
2500  FOR I=1 TO ND1
2510  IF I<>K THEN GOTO 2590
2520  K=K+INC
2530  PRINT I TAB(6)
2540  IF I>NS1 THEN GOTO 2580
2550  PRINT USING"     ###.#       ";HDMAX(I),HDMIN(I),HSMAX(I),HSMIN(I)
2560  GOTO 2590
2570  IF I>NS1 THEN GOTO 2590
2580  PRINT USING"     ###.#       ";HDMAX(I),HDMIN(I)
2590  NEXT I
2600  GOTO 2700
2610  REM PRINT MAX AND MIN PRESSURES FOR AIR VESSEL SYSTEM
2620  PRINT:PRINT"NODE     MAX PRESSURE   MIN PRESSURE"
2630  PRINT"              (m)              (m)"
2640  FOR I=1 TO ND1
2650  IF I<>K THEN GOTO 2690
2660  K=K+INC
2670  PRINT I TAB(6)
2680  PRINT USING"       +###.#   ";HDMAX(I),HDMIN(I)
2690  NEXT I
2700  PRINT:PRINT"DO YOU WISH TO PLOT THE PRESSURE TRANSIENT"
2710  INPUT"AT THE PUMP DELIVERY ON THE SCREEN (Y/N)";ANS$
2720  IF ANS$="N" THEN GOTO 2750
2730  IF ANS$<>"Y" GOTO 2700
2740  GOSUB 3360
2750  PRINT
2760  INPUT"DO YOU WISH TO RE-RUN THE PROGRAM (Y/N)";ANS$
```

```
2770  IF ANS$="N" GOTO 2790
2780  IF ANS$<>"Y" THEN GOTO 2760 ELSE GOTO 390
2790  END
2800  REM SUBROUTINE FRICTION FACTOR
2810  IF V<>0 THEN GOTO 2830 ELSE GOTO 2820
2820  X=0: RETURN
2830  R=ABS(V)*D/Y
2840  IF R>2000 THEN GOTO 2860
2850  X=64/R: RETURN
2860  TOP=.5: BOT=0
2870  X=(TOP+BOT)/2
2880  YOP=.5/SQR(X)
2890  XXX= (Z/(3.7*D))+(2.51/(R*SQR(X)))
2900  WW=YOP+LOG(XXX)/LOG(10)
2910  IF WW<0 THEN TOP=X
2920  IF WW>0 THEN BOT=X
2930  ZZ=(TOP+BOT)/2
2940  E=ABS((ZZ-X)/X)
2950  IF E<.005 THEN GOTO 2970
2960  X=ZZ: GOTO 2880
2970  X=ZZ
2980  RETURN
2990  REM COMPUTING TRANSIENTS AT AIR VESSEL NODE
3000  IF COUNT>1 GOTO 3030
3010  TAP=HD(1)-HPD-HSP0+10.329: TAV=VA0: HSP=HSP0: QDP(1)=QD(1): QD(1)=0
3020  CONST=TAP*VA0^BLF
3030  V=QD(2)/AD: D=DD: Z=ZD: GOSUB 2800
3040  RD=X*FD
3050  CM=HD(2)-BD*QD(2)+RD*QD(2)*ABS(QD(2))
3060  K1=0
3070  FOR I=1 TO 10
3080  DELV=DELT*(QD(1)+QDP(1))/2
3090  TTAV=TAV+DELV
3100  TAP=CONST/(TTAV^BLF)
3110  HSP=HSP-DELV/ASA
3120  FTAP=TAP-10.329+HSP+HPD
3130  IF QDP(1)>0 THEN C=CO ELSE C=-CI
3140  FQ=C*QDP(1)*QDP(1)+BD*QDP(1)+CM-FTAP
3150  DFQ=2*C*QDP(1)+BD
3160  QDP(1)=QDP(1)-FQ/DFQ
3170  NEXT I
3180  IF ABS(FQ/DFQ)<TOL GOTO 3220
3190  K1=K1+1
3200  IF K1<4 THEN GOTO 3070 ELSE PRINT "TROUBLE WITH AIR VESSEL"
3210  IF K1>4 THEN STOP
3220  HDP(1)=CM+BD*QDP(1)
3230  TAV=TTAV
3240  IF COUNT=1 THEN QD(1)=PQS
3250  GOTO 1740
3260  REM INPUT OF AIR VESSEL DATA
3270  PRINT
3280  INPUT"ENTER VOLUME OF AIR VESSEL (m^3)";VAV
3290  INPUT"ENTER INITIAL VOLUME OF AIR IN VESSEL (m^3)";VA0
3300  INPUT"ENTER AIR VESSEL X-SECT AREA (m^2)";ASA
3310  INPUT"ENTER LEVEL OF PUMP DISCHARGE PIPE AT AIR VESSEL (mOD)";HPD
3320  INPUT"HT OF WATER IN VESSEL ABOVE PUMP DISCHARGE PIPE (m)";HSP0
3330  INPUT"THROTTLE COEFF FOR OUTFLOW FROM VESSEL";CO
3340  INPUT"THROTTLE COEFF FOR INFLOW TO VESSEL";CI
3350  RETURN
3360  REM PLOTING RESULTS ON THE SCREEN
3370  SCREEN 8,3:CLS:KEY OFF
```

```
3380  LINE(80,16)-(80,176)
3390  LINE(80,144)-(480,144)
3400  Y=16
3410  FOR I=1 TO 11
3420  LINE(80,Y)-(88,Y)
3430  Y=Y+16
3440  NEXT I
3450  LOCATE 3,6:PRINT 1.6
3460  LOCATE 19,8:PRINT 0
3470  LOCATE 23,5:PRINT -.4
3480  LOCATE 7,4:PRINT"H/HS"
3490  LOCATE 9,6:PRINT 1!"
3500  X=88
3510  FOR I=1 TO 50
3520  LINE(X,142)-(X,146)
3530  IF I\5=I/5 THEN LINE(X,141)-(X,147)
3540  X=X+8
3550  NEXT I
3560  FOR I=1 TO 10
3570  XX=9+5*I
3580  LOCATE 20,XX
3590  PRINT 2*I
3600  NEXT I
3610  LOCATE 22,29:PRINT"TIME (s)"
3620  FOR COUNT=1 TO NITER
3630  PY=144-80*HP(COUNT)/PHD:PX=80+20*T(COUNT)
3640  PSET(PX,PY)
3650  NEXT COUNT
3660  LOCATE 24,10:PRINT "PRESS THE SPACE BAR TO RETURN TO PROGRAM CONTROL"
3670  Z$=INPUT$(1)
3680  SCREEN 0:KEY ON
3690  RETURN
```

Listing of program **OCSF.BAS**

```
10     REM      **** PROGRAM OCSF ****
20     PRINT"Program OCSF":PRINT
30     PRINT "THIS PROGRAM COMPUTES DISCHARGE, FLOW DEPTH AND FRICTION"
40     PRINT "SLOPE IN OPEN CHANNEL STEADY UNIFORM FLOW USING EITHER"
50     PRINT "THE MANNING OR DARCY-WEISBACH EQUATIONS.":PRINT:PRINT
60     PRINT "DO YOU WISH TO USE:  1 MANNING  OR   2 DARCY-WEISBACH ?"
70     INPUT "ENTER 1 OR 2, AS APPROPRIATE"; NUMB
80     IF NUMB =1 THEN FORM$="M": INPUT"ENTER MANNING N-VALUE";MN
90     IF NUMB=2 THEN FORM$="DW": INPUT"ENTER WALL ROUGHNESS (mm)";KS
100    KS=KS/1000
110    PRINT:PRINT"ENTER CHANNEL DATA:"
120    PRINT "IS SECTION   1 CIRCULAR  2 RECTANGULAR  3 TRAPEZOIDAL ?"
130    INPUT "ENTER 1, 2 OR 3, AS APPROPRIATE"; NO
140    IF NO=1 THEN SECT$="CIRC": INPUT"ENTER DIAMETER (mm)";D: D=D/1000
150    IF NO=2 THEN SECT$="RECT": INPUT"ENTER CHANNEL WIDTH (mm)";B: B=B/1000
160    IF NO=3 THEN SECT$="TRAP": INPUT"ENTER BOTTOM WIDTH (mm)";B: B=B/1000
170    IF NO=3 THEN INPUT"ENTER ANGLE OF SIDE TO HORL (deg)";FI
180    PRINT:PRINT"DO YOU WISH TO COMPUTE:  1 DISCHARGE"
190    PRINT"                              2 FRICTION SLOPE"
200    PRINT"                              3 FLOW DEPTH ?"
210    INPUT"ENTER 1, 2  OR  3, AS APPROPRIATE";NUM
```

```
220     IF NUM=1 THEN INPUT"ENTER FRICTION SLOPE";FS
230     IF NUM=1 THEN INPUT"ENTER FLOW DEPTH (mm)";Y: Y=Y/1000
240     IF NUM=2 THEN INPUT"ENTER DISCHARGE (m**3/s)";Q
250     IF NUM=2 THEN INPUT"ENTER FLOW DEPTH (mm)";Y: Y=Y/1000
260     IF NUM=3 THEN INPUT"ENTER DISCHARGE (m**3/s)";Q
270     IF NUM=3 THEN INPUT"ENTER FRICTION SLOPE";FS
280     PRINT:PRINT".....Computation in progress, please wait....."
290     IF NUM=1 THEN IF FORM$="M" THEN GOTO 350
300     IF NUM=1 THEN IF FORM$="DW" THEN GOTO 400
310     IF NUM=2 THEN IF FORM$="M" THEN GOTO 460
320     IF NUM=2 THEN IF FORM$="DW" THEN GOTO 510
330     IF NUM=3 THEN IF FORM$="M" THEN GOTO 570
340     IF NUM=3 THEN IF FORM$="DW" THEN GOTO 620
350     REM COMPUTE Q USING MANNING EQN
360     GOSUB 670
370     Q=(A/MN)*(RH^.67)*(FS^.5)
380     PRINT:PRINT "DISCHARGE (m**3/s) = ";Q
390     GOTO 1390
400     REM COMPUTE Q USING DARCY-WEISBACH EQN
410     GOSUB 670
420     BR=KS/(14.8*RH) + 2.2E-07/((9.810001*FS)^.5*RH^1.5)
430     Q=-A*(6.2*9.810001*RH*FS)^.5*LOG(BR)
440     PRINT:PRINT"DISCHARGE (m**3/s) = ";Q
450     GOTO 1390
460     REM COMPUTE FRICTION SLOPE USING MANNING EQN
470     GOSUB 670
480     FS=(MN*Q/A)^2*(RH^-1.33)
490     PRINT:PRINT"FRICTION SLOPE = ";FS
500     GOTO 1390
510     REM COMPUTE FRICTION SLOPE USING DARCY-WEISBACH EQN
520     GOSUB 670
530     V=Q/A:K=KS:GOSUB 1230
540     FS=F*Q*Q/(8*9.810001*A*A*RH)
550     PRINT:PRINT"FRICTION SLOPE = ";FS
560     GOTO 1390
570     REM COMPUTE FLOW DEPTH USING MANNING EQN
580     FSR=FS^.5
590     GOSUB 810
600     PRINT:PRINT"FLOW DEPTH (mm) = ";Y*1000
610     GOTO 1390
620     REM COMPUTE FLOW DEPTH USING DARCY-WEISBACH EQN
630     FSR=FS^.5
640     GOSUB 810
650     PRINT:PRINT"FLOW DEPTH (mm) = ";Y*1000
660     GOTO 1390
670     REM CALC OF A AND RH
680     IF SECT$="CIRC" THEN GOTO 720
690     IF SECT$="RECT" THEN A=B*Y:RH=B*Y/(B+2*Y):RETURN
700     IF SECT$="TRAP" THEN A=Y*(B+Y/TAN(FI*3.142/180))
710     RH=A/(B+(2*Y)/SIN(FI*3.142/180)):RETURN
720     HI=3.1416: LO=0!
730     TH=(HI+LO)/2
740     XR=1-(2*Y)/D-COS(TH)
750     IF XR<0 THEN LO=TH
760     IF XR>0 THEN HI=TH
770     Z=(HI+LO)/2
780     IF ABS(Z-TH)>.001 THEN GOTO 730
790     P=D*TH: A=.25*D*D*(TH-.5*SIN(2*TH)):RH=A/P
800     RETURN
810     REM COMPUTE FLOW DEPTH
820     HI=40:LO=.001
```

```
830     IF SECT$="CIRC" THEN GOTO 940
840     IF SECT$="RECT" THEN GOTO 1050
850     IF SECT$="TRAP" THEN GOTO 860
860     Y=(HI+LO)/2
870     A=Y*(B+Y/TAN(FI*3.142/180)): RH=A/(B+(2*Y)/SIN(FI*3.142/180))
880     IF FORM$="M" THEN FSH=MN*Q/(A*RH^.67):GOTO 910
890     IF FORM$="DW" THEN V=Q/A:K=KS:GOSUB 1230
900     FSH=(F/(8*9.810001*RH))^.5*V
910     GOSUB 1130
920     IF ABS(Z-Y)>.0002 THEN GOTO 860
930     RETURN
940     HI=3.1416:LO=.001
950     TH=(HI+LO)/2
960     A=.25*D*D*(TH-.5*SIN(2*TH))
970     P=D*TH:RH=A/P
980     IF FORM$="M" THEN FSH =MN*Q/(A*RH^.67):GOTO 1010
990     IF FORM$="DW" THEN V=Q/A:K=KS:GOSUB 1230
1000    FSH=(F/(8*9.810001*RH))^.5*V
1010    GOSUB 1190
1020    IF ABS(FSH-FSR)/FSR>.001 THEN GOTO 950
1030    Y=.5*D*(1-COS(TH))
1040    RETURN
1050    Y=(HI+LO)/2
1060    A=B*Y: RH=A/(B+2*Y)
1070    IF FORM$="M" THEN FSH=MN*Q/(A*RH^.67):GOTO 1100
1080    IF FORM$="DW" THEN V=Q/A:K=KS:GOSUB 1230
1090    FSH=(F/(8*9.810001*RH))^.5*V
1100    GOSUB 1130
1110    IF ABS(Z-Y)>.0002 THEN GOTO 1050
1120    RETURN
1130    REM INTERVAL-HALFING ROUTINE
1140    WW=FSH-FSR
1150    IF WW>0 THEN LO=Y
1160    IF WW<0 THEN HI=Y
1170    Z=(HI+LO)/2
1180    RETURN
1190    WW=FSH-FSR
1200    IF WW>0 THEN LO=TH
1210    IF WW<0 THEN HI=TH
1220    RETURN
1230    REM ********* SUBROUTINE FRICTION FACTOR *********
1240    KVISC=1.307E-06
1250    UPV=.5
1260    LOV=0
1270    F=(UPV+LOV)/2
1280    YY=1/SQR(F)
1290    X=K/(14.8*RH)+(2.51*KVISC)/(4*RH*V*SQR(F))
1300    W=YY+.88*LOG(X)
1310    IF W<0 THEN UPV=F
1320    IF W>0 THEN LOV=F
1330    Z=(UPV+LOV)/2
1340    E=ABS((Z-F)/F)
1350    IF E<.005 THEN GOTO 1370
1360    F=Z: GOTO 1280
1370    F=Z
1380    RETURN
1390    PRINT:INPUT"DO YOU WISH TO MAKE ANOTHER COMPUTATION (Y/N)";ANS$
1400    IF ANS$="Y" THEN GOTO 60
1410    END
```

Listing of program **HJUMP.BAS**

```
10      REM program hjump
20      PRINT:PRINT"Program HJUMP":PRINT
30      PRINT"THIS PROGRAM COMPUTES THE INCIDENT AND"
40      PRINT"SEQUENT DEPTHS FOR AN HYDRAULIC JUMP"
50      PRINT"IN RECTANGULAR, TRAPEZOIDAL AND CIRCULAR CHANNELS"
60      PRINT:PRINT"DO YOU WISH TO COMPUTE THE"
70      PRINT" 1  INCIDENT DEPTH  OR  2  SEQUENT DEPTH ?"
80      INPUT"ENTER 1 OR 2, AS APPROPRIATE";NUMB
90      IF NUMB=1 THEN INPUT"ENTER SEQUENT DEPTH (mm)";YS:YS=YS/1000
100     IF NUMB=2 THEN INPUT"ENTER INCIDENT DEPTH (mm)";YI:YI=YI/1000
110     PRINT:PRINT"ENTER CHANNEL DATA:"
120     PRINT"IS SECTION   1 CIRCULAR   2 RECTANGULAR   3 TRAPEZOIDAL ?"
130     INPUT"ENTER 1, 2 OR 3, AS APPROPRIATE"; NO
140     IF NO=1 THEN SECT$="CIRC":INPUT"DIAMETER (mm)";D:D=D/1000
150     IF NO=2 THEN SECT$="RECT":INPUT"BOTTOM WIDTH (mm)";B:B=B/1000
160     IF NO=3 THEN SECT$="TRAP":INPUT"BOTTOM WIDTH (mm)";B:B=B/1000
170     IF NO=3 THEN INPUT"ANGLE OF SIDE TO HORL (deg)";FI
180     PRINT:INPUT"DISCHARGE (m**3/s)";Q: G=9.810001
190     PRINT:PRINT"       Computation in progress; please wait ...."
200     GOSUB 600
210     PRINT:PRINT"CRITICAL DEPTH (mm) = ";YC*1000
220     IF NUMB=1 THEN Y=YS
230     IF NUMB=2 THEN Y=YI
240     GOSUB 440
250     SUM=A*YB+(Q*Q)/(G*A)
260     IF NUMB=1 THEN HI=YC:LO=0
270     IF NUMB=2 THEN LO=YC:HI=20
280     IF NUMB=2 THEN IF SECT$="CIRC" THEN HI=D
290     Y=(HI+LO)/2
300     GOSUB 440
310     SUMM=A*YB+(Q*Q)/(G*A)
320     IF (SUMM-SUM)>0 THEN IF NUMB=1 THEN LO=Y
330     IF (SUMM-SUM)<0 THEN IF NUMB=1 THEN HI=Y
340     IF (SUMM-SUM)>0 THEN IF NUMB=2 THEN HI=Y
350     IF (SUMM-SUM)<0 THEN IF NUMB=2 THEN LO=Y
360     Z=(HI+LO)/2
370     IF ABS(Z-Y)>.0002 THEN GOTO 290
380     IF NUMB=1 THEN PRINT"INCIDENT DEPTH (mm) = ";Y*1000
390     IF NUMB=2 THEN PRINT"SEQUENT DEPTH (mm)  = ";Y*1000
400     PRINT:INPUT"DO YOU WISH TO MAKE ANOTHER COMPUTATION (Y/N)";ANS$
410     IF ANS$="Y" THEN GOTO 60
420     END
430     REM SUBROUTINE TO COMPUTE AREA AND CENTROIDAL DEPTH
440     IF SECT$="CIRC" THEN GOTO 490
450     IF SECT$="RECT" THEN A=B*Y: YB=Y*.5:BBB=B:RETURN
460     IF SECT$="TRAP" THEN A=Y*(B+Y/TAN(FI*3.142/180))
470     BB=Y/TAN(FI*3.142/180):BBB=B+2*BB
480     YB=(B*Y*Y/2+BB*Y*Y/3)/A:RETURN
490     HII=3.142:LOO=0!
500     TH=(HII+LOO)/2
510     XR=1-(2*Y)/D-COS(TH)
520     IF XR<0 THEN LOO=TH
530     IF XR>0 THEN HII=TH
540     Z=(HII+LOO)/2
550     IF ABS(Z-TH)>.0002 THEN GOTO 500
560     A=.25*D*D*(TH-.5*SIN(2*TH)):BBB=D*SIN(TH)
570     YB=D^3*(SIN(TH)^3-1.5*COS(TH)*(TH-SIN(TH)*COS(TH)))/(12*A)
```

```
580     RETURN
590     REM SUBROUTINE TO COMPUTE CRITICAL DEPTH
600     IF SECT$="CIRC" THEN GOTO 720
610     IF SECT$="RECT" THEN YC=(Q^2/(B^2*G))^.3333:RETURN
620     IF SECT$="TRAP" THEN XC=Q^2/G
630     HI=20!:LO=0!
640     Y=(HI+LO)/2
650     BB=Y/TAN(FI*3.142/180):A=(B+BB)*Y:BBB=B+2*BB
660     XR=A^3/BBB-XC
670     IF XR<0 THEN LO=Y
680     IF XR>0 THEN HI=Y
690     Z=(HI+LO)/2
700     IF ABS(Z-Y)>.001 THEN GOTO 640
710     YC=Y:RETURN
720     HI=3.142:LO=0!
730     TH=(HI+LO)/2
740     A=.25*D*D*(TH-.5*SIN(2*TH)):BBB=D*SIN(TH)
750     XR=A^3/BBB-Q^2/G
760     IF XR<0 THEN LO=TH
770     IF XR>0 THEN HI=TH
780     Z=(HI+LO)/2
790     IF ABS(Z-TH)>.001 THEN GOTO 730
800     YC=.5*D*(1-COS(TH)):RETURN
```

Listing of program **GVF.BAS**

```
10      REM PROGRAM GVF: Analyses steady gradually varied flow in channels
20      PRINT:PRINT"Program GVF":PRINT
30      PRINT"THIS PROGRAM ANALYSES THREE CATEGORIES OF STEADY GRADUALLY"
40      PRINT"VARIED OPEN CHANNEL FLOW; IT CATERS FOR RECTANGULAR, TRAPEZOIDAL"
50      PRINT"AND CIRCULAR CHANNEL SECTIONS AND OFFERS A CHOICE BETWEEN THE"
60      PRINT"MANNING AND DARCY-WEISBACH FLOW EQUATIONS. THE ANALYSIS USES A"
70      PRINT"FOURTH ORDER RUNGE-KUTTA NUMERICAL COMPUTATIONAL SCHEME IN THE"
80      PRINT"THE SOLUTION OF THE RELEVANT WATER SURFACE SLOPE EQUATION."
90      PRINT:PRINT"THE THREE FLOW CATEGORIES ARE:"
100     PRINT
110     PRINT"       (1) GVF WIHOUT LATERAL INFLOW OR OUTFLOW"
120     PRINT"       (2) GVF WITH LATERAL INFLOW (COLLECTOR CHANNEL)"
130     PRINT"       (3) GVF WITH LATERAL OUTFLOW (SIDE WEIR CHANNEL)"
140     PRINT
150     INPUT"ENTER NUMBER OF YOUR CHOICE";NC
160     IF NC=1 THEN GOTO 210
170     IF NC=2 THEN GOTO 800
180     IF NC=3 THEN GOTO 1360
190     IF NC>3 THEN GOTO 150
200     REM*******FLOW CATEGORY 1********
210     CLS:PRINT"ANALYSIS OF GVF IN CHANNELS WITHOUT LATERAL INFLOW/OUTFLOW"
211     PRINT"The program computes the flow depth at specified intervals along"
212     PRINT"the channel, starting from a control point at which the depth"
213     PRINT"is specified. The programs outputs distance from the control"
214     PRINT"point and corresponding flow depth. Note that distances measured"
215     PRINT"upstream from the control point are printed as negative values."
216     PRINT:PRINT"DATA ENTRY:"
220     PRINT:PRINT"DO YOU WISH TO USE   1 MANNING OR   2 DARCY-WEISBACH ?"
230     INPUT"ENTER 1 OR 2, AS APPROPRIATE";NUMB
```

```
240      IF NUMB =1 THEN FORM$="M": INPUT"MANNING N-VALUE";MN
250      IF NUMB=2 THEN FORM$="DW": INPUT"WALL ROUGHNESS (mm)";KS:KS=KS/1000
260      PRINT:PRINT"ENTER CHANNEL DATA:"
270      PRINT "IS SECTION   1 CIRCULAR  2 RECTANGULAR  3 TRAPEZOIDAL ?"
280      INPUT "ENTER 1, 2 OR 3, AS APPROPRIATE"; NO
290      IF NO=1 THEN SECT$="CIRC": INPUT"DIAMETER (mm)";D: D=D/1000
300      IF NO=2 THEN SECT$="RECT": INPUT"CHANNEL WIDTH (mm)";B: B=B/1000
310      IF NO=3 THEN SECT$="TRAP": INPUT"BOTTOM WIDTH (mm)";B: B=B/1000
320      IF NO=3 THEN INPUT"ANGLE OF SIDE TO HORL (deg)";FI
330      PRINT:INPUT"DEPTH AT CONTROL SECTION (mm)";YO: YO=YO/1000
340      PRINT:INPUT"ENTER CHANNEL BED SLOPE";FO
350      PRINT:INPUT"ENTER DISCHARGE (m**3/s)";Q
360      PRINT:PRINT"IS COMPUTATION PROCEEDING UPSTREAM (UP) OR"
370      PRINT"DOWNSTREAM (DN) FROM CONTROL SECTION ?"
380      INPUT"ENTER UP OR DN, AS APPROPRIATE";ANS$
390      PRINT:INPUT"ENTER CHANNEL STEP COMPUTATION LENGTH (m)";DELX
400      PRINT:INPUT"ENTER NUMBER OF COMPUTATION STEPS";NS
410      PRINT:PRINT".....data input complete; computation now in progress....."
420      NNS=1: DIST=0: G=9.810001
430      IF ANS$="UP" THEN DELX=-DELX
440      REM *****COMPUTATION OF CRITICAL DEPTH*****
450      GOSUB 2290
460      PRINT:PRINT"CRITICAL DEPTH (mm) = ";YC*1000
470      REM *****COMPUTATION OF NORMAL DEPTH*****
480      GOSUB 2660
490      PRINT:PRINT"NORMAL DEPTH (mm) = ";Y*1000:PRINT
500      Y=YO
510      PRINT:PRINT" DISTANCE (m)  DEPTH (mm)"
520      PRINT USING "########.#";DIST, Y*1000
530      IF FORM$="M" THEN GOSUB 2040
540      IF FORM$="DW" THEN GOSUB 2080
550      A1=(FO-FS)/(1-Q*Q*TW/(G*A^3))
560      Y=YO+.5*A1*DELX
570      IF FORM$="M" THEN GOSUB 2040
580      IF FORM$="DW" THEN GOSUB 2080
590      A2=(FO-FS)/(1-Q*Q*TW/(G*A^3))
600      Y=YO+.5*A2*DELX
610      IF FORM$="M" THEN GOSUB 2040
620      IF FORM$="DW" THEN GOSUB 2080
630      A3=(FO-FS)/(1-Q*Q*TW/(G*A^3))
640      Y=YO+A3*DELX
650      IF FORM$="M" THEN GOSUB 2040
660      IF FORM$="DW" THEN GOSUB 2080
670      A4=(FO-FS)/(1-Q*Q*TW/(G*A^3))
680      DELY=(DELX/6)*(A1+2*A2+2*A3+A4)
690      Y=YO+DELY:YO=Y
700      NNS=NNS+1: DIST=DIST+DELX
710      PRINT USING "########.#";DIST, Y*1000
720      IF NNS\15=NNS/15 THEN GOTO 740
730      GOTO 760
740      PRINT:PRINT"Press the space bar to continue"
750      Z$=INPUT$(1)
760      IF NNS>NS THEN GOTO 770 ELSE GOTO 530
770      PRINT:INPUT"DO YOU WISH TO MAKE ANOTHER COMPUTATION (Y/N)";ANS$
780      IF ANS$="Y" THEN GOTO 60
790      END
800      REM*********FLOW CATEGORY 2*********
810      CLS:PRINT"ANALYSIS OF COLLECTOR CHANNEL FLOW":PRINT
820      PRINT"The analysis relates to collector channels having a uniform"
830      PRINT"lateral inflow over the channel length and a free overfall at"
840      PRINT"the outlet end. The analysis starts from the outlet end where"
```

```
850       PRINT"the flow depth is taken as the critical depth. For practical"
860       PRINT"computational reasons, as explained in Chapter 7, the starting"
870       PRINT"depth value is taken as 1.02 times the critical depth."
880       PRINT:PRINT"DATA ENTRY:"
890       PRINT:PRINT"DO YOU WISH TO USE   1 MANNING OR   2 DARCY-WEISBACH ?"
900       INPUT"ENTER 1 OR 2, AS APPROPRIATE";NUMB
910       IF NUMB =1 THEN FORM$="M": INPUT"MANNING N-VALUE";MN
920       IF NUMB=2 THEN FORM$="DW": INPUT"WALL ROUGHNESS (mm)";KS:KS=KS/1000
930       PRINT:PRINT"ENTER CHANNEL DATA:"
940       PRINT "IS SECTION   1 CIRCULAR  2 RECTANGULAR  3 TRAPEZOIDAL ?"
950       INPUT "ENTER 1, 2 OR 3, AS APPROPRIATE"; NO
960       IF NO=1 THEN SECT$="CIRC": INPUT"DIAMETER (mm)";D: D=D/1000
970       IF NO=2 THEN SECT$="RECT": INPUT"CHANNEL WIDTH (mm)";B: B=B/1000
980       IF NO=3 THEN SECT$="TRAP": INPUT"BOTTOM WIDTH (mm)";B: B=B/1000
990       IF NO=3 THEN INPUT"ANGLE OF SIDE TO HORL (deg)";FI
1000       PRINT:INPUT"ENTER CHANNEL BED SLOPE";FO
1010       PRINT:INPUT"ENTER DISCHARGE (m**3/s)";QO
1020       PRINT:INPUT"ENTER CHANNEL LENGTH (m)";XL
1030       PRINT:INPUT"ENTER CHANNEL STEP COMPUTATION LENGTH (m)";DELX
1040       PRINT:PRINT".....data input complete; computation now in progress....."
1050       DIST=0: G=9.810001:DELX=-DELX:Q=QO
1060       QL=QO/XL
1070       REM*******COMPUTE CRITICAL DEPTH*******
1080       GOSUB 2290
1090       PRINT:PRINT"CRITICAL DEPTH (mm) = ";YC*1000
1100       YO=1.02*YC:Y=YO
1110       PRINT:PRINT" DISTANCE (m)  DEPTH (mm)  Q (m**3/s)"
1120       PRINT USING "########.##";DIST, YO*1000,QO
1130       IF FORM$="M" THEN GOSUB 2040
1140       IF FORM$="DW" THEN GOSUB 2080
1150       A1=(FO-FS-QL*Q/(G*A*A))/(1-Q*Q*TW/(G*A^3))
1160       Y=YO+.5*A1*DELX:Q=Q+.5*DELX*QO/XL
1170       IF FORM$="M" THEN GOSUB 2040
1180       IF FORM$="DW" THEN GOSUB 2080
1190       A2=(FO-FS-QL*Q/(G*A*A))/(1-Q*Q*TW/(G*A^3))
1200       Y=YO+.5*A2*DELX
1210       IF FORM$="M" THEN GOSUB 2040
1220       IF FORM$="DW" THEN GOSUB 2080
1230       A3=(FO-FS-QL*Q/(G*A*A))/(1-Q*Q*TW/(G*A^3))
1240       Y=YO+A3*DELX:Q=Q+.5*DELX*QO/XL
1250       IF FORM$="M" THEN GOSUB 2040
1260       IF FORM$="DW" THEN GOSUB 2080
1270       A4=(FO-FS-QL*Q/(G*A*A))/(1-Q*Q*TW/(G*A^3))
1280       DELY=(DELX/6)*(A1+2*A2+2*A3+A4)
1290       Y=YO+DELY:YO=Y
1300       DIST=DIST+DELX
1310       PRINT USING "########.##";DIST, Y*1000,Q
1320       IF ABS(DIST+DELX)>XL THEN GOTO 1330 ELSE GOTO 1130
1330       PRINT:INPUT"DO YOU WISH TO MAKE ANOTHER COMPUTATION (Y/N)";ANS$
1340       IF ANS$="Y" THEN GOTO 850
1350       END
1360       REM*********FLOW CATEGORY 3*********
1365       CLS:PRINT"ANALYSIS OF GVF IN CHANNEL WITH SIDE OVERFLOW WEIRS":PRINT
1370       PRINT"The analysis relates to flow in a channel in which there is"
1380       PRINT"a lateral outflow over sharp-edged side weirs of specified crest"
1390       PRINT"level and crest length. The program computes the normal flow"
1400       PRINT"depth and the variation in depth over the weir length.It outputs"
1410       PRINT"the weir head and weir overflow rate, the channel flow rate and"
1420       PRINT"flow depth at the specified computational step intervals over"
1430       PRINT"the weir length."
1440       PRINT:PRINT"DATA ENTRY:"
```

```
1450    PRINT:PRINT"DO YOU WISH TO USE   1 MANNING OR   2 DARCY-WEISBACH ?"
1460    INPUT"ENTER 1 OR 2, AS APPROPRIATE";NUMB
1470    IF NUMB =1 THEN FORM$="M": INPUT"MANNING N-VALUE";MN
1480    IF NUMB=2 THEN FORM$="DW": INPUT"WALL ROUGHNESS (mm)";KS:KS=KS/1000
1490    PRINT:PRINT"ENTER CHANNEL DATA:"
1500    PRINT "IS SECTION   1 CIRCULAR  2 RECTANGULAR  3 TRAPEZOIDAL ?"
1510    INPUT "ENTER 1, 2 OR 3, AS APPROPRIATE"; NO
1520    IF NO=1 THEN SECT$="CIRC": INPUT"DIAMETER (mm)";D: D=D/1000
1530    IF NO=2 THEN SECT$="RECT": INPUT"CHANNEL WIDTH (mm)";B: B=B/1000
1540    IF NO=3 THEN SECT$="TRAP": INPUT"BOTTOM WIDTH (m)";B
1550    IF NO=3 THEN INPUT"ANGLE OF SIDE TO HORL (deg)";FI
1560    PRINT:INPUT"ENTER SINGLE SIDE WEIR LENGTH (m)";LW
1570    INPUT"ENTER NUMBER OF SIDE WEIRS (1 OR 2)";NW:PRINT
1580    INPUT"ENTER HT OF WEIR CREST ABOVE CHANNEL BED AT U/S END (mm)";HC
1590    HC=HC/1000
1600    PRINT:INPUT"ENTER CHANNEL BED SLOPE";FO
1610    PRINT:INPUT"ENTER DISCHARGE (m**3/s)";Q
1620    PRINT:INPUT"ENTER COMPUTATIONAL STEP LENGTH (m)";DELX
1630    PRINT:PRINT".....data input complete; computation now in progress....."
1640    NNS=1: DIST=0: G=9.810001
1650    REM COMPUTATION OF CRITICAL DEPTH
1660    GOSUB 2290
1670    PRINT:PRINT"CRITICAL DEPTH (mm) = ";YC*1000
1680    REM COMPUTATION OF NORMAL DEPTH
1690    GOSUB 2660
1700    PRINT:PRINT"NORMAL DEPTH (mm) = ";Y*1000:PRINT
1710    YO=Y:GOSUB 3090
1720    PRINT:PRINT"    DIST ALONG  WEIR HEAD  OVERFLOW  CHANNEL FLOW)"
1730    PRINT      "    WEIR (m)       (mm)    (m**3/s.m)   (m**3/s)"
1740    PRINT USING "########.##";DIST,HW*1000,QW,Q
1750    IF FORM$="M" THEN GOSUB 2040
1760    IF FORM$="DW" THEN GOSUB 2080
1780    IF HW<0 THEN GOTO 2010
1790    A1=(FO-FS+QW*Q/(G*A*A))/(1-Q*Q*TW/(G*A^3))
1800    Y=YO+.5*A1*DELX:Q=Q-.5*DELX*QW
1810    IF FORM$="M" THEN GOSUB 2040
1820    IF FORM$="DW" THEN GOSUB 2080
1830    A2=(FO-FS+QW*Q/(G*A*A))/(1-Q*Q*TW/(G*A^3))
1840    Y=YO+.5*A2*DELX
1850    IF FORM$="M" THEN GOSUB 2040
1860    IF FORM$="DW" THEN GOSUB 2080
1870    A3=(FO-FS+QW*Q/(G*A*A))/(1-Q*Q*TW/(G*A^3))
1880    Y=YO+A3*DELX:Q=Q-.5*DELX*QW
1890    IF FORM$="M" THEN GOSUB 2040
1900    IF FORM$="DW" THEN GOSUB 2080
1910    A4=(FO-FS+QW*Q/(G*A*A))/(1-Q*Q*TW/(G*A^3))
1920    DELY=(DELX/6)*(A1+2*A2+2*A3+A4)
1930    Y=YO+DELY:YO=Y
1940    DIST=DIST+DELX
1945    GOSUB 3090
1950    PRINT USING "########.##";DIST,HW*1000,QW,Q
1952    IF DIST>LW THEN GOTO 2010
1970    GOTO 1750
2010    PRINT:INPUT"DO YOU WISH TO MAKE ANOTHER COMPUTATION (Y/N)";ANS$
2020    IF ANS$="Y" THEN GOTO 1420
2030    END
2040    REM **SUBROUTINE TO COMPUTE FRICTION SLOPE USING MANNING EQN**
2050    GOSUB 2130
```

```
2060    FS=(MN*Q/A)^2*(RH^-1.33)
2070    RETURN
2080    REM **SUBROUTINE TO COMPUTE FRICTION SLOPE USING DARCY-WEISBACH EQN**
2090    GOSUB 2130
2100    V=Q/A:K=KS:GOSUB 2500
2110    FS=F*Q*Q/(8*9.810001*A*A*RH)
2120    RETURN
2130    REM **SUBROUTINE TO CALCULATE A,RH and TW**
2140    IF SECT$="CIRC" THEN GOTO 2180
2150    IF SECT$="RECT" THEN A=B*Y:RH=B*Y/(B+2*Y):TW=B:RETURN
2160    IF SECT$="TRAP" THEN A=Y*(B+Y/TAN(FI*3.142/180))
2170    RH=A/(B+(2*Y)/SIN(FI*3.142/180)):TW=B+(2*Y)/TAN(FI*3.142/180):RETURN
2180    HI=3.1416: LO=0!
2190    TH=(HI+LO)/2
2200    XR=1-(2*Y)/D-COS(TH)
2210    IF XR<0 THEN LO=TH
2220    IF XR>0 THEN HI=TH
2230    Z=(HI+LO)/2
2240    IF ABS(Z-TH)>.001 THEN GOTO 2190
2250    P=D*TH: A=.25*D*D*(TH-.5*SIN(2*TH)):RH=A/P
2260    TW=D*SIN(TH)
2270    RETURN
2280    REM **SUBROUTINE TO COMPUTE CRITICAL DEPTH**
2290    IF SECT$="CIRC" THEN GOTO 2410
2300    IF SECT$="RECT" THEN YC=(Q^2/(B^2*G))^.3333:RETURN
2310    IF SECT$="TRAP" THEN XC=Q^2/G
2320    HI=20!:LO=0!
2330    YY=(HI+LO)/2
2340    BB=YY/TAN(FI*3.142/180):A=(B+BB)*YY:BBB=B+2*BB
2350    XR=A^3/BBB-XC
2360    IF XR<0 THEN LO=YY
2370    IF XR>0 THEN HI=YY
2380    Z=(HI+LO)/2
2390    IF ABS(Z-YY)>.001 THEN GOTO 2330
2400    YC=YY:RETURN
2410    HI=3.142:LO=0!
2420    TH=(HI+LO)/2
2430    A=.25*D*D*(TH-.5*SIN(2*TH)):BBB=D*SIN(TH)
2440    XR=A^3/BBB-Q^2/G
2450    IF XR<0 THEN LO=TH
2460    IF XR>0 THEN HI=TH
2470    Z=(HI+LO)/2
2480    IF ABS(Z-TH)>.001 THEN GOTO 2420
2490    YC=.5*D*(1-COS(TH)):RETURN
2500    REM **SUBROUTINE FRICTION FACTOR**
2510    KVISC=1.307E-06
2520    UPV=.5
2530    LOV=0
2540    F=(UPV+LOV)/2
2550    YY=1/SQR(F)
2560    X=K/(14.8*RH)+(2.51*KVISC)/(4*RH*V*SQR(F))
2570    W=YY+.88*LOG(X)
2580    IF W<0 THEN UPV=F
2590    IF W>0 THEN LOV=F
2600    Z=(UPV+LOV)/2
2610    E=ABS((Z-F)/F)
2620    IF E<.005 THEN GOTO 2640
2630    F=Z: GOTO 2550
2640    F=Z
2650    RETURN
2660    REM **SUBROUTINE TO COMPUTE NORMAL DEPTH**
```

```
2670      FSR=FO^.5
2680      HI=40:LO=.001
2690      IF SECT$="CIRC" THEN GOTO 2800
2700      IF SECT$="RECT" THEN GOTO 2910
2710      IF SECT$="TRAP" THEN GOTO 2720
2720      Y=(HI+LO)/2
2730      A=Y*(B+Y/TAN(FI*3.142/180)): RH=A/(B+(2*Y)/SIN(FI*3.142/180))
2740      IF FORM$="M" THEN FSH=MN*Q/(A*RH^.67):GOTO 2770
2750      IF FORM$="DW" THEN V=Q/A:K=KS:GOSUB 2500
2760      FSH=(F/(8*9.810001*RH))^.5*V
2770      GOSUB 2990
2780      IF ABS(Z-Y)>.0002 THEN GOTO 2720
2790      RETURN
2800      HI=3.1416:LO=.001
2810      TH=(HI+LO)/2
2820      A=.25*D*D*(TH-.5*SIN(2*TH))
2830      P=D*TH:RH=A/P
2840      IF FORM$="M" THEN FSH =MN*Q/(A*RH^.67):GOTO 2870
2850      IF FORM$="DW" THEN V=Q/A:K=KS:GOSUB 2500
2860      FSH=(F/(8*9.810001*RH))^.5*V
2870      GOSUB 3050
2880      IF ABS(FSH-FSR)/FSR>.001 THEN GOTO 2810
2890      Y=.5*D*(1-COS(TH))
2900      RETURN
2910      Y=(HI+LO)/2
2920      A=B*Y: RH=A/(B+2*Y)
2930      IF FORM$="M" THEN FSH=MN*Q/(A*RH^.67):GOTO 2960
2940      IF FORM$-"DW" THEN V=Q/A:K=KS:GOSUB 2500
2950      FSH=(F/(8*9.810001*RH))^.5*V
2960      GOSUB 2990
2970      IF ABS(Z-Y)>.0002 THEN GOTO 2910
2980      RETURN
2990      REM INTERVAL-HALFING ROUTINE
3000      WW=FSH-FSR
3010      IF WW>0 THEN LO=Y
3020      IF WW<0 THEN HI=Y
3030      Z=(HI+LO)/2
3040      RETURN
3050      WW=FSH-FSR
3060      IF WW>0 THEN LO=TH
3070      IF WW<0 THEN HI=TH
3080      RETURN
3090      REM CALC OF WEIR DISCHARGE RATE QW
3100      CW=4.15-1.81*YC/Y-.14*YC/LW:HW=Y-(HC+DIST*FO)
3110      IF HW<0 THEN PRINT"WATER LEVEL BELOW WEIR CREST":GOTO 3130
3120      QW=NW*CW*HW^1.5
3130      RETURN
```

Listing of program **BROAD.BAS**

```
10  REM PROGRAM "BROAD.BAS"- RECTANGULAR BROAD-CRESTED WEIR DESIGN
20  CLS
30  PRINT"***DESIGN OF RECTANGULAR HORIZONTAL BROAD-CRESTED WEIR***"
40  PRINT
50  PRINT"Design parameters:"
60  PRINT"                          weir width b                 (m)
70  PRINT"                          weir length  L               (m)
80  PRINT"                          upstream step height  P1     (m)
90  PRINT"                          downstream step height  P2   (m)
100 PRINT"                          upstream nose radius  r      (m)
110 PRINT:PRINT"(above parameters are illustrated on Fig 8.1, Chap. 8)"
120 PRINT:PRINT:PRINT"Press the space bar to continue"
130 Y$=INPUT$(1)
140 CLS
150 PRINT "DATA INPUT":PRINT
160 INPUT "Enter maximum expected flow (m**3/s)";QMAX
170 INPUT "Enter minimum expected flow (m**3/s)";QMIN
180 INPUT "Enter channel slope (sin theta)";S
190 INPUT "Enter channel Manning n-value";N
200 INPUT "Enter channel bed width (m)";WID
210 INPUT "Enter channel side slope angle to horizontal (Degrees)=",ANG
220 INPUT "Enter value for b (weir width,m), not less than 0.3m";B
230 PRINT:PRINT
240 ANG=ANG*3.1416/180
250 REM*****Compute the tail water depth*****
260 FSR=S^.5
270 HI=40:LO=.001
280 HSD=(HI+LO)/2
290 A=HSD*(WID+HSD/TAN(ANG)):RH=A/(WID+(2*HSD)/SIN(ANG))
300 FSH=N*QMAX/(A*RH^.67)
310 WW=FSH-FSR
320 IF WW>0 THEN LO=HSD
330 IF WW<0 THEN HI=HSD
340 Z=(HI+LO)/2
350 IF ABS(Z-HSD)>.0002 THEN GOTO 280
360 PRINT "TAILWATER DEPTH (m) = ";HSD
370 HD=HSD+(QMAX/A)^2/19.62
380 PRINT:PRINT
390 PRINT "DESIGN FOR MAXIMUM FLOW CONDITION"
400 PRINT "================================="
410 PRINT:PRINT
420 PRINT "Cd and Cv values are assumed 0.95 & 1.1 initially to find h1 from"
430 PRINT "the formula   Q=Cd*Cv*(2/3)*(2*g/3)^0.5*b*h1^1.5"
440 CD=.95:CV=1.1
450 HS1=(QMAX/CD/CV/(2/3)/(2*9.810001/3)^.5/B)^(2/3)
460 H1=HS1*CV^(2/3)
470 PRINT:PRINT
480 PRINT "Selection of value for r ,the radius of the upstream rounded nose"
490 PRINT "-----------------------------------------------------------------"
500 PRINT "r(min)=0.11*H1(max)=";.11*H1
510 PRINT "r(max)=0.20*H1(max)=";.2*H1
520 INPUT "Enter r=",R
530 IF R<.11*H1 GOTO 500
540 IF R>.2*H1 GOTO 500
550 PRINT:PRINT
560 PRINT"Selection of value for L ,the crest length"
570 PRINT"------------------------------------------"
580 PRINT "L>=1.75*H1(max)+r and  H1/L between 0.05 & 0.5m    ie:"
```

```
590 PRINT "L>=";1.75*H1+R
600 PRINT "L is between ";H1/.05;" and ";H1/.5
610 INPUT "Enter L=";L
620 IF L<1.75*H1+R GOTO 580
630 IF L<H1/.5 GOTO 580
640 IF L>H1/.05 GOTO 580
650 PRINT:PRINT
660 PRINT"Selection of value for P1, height of crest over the upstream bed"
670 PRINT"----------------------------------------------------------------"
680 PRINT "Choose a value for P1; min P1=0.15m "
690 PRINT "Maximum H1/P1=1.5 ie   min P1=H1/1.5=";H1/1.5
700 INPUT "Enter P1=",P
710 IF P<.15 GOTO 680
720 IF P<H1/1.5 GOTO 680
730 PRINT:PRINT
740 PRINT"Selection of value for P2, height of crest over the downstream bed"
750 PRINT"------------------------------------------------------------------"
760 INPUT"Enter value for P2 (m, magnitude not less than P1)";P2
770 IF H1/P2<=1 THEN LIMIT= 2.718^(.0925*LOG(H1/P2)-.151)
780 IF H1/P2>1  THEN LIMIT= 2.718^(.057*LOG(H1/P2)-.151)
790 H2LIM=H1*LIMIT
800 IF P2<(HD-H2LIM) THEN PRINT"P2 is too small to satisfy modular flow
condition; try increased value":PRINT:GOTO 760
810 IF P2>(HD-H2LIM) THEN PRINT"This value of P2 satisfies modular flow
requirement"
820 PRINT:PRINT
830 PRINT"Selection of location for the measuring station from"
840 PRINT"----------------------------------------------------"
850 PRINT "Choose a distance of measuring station (should be of between two to"
860 PRINT "three H1(max) ie ";2*H1;"and";3*H1;"upstream of weir"
870 INPUT "Enter a value for Distance=",DIST
880 IF DIST<2*H1 GOTO 850
890 IF DIST>3*H1 GOTO 850
900 PRINT:PRINT
910 PRINT"Specifications for control section width b"
920 PRINT"------------------------------------------"
930 PRINT "Selected b should not be less than 0.3m ,H1(max)=";H1;" and L/5=";L/5
940 IF B<.3 THEN PRINT"Specifications are not satisfied ,increase b and try
again":GOTO 150
950 IF B<H1 THEN PRINT"Specifications are not satisfied ,increase b and try
again":GOTO 150
960 IF B<L/5 THEN PRINT"Specifications are not satisfied ,increase b or decrease
L and try again":GOTO 150
970 PRINT"You are within specifications"
980 A$="":PRINT:PRINT
990 PRINT"Check upstream water depth over weir crest"
1000 PRINT"------------------------------------------"
1010 PRINT "Lower limit of h1 is 0.06 or 0.05*L=";.05*L;"whichever is greater"
1020 PRINT "Actual h1=";HS1
1030 IF HS1<.06 THEN PRINT"Specifications are not satisfied ,decrease b and try
again":GOTO 150
1040 IF HS1<.05*L THEN PRINT"Specifications are not satisfied ,decrease b or L
and try again":GOTO 150
1050 PRINT"You are within specifications"
1060 PRINT:PRINT
1070 PRINT "DESIGN CHECK FOR MINIMUM FLOW"
1080 PRINT "============================="
1090 Q=QMIN:GOSUB 1920
1100 PRINT "Lower limit of h1 is 0.06m or 0.05*L=";.05*L;"whichever is greater"
1110 PRINT "H1/L should be between 0.05 & 0.5m"
1120 PRINT "h1=";HS1;"  H1/L=";H1/L
```

```
1130 IF HS1<.06 THEN PRINT"Specifications are not satisfied ,decrease b and try
again":GOTO 150
1140 IF HS1<.05*L THEN PRINT"Specifications are not satisfied ,decrease b or L
and try again":GOTO 150
1150 IF H1/L<.05 THEN PRINT"Specifications are not satisfied ,decrease b or L
and try again":GOTO 150
1160 PRINT "You are within specifications"
1170 A$="":PRINT:PRINT
1180 INPUT "Press the return key when you are ready",A$
1190 IF A$="" THEN CLS
1200 PRINT "DIMENSIONAL SUMMARY FOR BROAD-CRESTED WEIR"
1210 PRINT "========================================="
1220 PRINT "Crest width b = ";B;"(m)"
1230 PRINT "Crest length L = ";L;"(m)"
1240 PRINT "Radius of upstream rounded nose r = ";R;"(m)"
1250 PRINT "Upstream step-height P1 = ";P;"(m)"
1260 PRINT "Downstream step-height P2 = ";P2;"(m)"
1270 PRINT "Upstream distance of the head measurement station"
1280 PRINT "from control section = ";DIST;"(m)"
1290 PRINT:PRINT
1300 PRINT"Enter required form of head/discharge relationship"
1310 PRINT"1-Tabular form."
1320 PRINT"2-Graph form."
1330 INPUT FORM
1340 IF FORM=1 GOTO 1370
1350 IF FORM=2 GOTO 1510
1360 GOTO 1290
1370 CLS
1380 INPUT "Enter flow increment (m**3/s) for tabulation";DEQ
1390 PRINT:PRINT
1400 REM ***TABULATION OF HEAD/DISCHARGE VALUES***
1410 PRINT"Q(m3/s)      Cd        Cv          h1(m)"
1420 PRINT"======================================="
1430 Q=QMIN:GOSUB 1920
1440 PRINT USING"##.###       ";Q;
1450 PRINT USING"#.###     ";CD;
1460 PRINT USING"#.###        ";CV;
1470 PRINT USING"##.###       ";HS1
1480 Q=Q+DEQ
1490 IF Q<QMAX+DEQ THEN GOSUB 1920:GOTO 1440
1500 GOTO 1880
1510 REM ***SCREEN PLOT OF HEAD/DISCHARGE CORRELATION***
1520 SCREEN 8
1530 KEY OFF
1540 Q=QMIN:GOSUB 1920:HMIN=HS1
1550 Q=QMAX:GOSUB 1920:HMAX=HS1
1560 X1=HMIN-.3492*(HMAX-HMIN):X2=HMAX+.2381*(HMAX-HMIN)
1570 Y1=QMIN-.4918*(QMAX-QMIN):Y2=QMAX+.1475*(QMAX-QMIN)
1580 DX=X2-X1:DY=Y2-Y1
1590 CLS
1600 WINDOW (X1,Y1)-(X2,Y2)
1610 FOR I=0 TO 10
1620 LINE (X1+.22*DX,Y1+.3*DY+.061*DY*I)-(X1+.85*DX,Y1+.3*DY+.061*DY*I)
1630 LINE (X1+.22*DX+.063*DX*I,Y1+.3*DY)-(X1+.22*DX+.063*DX*I,Y1+.91*DY)
1640 NEXT I
1650 H=HMIN
1660 FOR I=16 TO 66 STEP 10
1670 LOCATE 19,I
1680 PRINT USING "##.###";H
1690 H=H+.2*(HMAX-HMIN)
1700 NEXT I
```

```
1710 LOCATE 21,40:PRINT"Head h1(m)"
1720 Q=QMAX
1730 FOR I=3 TO 18 STEP 3
1740 LOCATE I,10
1750 PRINT USING "##.###";Q
1760 Q=Q-.2*(QMAX-QMIN)
1770 NEXT I
1780 LOCATE 1,7:PRINT"Discharge (m3/s)"
1790 PX1=HMIN:PY1=QMIN
1800 DEQ=(QMAX-QMIN)/50
1810 PY2=PY1+DEQ
1820 Q=PY2:GOSUB 1920:PX2=HS1
1830 LINE (PX1,PY1)-(PX2,PY2)
1840 IF PY2<QMAX-DEQ THEN PX1=PX2:PY1=PY2:GOTO 1810
1850 A$="":LOCATE 23
1860 INPUT "Press the return key when you are ready",A$
1870 IF A$="" THEN SCREEN 0
1880 PRINT:PRINT
1890 INPUT"Enter (1) to run again;    enter (2) to quit"; A
1900 IF A=1 GOTO 10
1910 END
1920 REM  ***computation of CD and CV***
1930 CD=.95:CV=1.1:X=.005
1940 FOR I=1 TO 5
1950 HS1=(Q/CD/CV/(2/3)/(2*9.810001/3)^.5/B)^(2/3)
1960 CD=(1-2*X*(L-R)/B)*(1-X*(L-R)/HS1)^1.5
1970 A1=(HS1+P)*(WID+(HS1+P)/TAN(ANG))
1980 H1=HS1+Q*Q/(19.62*A1*A1)
1990 CV=(H1/HS1)^1.5
2000 NEXT I
2010 RETURN
```

Listing of program **FLUME.BAS**

```
10  REM  PROGRAM "FLUME.BAS"   -   LONG-THROATED FLUME DESIGN
20  PRINT"Program FLUME":PRINT
30  PRINT"****DESIGN OF TRAPEZOIDAL LONG-THROATED FLUMES****"
40  PRINT
50  PRINT"Design parameters:"
60  PRINT"                    throat bottom width                         "
70  PRINT"                    throat side wall inclination to horizontal  "
80  PRINT"                    throat length                               "
90  PRINT"                    throat step height                          "
100 PRINT:PRINT"(above parameters are illustrated on Fig 8.6, Chap. 8)"
110 PRINT:PRINT:PRINT"Press the space bar to continue"
120 Y$=INPUT$(1)
130 CLS
140 PRINT "INPUT OF FLOW AND CHANNEL DATA":PRINT
150 INPUT "Enter max expected flow (m**3/s)";QMAX
160 INPUT "Enter min expected flow (m**3/s)";QMIN
170 INPUT "Enter channel slope (sin theta)";S
180 INPUT "Enter channel bed width (m)";WID
190 INPUT "Enter channel side slope angle to horl. (deg)";ANG
200 INPUT "Enter channel Manning n-value";N
210 ANG=ANG*3.1416/180
220 PRINT:PRINT"INPUT OF CONTROL SECTION DATA":PRINT
230 INPUT"Enter value for b (throat bottom width,m)";B
240 INPUT"Enter value for angle of inclination to horl. of throat wall (deg)";FI
250 FI=FI*3.1416/180
260 INPUT"Enter value for L (throat length, m)";L
270 PRINT"Enter value for P, the throat step height (m)"
280 INPUT"Note:P value is not limited by any specifications.";P
290 PRINT:PRINT"INPUT OF DOWNSTREAM TRANSITION DATA":PRINT
300 PRINT"The expansion from the throat section to the downstream channel"
310 PRINT"should be gradual; a sidewall and bottom expansion in the range"
320 PRINT"of 1:4 to 1:6 is recommended but is not obligatory."
330 INPUT"Enter value of m, where expansion is expressed as 1:m";M
340 PRINT:PRINT"DATA INPUT IS NOW COMPLETE":PRINT
350 PRINT"Press the space bar to continue"
360 Y$=INPUT$(1):CLS
370 Q=QMAX
380 GOSUB 2120:REM ***to compute yc,H1,h1***
390 PRINT"DESIGN CHECK FOR MAX FLOW CONDITION"
400 PRINT"==================================="
410 PRINT"Specifications for h1"
420 PRINT"---------------------"
430 PRINT"Lower limit for h1 is 0.06m or 0.1*L=";.1*L;" whichever is greater"
440 PRINT"Actual value of h1=";HS1
450 IF HS1<.06 THEN PRINT"Specifications are not satisfied, reduce throat dims
and try again"
460 IF HS1<.06 GOTO 220
470 IF HS1<.1*L THEN PRINT"Specifications are not satisfied, reduce throat dims
and try again"
480 IF HS1<.1*L GOTO 220
490 PRINT"You are within specifications"
500 PRINT:PRINT
510 PRINT"Specifications for Froude number Fr."
520 PRINT"-----------------------------------"
530 PRINT"Fr in the approach channel shouldn't exceed 0.5"
540 A1=(HS1+P)*(WID+(HS1+P)/TAN(ANG)):V1=QMAX/A1
550 BC=WID+2*(HS1+P)/TAN(ANG)
560 FR=V1/(9.810001*A1/BC)^.5
```

```
570 PRINT"Fr = V1/(g*A1/BC)^0.5 = ";FR
580 PRINT"V1,A1,BC are velocity, x-sectional area and water surface width,"
590 PRINT"respectively, in the approach channel."
600 IF FR>.5 THEN PRINT"Specifications are not satisfied, reduce throat dims and
try again"
610 IF FR>.5 GOTO 220
620 PRINT"You are within specifications"
630 PRINT:PRINT"Press the space bar to continue":Y$=INPUT$(1)
640 PRINT:PRINT"Specifications for H1/L"
650 PRINT"--------------------"
660 PRINT"H1/L should be between 0.1 and 1.0"
670 PRINT"Actual H1/L=";H1/L
680 IF H1/L<.1 THEN PRINT"Specifications are not satisfied, reduce throat dims
and try again"
690 IF H1/L<.1 GOTO 220
700 IF H1/L>1! THEN PRINT"Specifications are not satisfied, increase throat dims
and try again"
710 IF H1/L>1! GOTO 220
720 PRINT"You are within specifications"
730 PRINT:PRINT
740 PRINT"Specifications for width of the water surface in the throat"
750 PRINT"-----------------------------------------------------------"
760 PRINT"Width of the water surface in the throat at the maximum stage should
not be"
770 PRINT"less than 0.3m,nor less than H1max=";H1;" nor less than L/5=";L/5
780 PRINT"Actual width = ";BBB
790 A$=""
800 IF BBB<.3 THEN A$="Y1"
810 IF BBB<H1 THEN A$="Y1"
820 IF BBB<L/5 THEN A$="Y2"
830 IF A$="Y1" THEN PRINT"Specifications are not satisfied; increase b or reduce
FI and try again"
840 IF A$="Y2" THEN PRINT"Specifications are not satisfied; increase b,L or
reduce FI and try again"
850 IF A$="Y1" GOTO 220
860 IF A$="Y2" GOTO 220
870 PRINT"You are within specifications"
880 PRINT:PRINT"Press the space bar to continue":Y$=INPUT$(1)
890 PRINT:PRINT"Specifications for the modular limit"
900 PRINT"------------------------------------"
910 PRINT"The modular ratio limit H2/H1 should not be exceeded"
920 MODL=.71+.01*M
930 PRINT"Modular limit H2/H1=";MODL
940 GOSUB 2400
950 PRINT"Actual H2/H1=";H2/H1
960 IF H2/H1>MODL THEN PRINT"Actual H2/H1 exceeds the modular limit decrease b
or increase m and try again"
970 IF H2/H1>MODL GOTO 220
980 PRINT"You are within specifications"
990 HMAX=H1
1000 PRINT:PRINT "Press the space bar to continue"
1010 Y$=INPUT$(1):PRINT
1020 PRINT "DESIGN CHECK AT MINIMUM FLOW"
1030 PRINT "==========================="
1040 PRINT
1050 Q=QMIN
1060 GOSUB 2120:REM---To find Cd,yc,H1,h1
1070 PRINT"Specifications for h1"
1080 PRINT"---------------------"
1090 PRINT"Lower limit for h1 is 0.06m or 0.1*L=";.1*L;" whichever is greater"
1100 PRINT"Actual value of h1=";HS1
```

```
1110 IF HS1<.06 THEN PRINT"Specifications are not satisfied, reduce throat dims
and try again"
1120 IF HS1<.06 THEN PRINT:PRINT:GOTO 220
1130 IF HS1<.1*L THEN PRINT"Specifications are not satisfied, reduce throat
dimsand try again"
1140 IF HS1<.1*L THEN PRINT:PRINT:GOTO 220
1150 PRINT"You are within specifications"
1160 PRINT
1170 PRINT"Specifications for H1/L"
1180 PRINT"--------------------"
1190 PRINT"H1/L should be between 0.1 and 1.0"
1200 PRINT"Actual H1/L=";H1/L
1210 IF H1/L<.1 THEN PRINT"Specifications are not satisfied, reduce throat dims
and try again"
1220 IF H1/L<.1 GOTO 220
1230 IF H1/L>1! THEN PRINT"Specifications are not satisfied, increase throat
dims and try again"
1240 IF H1/L>1! GOTO 220
1250 PRINT"You are within specifications"
1260 PRINT
1270 PRINT "Check for modular flow condition at min flow"
1280 PRINT "--------------------------------------------"
1290 PRINT"Modular limit H2/H1=";MODL
1300 GOSUB 2400
1310 PRINT"Actual H2/H1=";H2/H1
1320 IF H2/H1>MODL THEN PRINT"Actual H2/H1 exceeds the modular limit; reduce
throat dims and try again"
1330 IF H2/H1>MODL GOTO 220
1340 PRINT"You are within specifications"
1350 PRINT:PRINT"Press the space bar to continue"
1360 Y$=INPUT$(1)
1370 CLS:PRINT"SUMMARY OF DIMENSIONAL DATA":PRINT
1380 PRINT"throat width b = ";B;"(m)"
1390 PRINT"throat sidewall angle to horl. = ";FI*180/3.142;"(deg)"
1400 PRINT"throat length L = ";L;"(m)"
1410 PRINT"throat step height P = ";P;"(m)"
1420 PRINT"downstream divergence of sidewalls and bottom 1:";M:PRINT
1430 PRINT"The upstream convergence of side walls and bottom should be about 1:3
1440 PRINT"The floor of the entrance transition and of the approach channel
should be"
1450 PRINT"flat and level ,and at no point higher than the invert of the throat,
up to"
1460 PRINT"a distance 1.0*H1max=";HMAX;"(m) upstream of the head measurement
station."
1470 PRINT"This head measurement station should be located upstream of the flume
at a"
1480 PRINT"distance equal to between 2 & 3 H1max ie. ";2*HMAX;" and ";3*HMAX
1490 PRINT:PRINT
1500 PRINT"Enter required form of head/discharge relationship"
1510 PRINT"1-Tabular form."
1520 PRINT"2-Graph form."
1530 INPUT FORM
1540 IF FORM=1 GOTO 1570
1550 IF FORM=2 GOTO 1720
1560 GOTO 1500
1570 CLS
1580 INPUT "Enter flow increment (m**3/s) for tabulation";DEQ
1590 PRINT:PRINT
1600 PRINT"Q(m3/s)      yc(m)       h1(m)        H1(m)"
1610 PRINT"============================================"
1620 Q=QMIN
```

```
1630 GOSUB 2120
1640 PRINT USING"##.###       ";Q;
1650 PRINT USING"##.###       ";YC;
1660 PRINT USING"##.###        ";HS1;
1670 PRINT USING"##.###";H1
1680 Q=Q+DEQ
1690 IF Q<QMAX+DEQ GOTO 1630
1700 PRINT:PRINT"yc is the critical depth at the control section"
1710 GOTO 2080
1720 Q=QMIN:GOSUB 2120:HMIN=HS1
1730 Q=QMAX:GOSUB 2120:HMAX=HS1
1740 SCREEN 8
1750 KEY OFF
1760 X1=HMIN-.3492*(HMAX-HMIN):X2=HMAX+.2381*(HMAX-HMIN)
1770 Y1=QMIN-.4918*(QMAX-QMIN):Y2=QMAX+.1475*(QMAX-QMIN)
1780 DX=X2-X1:DY=Y2-Y1
1790 CLS
1800 WINDOW (X1,Y1)-(X2,Y2)
1810 FOR I=0 TO 10
1820 LINE (X1+.22*DX,Y1+.3*DY+.061*DY*I)-(X1+.85*DX,Y1+.3*DY+.061*DY*I)
1830 LINE (X1+.22*DX+.063*DX*I,Y1+.3*DY)-(X1+.22*DX+.063*DX*I,Y1+.91*DY)
1840 NEXT I
1850 H=HMIN
1860 FOR I=16 TO 66 STEP 10
1870 LOCATE 19,I
1880 PRINT USING "##.###";H
1890 H=H+.2*(HMAX-HMIN)
1900 NEXT I
1910 LOCATE 21,40:PRINT"Head h1(m)"
1920 Q=QMAX
1930 FOR I=3 TO 18 STEP 3
1940 LOCATE I,10
1950 PRINT USING "##.###";Q
1960 Q=Q-.2*(QMAX-QMIN)
1970 NEXT I
1980 LOCATE 1,7:PRINT"Discharge (m3/s)"
1990 PX1=HMIN:PY1=QMIN
2000 DEQ=(QMAX-QMIN)/50
2010 PY2=PY1+DEQ
2020 Q=PY2:GOSUB 2120:PX2=HS1
2030 LINE (PX1,PY1)-(PX2,PY2)
2040 IF PY2<QMAX-DEQ THEN PX1=PX2:PY1=PY2:GOTO 2010
2050 A$="":LOCATE 23
2060 INPUT "Press the return key when you are ready",A$
2070 IF A$="" THEN SCREEN 0 :CLS
2080 PRINT:PRINT
2090 PRINT"Enter (1) to run again;    enter (2) to quit":INPUT A
2100 IF A=1 THEN GOTO 140
2110 END
2120 REM---Subroutine to find yc, cd, H1 and h1 for a given value of Q
2130 REM---computation of yc
2140 XC=Q^2/9.810001
2150 HI=20!:LO=0
2160 Y=(HI+LO)/2
2170 BB=Y/TAN(FI): A=(B+BB)*Y:BBB=B+2*BB
2180 XR=A^3/BBB-XC
2190 IF XR<0 THEN LO=Y
2200 IF XR>0 THEN HI=Y
2210 Z=(HI+LO)/2
2220 IF ABS(Z-Y)>.001 THEN GOTO 2160
2230 YC=Y:VC=Q/A
```

```
2240 REM------COMPUTATION OF CD AND h1
2250 HI=20:LO=YC
2260 H1=(HI+LO)/2
2270 IF H1/L>.2 THEN CD=.95+.05*H1/L
2280 IF H1/L<=.2 THEN CD=.89+.2*H1/L
2290 QQ=CD*A*(19.62*(H1-YC))^.5
2300 IF QQ>Q THEN HI=H1
2310 IF QQ<Q THEN LO=H1
2320 Z=(HI+LO)/2
2330 IF ABS(Z-H1)>.0005 THEN GOTO 2260
2340 HS1=H1
2350 FOR I=1 TO 5
2360 A1=(HS1+P)*(WID+(HS1+P)/TAN(ANG))
2370 HS1=H1-(Q/A1)^2/19.62
2380 NEXT I
2390 RETURN
2400 REM ***Subroutine to compute tailwater depth***
2410 FSR=S^.5
2420 HI=40:LO=.001
2430 HSD=(HI+LO)/2
2440 A=HSD*(WID+HSD/TAN(ANG)):RH=A/(WID+(2*HSD)/SIN(ANG))
2450 FSH=N*Q/(A*RH^.67)
2460 WW=FSH-FSR
2470 IF WW>0 THEN LO=HSD
2480 IF WW<0 THEN HI=HSD
2490 Z=(HI+LO)/2
2500 IF ABS(Z-HSD)>.0002 THEN GOTO 2430
2510 PRINT"TAILWATER DEPTH (m) = ";HSD
2520 HD=HSD+(Q/A)^2/19.62:H2=HD-P
2530 RETURN
```

Listing of program **SHARP1.BAS**

```
10  REM PROGRAM "SHARP1.BAS" - RECTANGULAR SHARP-CRESTED WEIR
20  CLS
30  PRINT" ***DESIGN OF RECTANGULAR SHARP-CRESTED WEIR***"
40  PRINT
50  PRINT"Design parameters:"
60  PRINT"                      weir width  b               (m)"
70  PRINT"                      weir length  L              (m)"
80  PRINT"                      upstream crest height  P    (m)"
90  PRINT:PRINT"(above parameters are illustrated on Fig 8.3, Chap. 8)"
100 PRINT:PRINT"Press the space bar to continue":PRINT
110 Y$=INPUT$(1)
120 CLS
130 DIM X(20):DIM Y(20)
140 PRINT "DATA INPUT":PRINT
150 INPUT "Enter maximum expected flow (m**3/s)";QMAX
160 INPUT "Enter minimum expected flow (m**3/s)";QMIN
170 INPUT "Enter channel slope (sin theta)";S
180 INPUT "Enter channel Manning n-value";N
190 INPUT "Enter channel bed width (m)";WID
200 INPUT "Enter channel side slope (deg)";ANG
210 INPUT "Enter value for b (weir breadth) not less than 0.15m";BS
220 INPUT"Enter value for upstream step ht. P (not less than 0.1m)";P
230 IF P<.1 GOTO 220
240 ANG=ANG*3.1416/180:B=WID+2*P/TAN(ANG)
250 PRINT:PRINT"Channel width at weir crest B = ";B
260 IF BS>B THEN PRINT"b should be smaller than or equal to B"
270 IF BS>B GOTO 210
280 IF BS<.15 GOTO 210
290 PRINT
300 Q=QMAX
310 GOSUB 1400:REM---TO COMPUTE UPSTREAM HEAD AT MAX FLOW
320 HMAX=HS1
330 PRINT
340 PRINT"DESIGN CHECK FOR MAX FLOW CONDITION"
350 PRINT"==================================="
360 PRINT "Lower limit of h1 is 0.03m"
370 PRINT "Upper limit of h1/P =2.0"
380 PRINT "Actual h1=";HS1
390 PRINT "Actual h1/P=";HS1/P
400 IF HS1<.03 THEN PRINT"Specifications are not satisfied, reduce b and try
again"
410 IF HS1<.03 GOTO 140
420 IF HS1/P>2! THEN PRINT"Specifications are not satisfied ,increase b or P and
try again"
430 IF HS1/P>2 GOTO 140
440 PRINT"You are within specifications"
450 PRINT:PRINT"Check tailwater level to verify modular flow conditions."
460 PRINT"-------------------------------------------------------"
470 GOSUB 1770
480 PRINT"The tail water level should remain at least 0.05m below crest level"
490 PRINT"Actual freeboard (m) = "; (P-HSD)
500 IF (P-HSD)<.05 THEN PRINT "Freeboard inadequate, increase P":GOTO 140
510 PRINT:PRINT"Press the space bar to continue"
520 Y$=INPUT$(1)
530 CLS
540 PRINT "DESIGN CHECK FOR MIN FLOW"
550 PRINT "========================="
```

```
560 PRINT
570 Q=QMIN
580 GOSUB 1400:REM---To find h1 which is HS1 ,and Ce
590 PRINT:PRINT "Lower limit of h1 is 0.03m"
600 PRINT "Actual h1=";HS1
610 IF HS1<.03 THEN PRINT"Specifications are not satisfied, reduce b and try
again"
620 IF HS1<.03 GOTO 140
630 PRINT"You are within specifications"
640 A$="":PRINT
650 INPUT "Press the return key when you are ready",A$
660 IF A$="" THEN CLS
670 PRINT "DIMENSIONAL SUMMARY FOR RECTANGULAR WEIR"
680 PRINT "======================================="
690 PRINT"Channel width B at weir crest level = ";B;"(m)"
700 PRINT"Notch width b = ";BS;"(m)"
710 PRINT"Height P of weir crest above channel floor = ";P;"(m)"
720 PRINT:PRINT
730 PRINT"Enter required form of head/discharge relationship"
740 PRINT"1-Tabular form."
750 PRINT"2-Graph form."
760 INPUT FORM
770 IF FORM=1 GOTO 800
780 IF FORM=2 GOTO 950
790 GOTO 730
800 CLS
810 INPUT "Enter flow increment (m**3/s) for tabulation";DEQ
820 PRINT
830 PRINT" Q(m3/s)      Ce        be(m)     he(m)      h1(m)"
840 PRINT"================================================"
850 Q=QMIN
860 GOSUB 1400
870 PRINT USING"##.####      ";Q;
880 PRINT USING"#.####    ";CE;
890 PRINT USING"##.####   ";BE;
900 PRINT USING"##.####    ";HE;
910 PRINT USING"##.####      ";HS1
920 Q=Q+DEQ
930 IF Q<QMAX+DEQ GOTO 860
940 GOTO 1320
950 Q=QMIN:GOSUB 1400:HMIN=HS1
960 Q=QMAX:GOSUB 1400:HMAX=HS1
970 SCREEN 8
980 KEY OFF
990 X1=HMIN-.3492*(HMAX-HMIN):X2=HMAX+.2381*(HMAX-HMIN)
1000 Y1=QMIN-.4918*(QMAX-QMIN):Y2=QMAX+.1475*(QMAX-QMIN)
1010 DX=X2-X1:DY=Y2-Y1
1020 CLS
1030 WINDOW (X1,Y1)-(X2,Y2)
1040 FOR I=0 TO 10
1050 LINE (X1+.22*DX,Y1+.3*DY+.061*DY*I)-(X1+.85*DX,Y1+.3*DY+.061*DY*I)
1060 LINE (X1+.22*DX+.063*DX*I,Y1+.3*DY)-(X1+.22*DX+.063*DX*I,Y1+.91*DY)
1070 NEXT I
1080 H=HMIN
1090 FOR I=16 TO 66 STEP 10
1100 LOCATE 19,I
1110 PRINT USING "##.###";H
1120 H=H+.2*(HMAX-HMIN)
1130 NEXT I
1140 LOCATE 21,40:PRINT"Head h1(m)"
1150 Q=QMAX
```

```
1160 FOR I=3 TO 18 STEP 3
1170 LOCATE I,10
1180 PRINT USING "##.###";Q
1190 Q=Q-.2*(QMAX-QMIN)
1200 NEXT I
1210 LOCATE 1,7:PRINT"Discharge (m3/s)"
1220 PX1=HMIN:PY1=QMIN
1230 DEQ=(QMAX-QMIN)/50
1240 PY2=PY1+DEQ
1250 Q=PY2:GOSUB 1400:PX2=HS1
1260 LINE (PX1,PY1)-(PX2,PY2)
1270 IF PY2<QMAX-DEQ THEN PX1=PX2:PY1=PY2:GOTO 1240
1280 A$="":LOCATE 23
1290 INPUT "Press the return key when you are ready",A$
1300 IF A$="" THEN SCREEN 0 :CLS
1310 PRINT"Ce is the effective discharge coefficient"
1320 PRINT:PRINT"        Q=Ce*(2/3)*(2*g)^0.5*be*he^1.5"
1330 PRINT"        be=b+Kb, be is the effective breadth"
1340 PRINT"        he=h1+Kh, he is the effective head"
1350 PRINT"        Kb, Kh represent the effects of fluid properties"
1360 PRINT:PRINT
1370 PRINT"Enter (1) to run again;    enter (2) to quit":INPUT A
1380 IF A=1 THEN GOTO 20
1390 END
1400 REM ***SUBROUTINE TO COMPUTE DISCHARGE***
1410 REM   Ce is the effective discharge coefficient ;it is used in the
formula:"
1420 REM  Q=Ce*(2/3)*(2*g)^0.5*be*he^1.5
1430 REM  be=b+Kb  ,be is the effective breadth
1440 REM  he=h1+Kh ,he is the effective head
1450 REM  Kb , Kh represent the effects of fluid properties
1460 REM ***TABLE 8.2 COEFFICIENT VALUES***
1470 X(0)=0:Y(0)=.0024:K1(0)=.587:K2(0)=-.0023
1480 X(1)=.1:Y(1)=.00238:K1(1)=.588:K2(1)=-.0021
1490 X(2)=.2:Y(2)=.0024:K1(2)=.589:K2(2)=-.0018
1500 X(3)=.3:Y(3)=.00253:K1(3)=.59:K2(3)=.002
1510 X(4)=.4:Y(4)=.0027:K1(4)=.591:K2(4)=.0058
1520 X(5)=.5:Y(5)=.003:K1(5)=.592:K2(5)=.011
1530 X(6)=.6:Y(6)=.0037:K1(6)=.593:K2(6)=.018
1540 X(7)=.7:Y(7)=.00407:K1(7)=.595:K2(7)=.03
1550 X(8)=.8:Y(8)=.0043:K1(8)=.597:K2(8)=.045
1560 X(9)=.9:Y(9)=.0037:K1(9)=.599:K2(9)=.064
1570 X(10)=1!:Y(10)=-.0009:K1(10)=.602:K2(10)=.075
1580 FOR I=0 TO 9
1590 IF BS/B=X(I) THEN KB=Y(I):GOTO 1630
1600 IF BS/B=X(I+1) THEN KB=Y(I+1):GOTO 1630
1610 IF BS/B>X(I) THEN IF BS/B<X(I+1) THEN
KB=Y(I)+(BS/B-X(I))*(Y(I+1)-Y(I))/(X(I+1)-X(I)):GOTO 1630
1620 NEXT I
1630 BE=BS+KB
1640 REM***Ce is initially assumed 0.65 ,but the exact value is then calculated.
1650 CE=.65:KH=.001
1660 FOR CYCLE=1 TO 5
1670 HE=(Q/CE/(2/3)/(2*9.810001)^.5/BE)^(2/3)
1680 HS1=HE-KH
1690 REM***Computing the value of Ce as a function of b/B  &  h1/P
1700 FOR I=0 TO 9
1710 IF BS/B=X(I) THEN CE=K1(I)+K2(I)*(HS1/P):GOTO 1760
1720 IF BS/B=X(I+1) THEN CE=K1(I+1)+K2(I+1)*(HS1/P):GOTO 1760
1730 IF BS/B>X(I) THEN IF BS/B<X(I+1) THEN
CE=.5*(K1(I)+K2(I)*HS1/P+K1(I+1)+K2(I+1)*HS1/P)
```

```
1740 NEXT I
1750 NEXT CYCLE
1760 RETURN
1770 REM ***Computing the tailwater depth***
1780 FSR=S^.5
1790 HI=40:LO=.001
1800 HSD=(HI+LO)/2
1810 A=HSD*(WID+HSD/TAN(ANG)):RH=A/(WID+(2*HSD)/SIN(ANG))
1820 FSH=N*QMAX/(A*RH^.67)
1830 WW=FSH-FSR
1840 IF WW>0 THEN LO=HSD
1850 IF WW<0 THEN HI=HSD
1860 Z=(HI+LO)/2
1870 IF ABS(Z-HSD)>.0002 THEN GOTO 1800
1880 PRINT"TAILWATER DEPTH (m) = ";HSD
1890 RETURN
```

Listing of program **SHARP2.BAS**

```
10  REM PROGRAM "SHARP2.BAS"  -  V-NOTCH SHARP-CRESTED WEIR
20  CLS
30  PRINT" ***DESIGN OF V-NOTCH SHARP-CRESTED WEIR***"
40  PRINT
50  PRINT"Design parameters:"
60  PRINT"                    V-notch angle  theta         (deg)"
70  PRINT"                    Upstream crest height  P      (m)"
80  PRINT:PRINT"    (above parameters are illustrated on Fig 8.4, Chap. 8)"
90  PRINT:PRINT"Press the space bar to continue":PRINT
100 Y$=INPUT$(1)
110 DIM X(10):DIM Y(10):DIM Z(10)
120 PRINT "DATA INPUT":PRINT
130 INPUT "Enter maximum expected flow (m**3/s)";QMAX
140 INPUT "Enter minimum expected flow (m**3/s)";QMIN
150 INPUT "Enter channel slope (sin theta)";S
160 INPUT "Enter channel Manning n-value";N
170 INPUT "Enter channel bed width (m)";WID
180 INPUT "Enter channel side slope (deg)";ANG
190 PRINT:PRINT"NOTE THAT QMAX SHOULD NOT EXCEED 0.450 m**3/s":PRINT
200 INPUT "Choose a value for θ (notch angle) between 25-100 degrees  ";THETA
210 PRINT:IF THETA<25 GOTO 200
220 IF THETA>100 GOTO 200
230 PRINT"Choose a value for P (ht. of vertex over the approach channel bed,m)"
240 PRINT"Vertex should be higher than the expected tailwater depth by not less
than 0.05m."
250 PRINT"Lower limit of P is 0.10 m. ";
260 INPUT P
270 IF P<.10 GOTO 230
280 PRINT:PRINT
290 Q=QMAX
300 GOSUB 1640:REM***To find h1 which is HS1 ,and Ce***
310 HMAX=HS1
320 ANG=ANG*3.1416/180:B=WID+2*(P+HS1)/TAN(ANG)
330 CLS
340 PRINT"DESIGN CHECK FOR MAX FLOW CONDITION"
350 PRINT"==================================="
360 PRINT"h1/P should not be greater than 1.2"
370 PRINT"Actual h1/P=";HS1/P
380 IF HS1/P>1.2 THEN PRINT"Specifications are not satisfied ,increase P or
decrease h1 by increasing θ ":PRINT"and try again with new data.":PRINT
390 IF HS1/P>1.2 GOTO 120
400 PRINT"h1/P satisfies specifications"
410 PRINT
420 PRINT"h1/B should not be greater than 0.4"
430 PRINT"Actual h1/B=";HS1/B
440 IF HS1/B>.4 THEN PRINT"Specifications are not satisfied ,increase B or
decrease h1 by increasing θ ":PRINT"and try again with new data":PRINT
450 IF HS1/B>.4 GOTO 120
460 PRINT"h1/B satisfies specifications"
470 PRINT
480 PRINT"h1 should be less than 0.60"
490 PRINT"Actual h1=";HS1
500 IF HS1>.60 THEN PRINT"Specifications are not satisfied ,decrease h1 by
increasing θ and try again.":PRINT
510 IF HS1>.60 GOTO 120
520 PRINT"h1 satisfies specifications"
530 PRINT
540 PRINT"h1 should be greater than 0.05"
550 PRINT"Actual h1=";HS1
```

```
560 IF HS1<.05 THEN PRINT"Specifications are not satisfied ,increase h1 by
decreasing θ and try again.":PRINT
570 IF HS1<.05 GOTO 120
580 PRINT"h1 satisfies specifications"
590 PRINT:PRINT "Press the space bar to continue"
600 Y$=INPUT$(1)
610 PRINT "Check tailwater level to verify modular flow conditions."
620 PRINT "-------------------------------------------------------"
630 GOSUB 1800
640 PRINT "The tailwater level shouldremain at least 0.05m below crest level"
650 PRINT "Actual freeboard (m) = "; (P-HSD)
660 IF (P-HSD)<.05 THEN PRINT "Freeboard inadequate, increase P":GOTO 120
670 PRINT:PRINT "Press the space bar to continue"
680 Y$=INPUT$(1)
690 PRINT "DESIGN CHECK FOR MIN FLOW CONDITION"
700 PRINT "=================================="
710 Q=QMIN
720 GOSUB 1640:REM***To find h1 which is HS1 ,and Ce***
730 PRINT "Lower limit of h1 is 0.05m"
740 PRINT "Actual h1=";HS1
750 IF HS1<.05 THEN PRINT"Specifications are not satisfied, increase h1 by
decreasing θ and try again.":PRINT
760 IF HS1<.05 GOTO 120
770 PRINT"You are within specifications"
780 A$="":PRINT:PRINT
790 INPUT "Press the return key when you are ready",A$
800 IF A$="" THEN CLS
810 PRINT "DIMENSIONAL SUMMARY FOR V-NOTCH WEIR"
820 PRINT "===================================="
830 PRINT"Notch angle θ = ";THETA;"(degrees)"
840 PRINT"Vertex height P = ";P;"(m)"
850 PRINT"Channel width B at upstream max water level = ";B;"(m)":PRINT
860 PRINT "The distance upstream from the weir of the head measurement station"
870 PRINT"should be in the region of 3 to 4 times the head on the weir at"
880 PRINT"the maximum expected flow  i.e. ";3*HMAX;" - ";4*HMAX;"(m)"
890 PRINT:PRINT
900 PRINT"Enter required form of head/discharge relationship"
910 PRINT"1-Tabular form."
920 PRINT"2-Graph form."
930 INPUT FORM
940 IF FORM=1 GOTO 970
950 IF FORM=2 GOTO 1130
960 GOTO 900
970 CLS
980 PRINT "Enter required interval between consecutive flow values to give
upstream"
990 PRINT "water level values in tabular form ";
1000 INPUT DEQ
1010 PRINT:PRINT
1020 PRINT"Q(m3/s)       Ce       Kh(m)        h1(m)"
1030 PRINT"===================================="
1040 Q=QMIN
1050 GOSUB 1640
1060 PRINT USING"##.####      ";Q;
1070 PRINT USING"#.####    ";CE;
1080 PRINT USING"#.####     ";KH;
1090 PRINT USING"##.####";HS1
1100 Q=Q+DEQ
1110 IF Q<QMAX+DEQ GOTO 1050
1120 GOTO 1490
1130 Q=QMIN:GOSUB 1640:HMIN=HS1
```

```
1140 Q=QMAX:GOSUB 1640:HMAX=HS1
1150 SCREEN 8
1160 KEY OFF
1170 X1=HMIN-.3492*(HMAX-HMIN):X2=HMAX+.2381*(HMAX-HMIN)
1180 Y1=QMIN-.4918*(QMAX-QMIN):Y2=QMAX+.1475*(QMAX-QMIN)
1190 DX=X2-X1:DY=Y2-Y1
1200 CLS
1210 WINDOW (X1,Y1)-(X2,Y2)
1220 FOR I=0 TO 10
1230 LINE (X1+.22*DX,Y1+.3*DY+.061*DY*I)-(X1+.85*DX,Y1+.3*DY+.061*DY*I)
1240 LINE (X1+.22*DX+.063*DX*I,Y1+.3*DY)-(X1+.22*DX+.063*DX*I,Y1+.91*DY)
1250 NEXT I
1260 H=HMIN
1270 FOR I=16 TO 66 STEP 10
1280 LOCATE 19,I
1290 PRINT USING "##.###";H
1300 H=H+.2*(HMAX-HMIN)
1310 NEXT I
1320 LOCATE 21,40:PRINT"Head h1(m)"
1330 Q=QMAX
1340 FOR I=3 TO 18 STEP 3
1350 LOCATE I,10
1360 PRINT USING "##.###";Q
1370 Q=Q-.2*(QMAX-QMIN)
1380 NEXT I
1390 LOCATE 1,7:PRINT"Discharge (m3/s)"
1400 PX1=HMIN:PY1=QMIN
1410 DEQ=(QMAX-QMIN)/50
1420 PY2=PY1+DEQ
1430 Q=PY2:GOSUB 1640:PX2=HS1
1440 LINE (PX1,PY1)-(PX2,PY2)
1450 IF PY2<QMAX-DEQ THEN PX1=PX2:PY1=PY2:GOTO 1420
1460 A$="":LOCATE 23
1470 INPUT "Press the return key when you are ready",A$
1480 IF A$="" THEN SCREEN 0 :CLS
1490 PRINT"Ce is the effective discharge coefficient ;it is used in the
formula:"
1500 PRINT"Q=Ce*(8/15)*(2*g)^0.5*tan(θ/2)*he^2.5"
1510 PRINT"he=h1+Kh ,he is the effective head over the notch vertex"
1520 PRINT"h1 is the upstream water depth over the notch vertex"
1530 PRINT"Kh represents the effects of fluid properties"
1540 PRINT:PRINT
1550 PRINT"Enter (1) to run again          Enter (2) to quit":INPUT A
1560 IF A=1 THEN RUN "SHARP2"
1570 END
1580 REM  Ce is the effective discharge coefficient; it is used in the formula:
1590 REM   Q=Ce*(8/15)*(2*g)^0.5*tan(θ/2)*he^2.5
1600 REM   he=h1+Kh ,he is the effevtive head over the notch vertex
1610 REM   h1 is the upstream water depth over the notch vertex
1620 REM   Kh represents the effects of fluid properties
1630 REM   X,Y,Z are the empirical values from Table 8.3
1640 X(1)=20:Y(1)=.0029:Z(1)=.595
1650 X(2)=30:Y(2)=.00235:Z(2)=.586
1660 X(3)=40:Y(3)=.0018:Z(3)=.581
1670 X(4)=50:Y(4)=.0015:Z(4)=.579
1680 X(5)=60:Y(5)=.0012:Z(5)=.577
1690 X(6)=70:Y(6)=.00105:Z(6)=.577
1700 X(7)=80:Y(7)=.0009:Z(7)=.578
1710 X(8)=90:Y(8)=.00085:Z(8)=.579
1720 X(9)=100:Y(9)=.0008:Z(9)=.581
1730 FOR I=1 TO 8
```

```
1740 IF THETA=X(I) THEN KH=Y(I):CE=Z(I)
1750 IF THETA=X(I+1) THEN KH=Y(I+1):CE=Z(I+1)
1760 IF THETA>X(I) THEN IF THETA<X(I+1) THEN
KH=Y(I)+(THETA-X(I))*(Y(I+1)-Y(I))/(X(I+1)-X(I))
:CE=Z(I)+(THETA-X(I))*(Z(I+1)-Z(I))/(X(I+1)-X(I))
1770 NEXT I
1780 HS1=(Q/CE/(8/15)/(2*9.810001)^.5/TAN(THETA*3.1416/180/2))^(1/2.5)-KH
1790 RETURN
1800 REM ***Computing tailwater depth***"
1810 FSR=S^.5
1820 HI=40:LO=.001
1830 HSD=(HI+LO)/2
1840 A=HSD*(WID+HSD/TAN(ANG)):RH=A/(WID+(2*HSD)/SIN(ANG))
1850 FSH=N*QMAX/(A*RH^.67)
1860 WW=FSH-FSR
1870 IF WW>0 THEN LO=HSD
1880 IF WW<0 THEN HI=HSD
1890 Z=(HI+LO)/2
1900 IF ABS(Z-HSD)>.0002 THEN GOTO 1830
1910 PRINT"TAILWATER DEPTH (m) = ";HSD
1920 RETURN
```

Listing of program **SUTRO.BAS**

```
10  REM PROGRAM "SUTRO.BAS" - PROPORTIONAL-FLOW WEIR
20  CLS
30  PRINT" ***DESIGN OF SUTRO (PROPORTIONAL FLOW) WEIR***"
40  PRINT
50  PRINT"Design parameters:"
60  PRINT"                    weir bottom width  b          (m)"
70  PRINT"                    weir bottom height  a         (m)"
80  PRINT"                    upstream crest height  P1     (m)"
90  PRINT"                    downstream crest height  P2   (m)"
100 PRINT:PRINT"(above parameters are illustrated on Fig 8.5, Chap. 8)"
110 PRINT:PRINT"Press the space bar to continue":PRINT
120 Y$=INPUT$(1)
130 CLS
140 DIM A(10),B(10),CDV(10,10)
150 GOSUB 1420: REM----To read Cd values (refer Table 8.4 in text)----
160 PRINT "DATA INPUT":PRINT
170 INPUT "Enter maximum expected flow (m**3/s)";QMAX
180 INPUT "Enter minimum expected flow (m**3/s)";QMIN
190 INPUT "Enter channel slope (sin theta)";S
200 INPUT "Enter channel Manning n-value";N
210 INPUT "Enter channel bed width (m)";WID
220 INPUT "Enter channel sidewall slope (deg)";ANG
230 INPUT "Enter value for b (weir bottom width, design range 0.15-0.45m)";BS
240 INPUT "Enter value for a (weir bottom ht., design range 0.006-0.09m)";AS
250 INPUT "Enter value for upstream step ht. P1 (not greater than b)";P1
260 INPUT "Enter value for downstream step ht. P2 (not less than P1)";P2
270 ANG=ANG*3.1416/180:B=WID+2*P1/TAN(ANG)
280 PRINT:PRINT"Channel width at weir crest B = ";B
290 IF B/BS<3 THEN PRINT"B/b should be >= 3; this requirement is not satisfied;"
300 IF B/BS<3 THEN PRINT"reduce value of b and try again."
310 IF B/BS<3 THEN GOTO 230
320 PRINT
330 Q=QMAX
340 GOSUB 1550:REM***TO COMPUTE THE DISCHARGE COEFFICIENT Cd***
350 GOSUB 1670:REM***TO COMPUTE UPSTREAM HEAD h1***
360 HMAX=HS1
370 PRINT
380 PRINT"DESIGN CHECK FOR MAX FLOW CONDITION"
390 PRINT"==================================="
400 PRINT "Lower limit of h1 is 2a OR 0.03m, whichever is greater"
410 PRINT "Actual h1 = ";HS1
420 PRINT "2a = ";2*AS
430 IF HS1<.03 THEN PRINT"Specifications are not satisfied, reduce weir dims and
try again"
440 IF HS1<.03 GOTO 230
450 IF HS1<2*AS THEN PRINT"Specifications are not satisfied, reduce a and try
again"
460 IF HS1<2*AS THEN GOTO 230
470 PRINT"You are within specifications"
480 PRINT:PRINT"Check tailwater level to verify modular flow conditions."
490 PRINT"-------------------------------------------------------"
500 GOSUB 1780
510 PRINT"The tail water level should remain at least 0.05m below crest level"
520 PRINT"Actual freeboard (m) = "; (P2-HSD)
530 IF (P2-HSD)<.05 THEN PRINT "Freeboard inadequate, increase P2":GOTO 230
540 PRINT:PRINT"Press the space bar to continue"
550 Y$=INPUT$(1)
```

```
560 CLS
570 PRINT "DESIGN CHECK FOR MIN FLOW"
580 PRINT "========================"
590 PRINT
600 Q=QMIN
610 GOSUB 1670:REM---To compute h1---
620 PRINT:PRINT "Lower limit of h1 is 2a or 0.03m, whichever is greater"
630 PRINT "Actual h1=";HS1
640 PRINT "2a = ";2*AS
650 IF HS1<.03 THEN PRINT"Specifications are not satisfied, reduce weir dims and
 try again"
660 IF HS1<.03 GOTO 230
670 IF HS1<2*AS THEN PRINT"Specifications are not satisfied, reduce value of a
and try again"
680 IF HS1<2*AS GOTO 230
690 PRINT"You are within specifications"
700 PRINT
710 PRINT "Press the space bar to continue":A$=INPUT$(1)
720 CLS
730 PRINT "DIMENSIONAL SUMMARY FOR RECTANGULAR WEIR"
740 PRINT "========================================"
750 PRINT"Channel width B at weir crest level = ";B;"(m)"
760 PRINT"Notch width b = ";BS;"(m)"
770 PRINT"Height dimension a = ";AS;"(m)"
780 PRINT"Upstream step height P1 = ";P1;"(m)"
790 PRINT"Downstream step height P2 = ";P2;"(m)"
800 PRINT:PRINT
810 PRINT"Enter required form of head/discharge relationship"
820 PRINT"1-Tabular form."
830 PRINT"2-Graph form."
840 INPUT FORM
850 IF FORM=1 GOTO 880
860 IF FORM=2 GOTO 1000
870 GOTO 810
880 CLS
890 INPUT "Enter flow increment (m**3/s) for tabulation";DEQ
900 PRINT
910 PRINT" Q(m3/s)        h1(m)"
920 PRINT"==================="
930 Q=QMIN
940 GOSUB 1670
950 PRINT USING"##.####      ";Q;
960 PRINT USING"##.####      ";HS1
970 Q=Q+DEQ
980 IF Q<QMAX+DEQ GOTO 940
990 GOTO 1360
1000 Q=QMIN:GOSUB 1670:HMIN=HS1
1010 Q=QMAX:GOSUB 1670:HMAX=HS1
1020 SCREEN 8
1030 KEY OFF
1040 X1=HMIN-.3492*(HMAX-HMIN):X2=HMAX+.2381*(HMAX-HMIN)
1050 Y1=QMIN-.4918*(QMAX-QMIN):Y2=QMAX+.1475*(QMAX-QMIN)
1060 DX=X2-X1:DY=Y2-Y1
1070 CLS
1080 WINDOW (X1,Y1)-(X2,Y2)
1090 FOR I=0 TO 10
1100 LINE (X1+.22*DX,Y1+.3*DY+.061*DY*I)-(X1+.85*DX,Y1+.3*DY+.061*DY*I)
1110 LINE (X1+.22*DX+.063*DX*I,Y1+.3*DY)-(X1+.22*DX+.063*DX*I,Y1+.91*DY)
1120 NEXT I
1130 H=HMIN
1140 FOR I=16 TO 66 STEP 10
```

```
1150 LOCATE 19,I
1160 PRINT USING "##.###";H
1170 H=H+.2*(HMAX-HMIN)
1180 NEXT I
1190 LOCATE 21,40:PRINT"Head h1(m)"
1200 Q=QMAX
1210 FOR I=3 TO 18 STEP 3
1220 LOCATE I,10
1230 PRINT USING "##.###";Q
1240 Q=Q-.2*(QMAX-QMIN)
1250 NEXT I
1260 LOCATE 1,7:PRINT"Discharge (m3/s)"
1270 PX1=HMIN:PY1=QMIN
1280 DEQ=(QMAX-QMIN)/50
1290 PY2=PY1+DEQ
1300 Q=PY2:GOSUB 1670:PX2=HS1
1310 LINE (PX1,PY1)-(PX2,PY2)
1320 IF PY2<QMAX-DEQ THEN PX1=PX2:PY1=PY2:GOTO 1290
1330 A$="":LOCATE 23
1340 INPUT "Press the return key when you are ready",A$
1350 IF A$="" THEN SCREEN 0 :CLS
1360 PRINT: PRINT"discharge coefficient Cd = ";CD
1370 PRINT:PRINT
1380 PRINT"Enter (1) to run again           Enter (2) to quit":INPUT A
1390 IF A=1 THEN CLS:GOTO 170
1400 IF A=2 THEN GOTO 2000
1410 GOTO 1380
1420 REM ***SUBROUTINE TO COMPUTE THE DISCHARGE COEFFICIENT Cd***
1430 REM ***Cd IS COMPUTED FROM THE DATA IN TABLE 8.4***
1440 REM ***READING DATA FROM TABLE 8.4:***
1450 FOR I=1 TO 7:READ A(I):NEXT I
1460 FOR I=1 TO 5:READ B(I):NEXT I
1470 I=1:FOR J=1 TO 5:READ CDV(I,J):NEXT J
1480 I=2:FOR J=1 TO 5:READ CDV(I,J):NEXT J
1490 I=3:FOR J=1 TO 5:READ CDV(I,J):NEXT J
1500 I=4:FOR J=1 TO 5:READ CDV(I,J):NEXT J
1510 I=5:FOR J=1 TO 5:READ CDV(I,J):NEXT J
1520 I=6:FOR J=1 TO 5:READ CDV(I,J):NEXT J
1530 I=7:FOR J=1 TO 5:READ CDV(I,J):NEXT J
1540 RETURN
1550 REM ***COMPUTATION OF DISCHARGE COEFFICIENT Cd***
1560 FOR I=1 TO 7
1570 IF AS<A(I) THEN GOTO 1590
1580 NEXT I
1590 II=I
1600 FOR J=1 TO 5
1610 IF BS<B(J) THEN GOTO 1630
1620 NEXT J
1630 JJ=J
1640 CD=CDV(II,JJ)
1650 PRINT"cd=";CD
1660 RETURN
1670 REM ***COMPUTATION OF UPSTREAM HEAD h1***
1680 HI=20:LO=0
1690 HS1=(HI+LO)/2
1700 FUNC=CD*BS*(19.62*AS)^.5*(HS1-AS/3)
1710 IF (FUNC-Q)>0 THEN HI=HS1
1720 IF (FUNC-Q)<0 THEN LO=HS1
1730 Z=(HI+LO)/2
1740 IF ABS(Z-HS1)>.0002 THEN GOTO 1690
1750 HS1=Z
```

```
1760 RETURN
1770 RETURN
1780 REM ***Computing the tailwater depth***
1790 FSR=S^.5
1800 HI=40:LO=.001
1810 HSD=(HI+LO)/2
1820 A=HSD*(WID+HSD/TAN(ANG)):RH=A/(WID+(2*HSD)/SIN(ANG))
1830 FSH=N*QMAX/(A*RH^.67)
1840 WW=FSH-FSR
1850 IF WW>0 THEN LO=HSD
1860 IF WW<0 THEN HI=HSD
1870 Z=(HI+LO)/2
1880 IF ABS(Z-HSD)>.0002 THEN GOTO 1810
1890 PRINT"TAILWATER DEPTH (m) = ";HSD
1900 RETURN
1910 DATA 0.006,0.015,0.030,0.046,0.061,0.076,0.091
1920 DATA 0.15,0.23,0.30,0.38,0.46
1930 DATA 0.608,0.613,0.617,0.618,0.619
1940 DATA 0.606,0.611,0.615,0.617,0.617
1950 DATA 0.603,0.608,0.612,0.613,0.614
1960 DATA 0.601,0.606,0.610,0.612,0.612
1970 DATA 0.599,0.604,0.608,0.610,0.610
1980 DATA 0.598,0.603,0.607,0.608,0.609
1990 DATA 0.597,0.602,0.606,0.608,0.608
2000 END
```

Listing of program **OCUSF.BAS**

```
10  REM PROGRAM OCUSF
20  PRINT"                            PROGRAM OCUSF                    "
30  PRINT
40  PRINT"     This program computes the transient flow and water depth in"
50  PRINT"     open channels of rectangular, trapezoidal and circular cross-"
60  PRINT"     sections, using a numerical computation procedure based on the"
70  PRINT"     method of characteristics, as outlined in this chapter. The"
80  PRINT"     computation of frictional resistance is based on the Manning"
90  PRINT"     equation."
100 PRINT
110 PRINT"     Computation starts from a specified steady state at time zero."
120 PRINT"     The program offers a choice of two initial steady states viz."
130 PRINT"     steady uniform flow and zero flow."
140 PRINT
150 PRINT"     Boundary conditions:  The program caters for the following"
160 PRINT"     parameter variations at both ends of the channel:"
170 PRINT"        (a) linear variation of flow depth with time"
180 PRINT"        (b) linear variation of discharge rate with time"
190 PRINT
200 PRINT"     A constant value for either boundary parameter is obtained"
210 PRINT"     by specifying a zero rate for the parameter variation."
220 PRINT:PRINT "Press the space bar to continue"
230 Y$=INPUT$(1):CLS
240 DIM Y(100),V(100),YP(100),VP(100),C(100),Q(100),QP(100),SF(100)
250 PRINT:PRINT"ENTER CHANNEL DATA:":PRINT
260 PRINT "Is section:   1 CIRCULAR    2 RECTANGULAR   3 TRAPEZOIDAL ?"
270 PRINT:INPUT "Enter 1, 2 or 3, as appropriate";NO:PRINT
280 IF NO=1 THEN SECT$="CIRC":INPUT"Diameter (m)";D:
290 IF NO=2 THEN SECT$="RECT": INPUT "Channel width (m)";B
300 IF NO=3 THEN SECT$="TRAP": INPUT "Bottom width (m)";B
310 IF NO=3 THEN INPUT "Angle of side wall to horl (deg)";FI
320 PRINT:INPUT"Enter channel length (m)";XL
330 INPUT"Enter Manning's n-value";MN
340 INPUT"Enter channel bed slope (sin theta)";S0:PRINT
350 INPUT"Enter initial steady flow rate (m**3/s)";Q0
360 IF Q0=0 THEN INPUT"Enter water depth at upstream end of channel (m)";Y0
370 PRINT:PRINT"Select upstream boundary condition:"
380 PRINT"                     1. linear variation of discharge with time"
390 PRINT"                     2. linear variation of depth with time"
400 INPUT"Enter 1 or 2 as appropriate";UN
410 IF UN=1 THEN INPUT"Enter rate of discharge variation (m**3/s/s)";QURATE
420 IF UN=1 THEN INPUT"Enter final upstream discharge rate (m**3/s)";QUF
430 IF UN=2 THEN INPUT"Enter rate of depth variation with time (m/s)";HURATE
440 IF HURATE=0 THEN GOTO 460
450 IF UN=2 THEN INPUT"Enter final u/s depth (m)";YUF
460 PRINT:PRINT"Select downstream boundary condition:"
470 PRINT"                     1. linear variation of discharge with time"
480 PRINT"                     2. linear variation of depth with time"
490 INPUT"Enter 1 or 2 as appropriate";DN
500 IF DN=1 THEN INPUT"Enter rate of discharge variation (m**3/s/s)";QDRATE
510 IF DN=1 THEN INPUT"Enter final d/s discharge rate (m**3/s)";QDF
520 IF DN=2 THEN INPUT"Enter rate of depth variation with time (m/s)";HDRATE
530 IF HDRATE=0 THEN GOTO 550
540 IF DN=2 THEN INPUT"Enter final d/s depth (m)";YDF
550 PRINT
560 PRINT"Enter number of reaches into which the channel length is divided"
570 INPUT"for computational purposes (multiple of 10)";N
580 PRINT:INPUT"Enter number of computation iterations";NITER
```

```
590 PRINT:PRINT"DATA INPUT COMPLETED; COMPUTATION IN PROGRESS"
600 G=9.806:NS=N+1:NP=N/5:FI=FI*3.142/180
610 REM COMPUTATION OF STEADY FLOW DEPTH YN AND VELOCITY VN
620 IF Q0<>0 THEN Q=ABS(Q0):GOSUB 2060
630 REM ASSIGNMENT OF INITIAL VALUES TO VARIABLES
640 FOR I=1 TO NS
650 IF Q0=0 THEN V(I)=0:Y(I)=Y0+(I-1)*XL/N*S0:GOTO 670
660 IF Q0<>0 THEN V(I)=VN:Y(I)=YN
670 YY=Y(I):GOSUB 1860
680 VV=V(I):GOSUB 2010
690 SF(I)=SF
700 Q(I)=Q0
710 C(I)=C
720 NEXT I
730 DX=XL/N
740 DT=DX/(V(NS)+C(NS))
750 T=0
760 PRINT:PRINT:PRINT"           **** Tabulation of computed values follows ****"
770 PRINT:PRINT"   TIME                  DISTANCE ALONG CHANNEL"
780 PRINT"    (MIN)   0.0L     0.2L     0.4L     0.6L     0.8L     1.0L":PRINT
790 PRINT USING"  ##.##  ";TM;
800 FOR I=1 TO N STEP N/5
810 PRINT USING"###.###  ";Q(I);:NEXT I
820 PRINT USING"###.###  ";Q(NS);:PRINT"  Q (m**3/s)"
830 PRINT"          ";
840 FOR I=1 TO N STEP N/5
850 PRINT USING"###.###  ";Y(I);:NEXT I
860 PRINT USING"###.###  ";Y(NS);:PRINT"  DEPTH (m)"
870 PRINT
880 FOR COUNT=1 TO NITER
890 IF UN=1 THEN QP(1)=Q(1)+DT*QURATE:GOSUB 2210
900 IF UN=2 THEN YP(1)=Y(1)+DT*HURATE:GOSUB 2270
910 IF DN=1 THEN QP(NS)=Q(NS)+DT*QDRATE:GOSUB 2340
920 IF DN=2 THEN YP(NS)=Y(NS)+DT*HDRATE:GOSUB 2400
930 TM=T/60
940 DXX=0
950 FOR I=1 TO NS
960 DXI=(ABS(V(I))+C(I))*DT
970 IF (DXI>DXX) THEN DXX=DXI
980 NEXT I
990 ZETA=DXX/DX
1000 DT=DT/ZETA
1010 T=T+DT:TM=T/60
1020 TH=DT/DX
1030 REM:INTERIOR POINTS
1040 FOR I=2 TO N
1050 CA=C(I)-C(I-1)
1060 VR=(V(I)+TH*(C(I)*V(I-1)-V(I)*C(I-1)))/(1!+TH*(V(I)-V(I-1)+CA))
1070 CR=(C(I)-VR*TH*CA)/(1!+TH*CA)
1080 YR=Y(I)-TH*(VR+CR)*(Y(I)-Y(I-1))
1090 YY=YR:GOSUB 1860
1100 VV=VR:GOSUB 2010
1110 CR=C:SR=SF
1120 CB=C(I)-C(I+1)
1130 VS=(V(I)-TH*(V(I)*C(I+1)-C(I)*V(I+1)))/(1!-TH*(V(I)-V(I+1)-CB))
1140 CS=(C(I)+VS*TH*CB)/(1!+TH*CB)
1150 YS=Y(I)+TH*(VS-CS)*(Y(I)-Y(I+1))
1160 YY=YS:GOSUB 1860
1170 VV=VS:GOSUB 2010
1180 CS=C:SS=SF
1190 YP(I)=(YS*CR+YR*CS+CR*CS*((VR-VS)/G-DT*(SR-SS)))/(CR+CS)
```

```
1200 VP(I)=VR-G*((YP(I)-YR)/CR+DT*(SR-S0))
1210 NEXT I
1220 REM:UPSTREAM BOUNDARY CONDITIONS
1230 CB=C(1)-C(2)
1240 VS=(V(1)-TH*(V(1)*C(2)-C(1)*V(2)))/(1-TH*(V(1)-V(2)-CB))
1250 CS=(C(1)+VS*TH*CB)/(1+TH*CB)
1260 YS=Y(1)+TH*(VS-CS)*(Y(1)-Y(2))
1270 YY=YS:GOSUB 1860
1280 VV=VS:GOSUB 2010
1290 C2=G/C
1300 CM=VS-C2*YS-G*DT*(SF-S0)
1310 IF UN=2 THEN GOTO 1410
1320 REM SPECIFIED VARIATION IN Q AT US BOUNDARY
1330 YY=Y(1)
1340 GOSUB 1860
1350 FY=QP(1)/A-C2*YY-CM
1360 FDY=-(QP(1)/A^2)*TW-C2
1370 DELY=-FY/FDY
1380 YY=YY+DELY
1390 IF ABS(DELY)>.001 THEN GOTO 1340
1400 YP(1)=YY:VP(1)=C2*YP(1)+CM
1410 REM SPECIFIED VARIATION IN Y AT US BOUNDARY
1420 VP(1)=C2*YP(1)+CM
1430 REM:DOWNSTREAM BOUNDARY
1440 CA=C(NS)-C(N)
1450 VR=(V(NS)+TH*(C(NS)*V(N)-V(NS)*C(N)))/(1+TH*(V(NS)-V(N)+CA))
1460 CR=(C(NS)-VR*TH*CA)/(1+TH*CA)
1470 YR=Y(NS)-TH*(VR+CR)*(Y(NS)-Y(N))
1480 YY=YR:GOSUB 1860
1490 VV=VR:GOSUB 2010
1500 C4=G/C
1510 CP=VR+C4*YR-G*DT*(SF-S0)
1520 IF DN=2 THEN GOTO 1620
1530 REM SPECIFIED VARIATION IN DS DISCHARGE Q
1540 YY=Y(NS)
1550 GOSUB 1860
1560 FY=QP(NS)/A+C4*YY-CP
1570 FDY=-(QP(NS)/A^2)*TW+C4
1580 DELY=-FY/FDY
1590 YY=YY+DELY
1600 IF ABS(DELY)>.001 THEN GOTO 1550
1610 YP(NS)=YY:VP(NS)=CP-C4*YP(NS)
1620 REM SPECIFIED VARIATION IN DOWNSTREAM Y
1630 VP(NS)=CP-C4*YP(NS)
1640 REM UPDATE VARIABLE VALUES FOR CURRENT TIME STEP
1650 FOR I=1 TO NS
1660 YY=YP(I):GOSUB 1860
1670 VV=VP(I):GOSUB 2010
1680 V(I)=VP(I)
1690 Q(I)=V(I)*A
1700 C(I)=C
1710 Y(I)=YP(I)
1720 SF(I)=SF
1730 NEXT I
1740 IF COUNT\2=COUNT/2 THEN GOTO 1750 ELSE GOTO 1840
1750 PRINT USING"  ##.##  ";TM;
1760 FOR I=1 TO N STEP N/5
1770 PRINT USING"###.###  ";Q(I);:NEXT I
1780 PRINT USING"###.###  ";Q(NS);:PRINT"  Q (m**3/s)"
1790 PRINT"          ";
1800 FOR I=1 TO N STEP N/5
```

```
1810 PRINT USING"###.###  ";Y(I);:NEXT I
1820 PRINT USING"###.###  ";Y(NS);:PRINT"  DEPTH (m)"
1830 PRINT
1840 NEXT COUNT
1850 END
1860 REM SUBROUTINE TO CALCULATE SECTION FLOW PARAMETERS
1870 IF SECT$="CIRC" THEN GOTO 1920
1880 IF SECT$="RECT" THEN A=B*YY:P=(B+2*YY):TW=B:RETURN
1890 IF SECT$="TRAP" THEN A=YY*(B+YY/TAN(FI))
1900 P=B+2*YY/SIN(FI):TW=B+2*YY/TAN(FI)
1910 RETURN
1920 HI=3.1416: LO=0!
1930 EST=(HI+LO)/2
1940 XR=1-(2*YY)/D-COS(EST)
1950 IF XR<0 THEN LO=EST
1960 IF XR>0 THEN HI=EST
1970 Z=(HI+LO)/2
1980 IF ABS(Z-EST)>.001 THEN GOTO 1930
1990 P=D*EST: A=.25*D*D*(EST-.5*SIN(2*EST))
2000 TW=D*SIN(EST):RETURN
2010 REM SUBROUTINE TO CALCULATE Q,C,SF
2020 Q=VV*A
2030 C=(G*A/TW)^.5
2040 SF=MN^2*(P/A)^1.3333*VV*ABS(VV)
2050 RETURN
2060 REM SUBROUTINE TO CALCULATE STEADY FLOW DEPTH
2070 FSR=S0^.5
2080 HII=40:LOO=.001
2090 IF SECT$="CIRC" THEN HII=D
2100 YY=(HII+LOO)/2
2110 GOSUB 1860
2120 RH=A/P
2130 FSH=MN*Q/(A*RH^.67)
2140 WW=FSH-FSR
2150 IF WW>0 THEN LOO=YY
2160 IF WW<0 THEN HII=YY
2170 Z=(HII+LOO)/2
2180 IF ABS(Z-YY)>.0002 THEN GOTO 2100
2190 YN=YY:VN=Q0/A
2200 RETURN
2210 REM UPSTREAM BOUNDARY - FINAL DISCHARGE CHECK
2220 IF QURATE <0 THEN GOTO 2250
2230 IF QP(1)>=QUF THEN QP(1)=QUF
2240 RETURN
2250 IF QP(1)<=QUF THEN QP(1)=QUF
2260 RETURN
2270 REM UPSTREAM BOUNDARY - FINAL DEPTH CHECK
2280 IF HURATE=0 THEN YP(1)=Y(1):RETURN
2290 IF HURATE <0 THEN GOTO 2320
2300 IF YP(1)>=YUF THEN YP(1)=YUF
2310 RETURN
2320 IF YP(1)<=YUF THEN YP(1)=YUF
2330 RETURN
2340 REM DOWNSTREAM BOUNDARY - FINAL DISCHARGE CHECK
2350 IF QDRATE <0 THEN GOTO 2380
2360 IF QP(NS)>=QDF THEN QP(NS)=QDF
2370 RETURN
2380 IF QP(NS)<=QDF THEN QP(NS)=QDF
2390 RETURN
2400 REM DOWNSTREAM BOUNDARY - FINAL DEPTH CHECK
2410 IF HDRATE <0 THEN GOTO 2450
```

```
2420 IF HDRATE=0 THEN YP(NS)=Y(NS):RETURN
2430 IF YP(NS)>=YDF THEN YP(NS)=YDF
2440 RETURN
2450 IF YP(NS)<=YDF THEN YP(NS)=YDF
2460 RETURN
```

Listing of program **AIRLIFT.BAS**

```
10  REM PROGRAM AIRLIFT
20  PRINT "This program computes the following airlift pump design"
30  PRINT"parameters, given the values for the remaining parameters"
40  PRINT"as input data:"
50  PRINT:PRINT"   1.  REQUIRED AIR INPUT "
60  PRINT"   2.  REQUIRED SUBMERGENCE"
70  PRINT:INPUT"ENTER 1 OR 2, AS APPROPRIATE";NUMB
80  PRINT:INPUT"ENTER PIPE DIAMETER (mm)";PD:PD=PD/1000
90  IF NUMB=1 THEN INPUT"ENTER SUBMERGENCE OF AIR INJECTION PT. (m)";HS
100 INPUT"ENTER STATIC LIFT (m)";HL
110 INPUT"ENTER LENGTH OF SUCTION PIPE U/S OF INJECTION PT. (m)";LS
120 INPUT "ENTER PIPE WALL ROUGHNESS (mm)";WR:WR=WR/1000
130 IF NUMB=2 THEN INPUT"ENTER FREE AIR INPUT RATE (m**3/s)";QA
140 KG=1.2:N=1.5:ITER=1
150 INPUT"ENTER REQUIRED LIQUID PUMPING RATE (m**3/s)";QL
160 INPUT"ENTER LIQUID DENSITY (kg/m**3)";RHOL
170 INPUT"ENTER LIQUID VISCOSITY (Ns/m**2)";MU
180 VL=QL/(.785*PD*PD):GOSUB 660
190 KF=F*VL*VL*RHOL/(2*PD)
200 PA=9810*10.3:VD=.35*(9.810001*PD)^.5
210 IF NUMB=2 THEN GOTO 440
220 P0=RHOL*9.810001*(HS-.5*VL*VL/19.6-F*LS*VL*VL/(19.6*PD)+10.3)
230 QAU=QL*5
240 QAL=QL/5
250 QA=(QAL+QAU)/2:VGA=QA/(.785*PD*PD)
260 R=RHOL*9.810001*VGA*PA*(KG-1)+KF*VGA*PA*(KG+N)
270 S=KG*VL+VD
280 W=(PA-P0)/(RHOL*9.810001+KF)-(((KF*N-RHOL*9.810001)*VGA*PA)/((RHOL*9.810001
+KF)^2*S))*LOG((R+(RHOL*9.810001+KF)*S*PA)/(R+(RHOL*9.810001+KF)*S*P0))+HS+HL
290 IF W>0 THEN QAL=QA
300 IF W<0 THEN QAU=QA
310 Z=(QAU+QAL)/2:E=ABS((Z-QA)/QA)
320 IF E>.01 THEN GOTO 250
330 ITER=ITER+1
340 IF ITER>100 THEN PRINT "SOLUTION NOT CONVERGING":GOTO 650
350 IF ABS(W)>.05 THEN GOTO 250
360 EFF=(RHOL*9.810001*HL*QL)/(PA*QA*LOG(P0/PA)):SR=HS/(HS+HL)
370 PRINT:PRINT"Computed output values:"
380 PRINT:PRINT"REQUIRED AIR INPUT RATE (m**3/s) = ";QA
390 PRINT"SUPERFICIAL AIR VEL. (m/s) = ";VGA
400 PRINT"SUPERFICIAL WATER VELOCITY (m/s) = ";VL
410 PRINT"SUBMERGENCE RATIO = ";SR
420 PRINT"EFFICIENCY = ";EFF
430 GOTO 650
440 VGA=QA/(.785*PD*PD)
450 R=RHOL*9.810001*VGA*PA*(KG-1)+KF*VGA*PA*(KG+N)
460 HSL=HL:HSU=40*HL
```

```
470 HS=(HSU+HSL)/2
480 P0=RHOL*9.810001*(HS-.5*VL*VL/19.6-F*LS*VL*VL/(19.6*PD)+10.3)
490 S=KG*VL+VD
500 W=(PA-P0)/(RHOL*9.810001+KF)-(((KF*N-RHOL*9.810001)*VGA*PA)/((RHOL*9.810001
+KF)^2*S))*LOG((R+(RHOL*9.810001+KF)*S*PA)/(R+(RHOL*9.810001+KF)*S*P0))+HS+HL
510 IF W<0 THEN HSU=HS
520 IF W>0 THEN HSL=HS
530 Z=(HSU+HSL)/2:E=ABS((Z-HS)/HS)
540 IF E>.01 THEN GOTO 470
550 ITER=ITER+1
560 IF ITER>100 THEN PRINT"SOLUTION NOT CONVERGING":GOTO 650
570 IF ABS(W)>.05 THEN GOTO 470
580 EFF=(RHOL*9.810001*HL*QL)/(PA*QA*LOG(P0/PA)):SR=HS/(HS+HL)
590 PRINT:PRINT"Computed output values:"
600 PRINT:PRINT"REQUIRED SUBMERGENCE (m) = ";HS
610 PRINT"SUPERFICIAL AIR VELOCITY (m/s) = ";VGA
620 PRINT"SUPERFICIAL WATER VELOCITY (m/s) = ";VL
630 PRINT"SUBMERGENCE RATIO = ";SR
640 PRINT"EFFICIENCY = ";EFF
650 END
660 REM SUBROUTINE TO COMPUTE THE FRICTION FACTOR F
670 KVISC=MU/RHOL:V=VL:D=PD:K=WR
680 RE=ABS(V)*D/KVISC
690 IF RE>2000 THEN GOTO 700 ELSE F=64/RE: GOTO 830
700 UPV=.5
710 LOV=0
720 F=(UPV+LOV)/2
730 Y=.5/SQR(F)
740 X=((K/(3.7*D))+2.51/(RE*SQR(F)))
750 W=Y+LOG(X)/LOG(10)
760 IF W<0 THEN UPV=F
770 IF W>0 THEN LOV=F
780 Z=(UPV+LOV)/2
790 E=ABS(Z-F)
800 IF E<.0001 THEN GOTO 820
810 F=Z: GOTO 730
820 F=Z
830 RETURN
```

Listing of program **PUMP.BAS**

```
10  REM PROGRAM PUMP
20  PRINT"This program computes the coefficients A1 and A2 in the rotodynamic"
30  PRINT"pump equation: H = A0+A1*Q+A2*Q*Q, given 3 points on the H/Q standard"
40  PRINT"speed curve, including the shut-off head A0." :PRINT
50  INPUT"ENTER THE VALUE OF A0 (m)"; A0
60  PRINT"ENTER VALUES FOR H (m) AND Q(m**/s), SEPARATED BY A COMMA"
70  INPUT"ENTER FIRST PAIR OF VALUES";H1,Q1
80  INPUT"ENTER SECOND PAIR OF VALUES";H2,Q2
90  INPUT"ENTER VALUE OF STANDARD PUMP SPEED (rpm)";PS0
100 INPUT"ENTER VALUE OF DUTY SPEED (rpm)";PS
110 A2=((H1*Q2-H2*Q1)-A0*(Q2-Q1))/(Q1*Q2*(Q1-Q2))
120 A1=(H1-A0-A2*Q1*Q1)/Q1
130 PRINT:PRINT"Coefficient values at standard speed";PS0;" rpm are:"
```

```
140 PRINT:PRINT"       A0 = ";A0
150 PRINT"       A1 = ";A1
160 PRINT"       A2 = ";A2:PRINT
170 REM COMPUTATION OF PUMP DUTY POINT
180 FR=PS/PS0
190 A0=A0*FR*FR:A1=A1*FR
200 INPUT"ENTER PUMP SUMP WATER LEVEL (mOD)";SRL
210 INPUT"ENTER RISING MAIN DISCHARGE LEVEL (mOD)";DRL
220 INPUT"ENTER SUCTION MAIN LENGTH (m)";LS
230 INPUT"ENTER SUCTION MAIN DIAMETER (mm)";DS:DS=DS/1000
240 INPUT"ENTER SUCTION MAIN WALL ROUGHNESS (mm)";KS:KS=KS/1000
250 INPUT"ENTER DELIVERY MAIN LENGTH (m)";LD
260 INPUT"ENTER DELIVERY MAIN DIAMETER (mm)";DD:DD=DD/1000
270 INPUT"ENTER DELIVERY MAIN WALL ROUGHNESS (mm)";KD:KD=KD/1000
280 INPUT"ENTER ESTIMATE OF DUTY POINT DISCHARGE (m**3/S)";PQ
290 FOR COUNT=1 TO 10
300 AS=.785*DS*DS:AD=.785*DD*DD
310 V=PQ/AS:D=DS:K=KS:GOSUB 450
320 KP=F*LS/(19.62*DS*AS*AS)
330 V=PQ/AD:D=DD:K=KD:GOSUB 450
340 KP=KP+F*LD/(19.62*DD*AD*AD)
350 F=KP*PQ*PQ+DRL-SRL-A0-A1*PQ-A2*PQ*PQ
360 DF=2*KP*PQ-A1-2*A2*PQ
370 PQ=PQ-F/DF
380 NEXT COUNT
390 PH=A0+A1*PQ+A2*PQ*PQ
400 PRINT:PRINT"Computed duty point values are:"
410 PRINT:PRINT"       DUTY POINT DISCHARGE (m**3/s) = ";PQ
420 PRINT"       DUTY POINT HEAD (m) = ";PH
430 PRINT"       VALUES RELATE TO PUMP SPEED (rpm) = ";PS
440 END
450 REM SUBROUTINE TO COMPUTE THE FRICTION FACTOR F
460 KVISC=.000001
470 RE=ABS(V)*D/KVISC
480 IF RE>2000 THEN GOTO 490 ELSE F=64/RE: GOTO 620
490 UPV=.5
500 LOV=0
510 F=(UPV+LOV)/2
520 Y=.5/SQR(F)
530 X=((K/(3.7*D))+2.51/(RE*SQR(F)))
540 W=Y+LOG(X)/LOG(10)
550 IF W<0 THEN UPV=F
560 IF W>0 THEN LOV=F
570 Z=(UPV+LOV)/2
580 E=ABS(Z-F)
590 IF E<.0001 THEN GOTO 610
600 F=Z: GOTO 520
610 F=Z
620 RETURN
```

Appendix B
Computational algorithms

B.1 The interval-halving method

The method of interval halving (sometimes called the bisection method) has been used in a number of the computer programs presented in earlier chapters. It is used to determine, by systematic successive approximations, the root of a single-variable equation of the form $f(x) = 0$ within a specified variable interval. As a starting point the variable interval within which the required root lies is specified by an upper boundary value x_U and a lower boundary value x_L, as illustrated in Fig. B.1. Clearly, since a root lies within the reach defined by x_U and x_L, $f(x_U)$ and $f(x_L)$ must be of opposite sign. The root is approximated as the midpoint of the interval:

$$x_i = \frac{x_U + x_L}{2}. \tag{B.1}$$

If $f(x_U) \times f(x_i) > 0$, that is, are of the same sign, then it follows that the root does not lie in the upper interval half bounded by x_U and x_i but is within the lower interval half defined by x_i and x_L. By the same reasoning, if $f(x_U) \times f(x_i) < 0$, the root must lie in the interval defined by x_i and x_U. Thus a new interval, which is half the preceding interval, is defined by each succesive computation. Computation is continued until the specified tolerance is reached, that is, $|x_U - x_L| \leq$ tolerance or $f(x_i) = 0$. The progress of the computation is illustrated in Fig. B.1.

B.2 The Newton–Raphson method

The Newton–Raphson method has also been used in a number of the programs presented in earlier chapters. Like the interval-halving method, it has been used to determine the root of a single-variable equation of the form $f(x) = 0$. Computation starts from an initial estimate of the root x_i. The point where the tangent at x_i cuts the x-axis is used as the next approximation of the root, as illustrated in Fig. B.2. In computational terms this procedure is

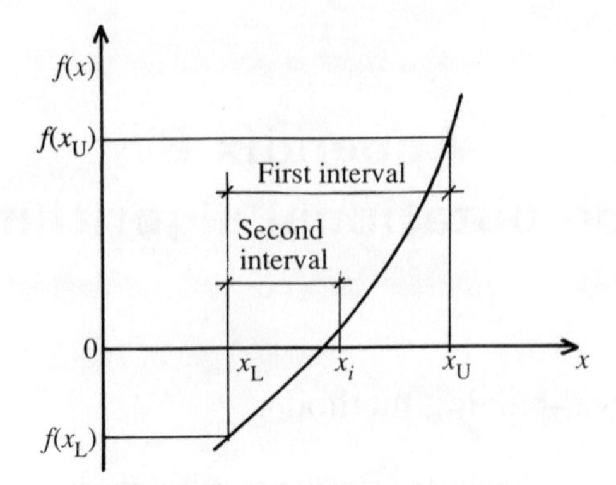

Fig. B.1 Interval-halving algorithm.

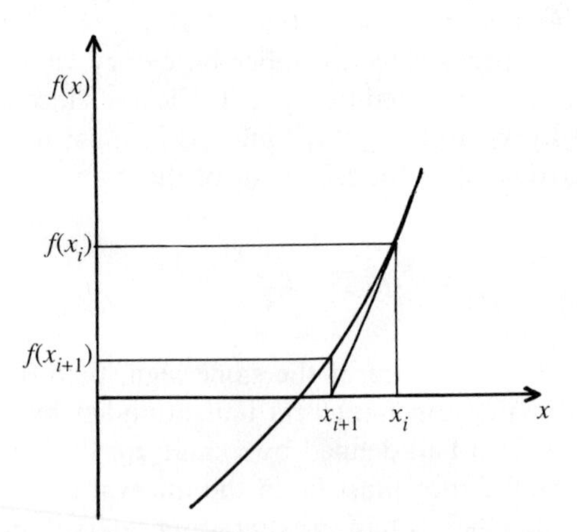

Fig. B.2 Newton–Raphson algorithm.

written as follows:

$$x_{i+1} = x_i - \frac{f(x_i)}{f'(x_i)} \tag{B.2}$$

where $f'(x_i)$ denotes the derivative of $f(x)$ at $x = x_i$. Expression (B.2) is the Newton–Raphson formula. The procedure is repeated until $|x_i - x_{i+1}| \leq$ specified tolerance.

It is worth noting that the loop flow correction algorithm used in the pipe network analysis procedure presented in Chapter 5 is based on the Newton–Raphson formula.

References

Chapra, S. C. and Canale, R. P. (1985). *Numerical Methods for Engineers*. McGraw-Hill, New York.

Curtis, F. G. and Wheatley, P. O. (1989). *Applied Numerical Analysis*, (4th edn), Addison-Wesley, New York.

Index